Drug Delivery Systems Using Quantum Computing

Scrivener Publishing
100 Cummings Center, Suite 541J
Beverly, MA 01915-6106

Publishers at Scrivener
Martin Scrivener (martin@scrivenerpublishing.com)
Phillip Carmical (pcarmical@scrivenerpublishing.com)

Drug Delivery Systems Using Quantum Computing

Edited by

Rishabha Malviya

School of Medical and Allied Sciences, Galgotias University, India

Sonali Sundram

School of Medical and Allied Sciences, Galgotias University, India

and

Dhanalekshmi Unnikrishnan Meenakshi

College of Pharmacy, National University of Science and Technology, Muscat, Sultanate of Oman

Scrivener Publishing

WILEY

This edition first published 2024 by John Wiley & Sons, Inc., 111 River Street, Hoboken, NJ 07030, USA and Scrivener Publishing LLC, 100 Cummings Center, Suite 541J, Beverly, MA 01915, USA

For more information about Scrivener publications please visit www.scrivenerpublishing.com.

Wiley Global Headquarters
111 River Street, Hoboken, NJ 07030, USA

For details of our global editorial offices, customer services, and more information about Wiley products visit us at www.wiley.com.

Library of Congress Cataloging-in-Publication Data

ISBN 978-1-394-15914-7

Cover image: Pixabay.Com
Cover design by Russell Richardson

Set in size of 11pt and Minion Pro by Manila Typesetting Company, Makati, Philippines

Printed in the USA

10 9 8 7 6 5 4 3 2 1

Dear healthcare professionals,
We are dedicating this book to you. Our love for the profession shall live forever.

Contents

Foreword

The origin of this unique book started in a conversation I had with Rishabha about ten years ago. He had just started his career in research and wished to make a contribution to pharmacy professionals. I knew he was going to be a star in his field as he had all the qualities of a high-flyer. Every story and every achievement were a performance. Today, I am writing this Foreword for his dream project.

Identifying and developing small molecules and macromolecules that might help cure illnesses and diseases is the core activity of pharmaceutical companies. Given its focus on molecular formations, the pharmaceutical industry is a natural candidate for quantum computing. Today's trendiest issues in technology are all related to quantum computing. Thanks to technology, computing issues that were once thought to be insurmountable are now being solved by individuals and businesses. This technology has had a substantial impact on many different domains, including healthcare, the pharmaceutical industry, drug discovery, and many more.

The development of this emerging technology, both by itself and in concert with other technologies, has the potential to make a significantly positive impact on society. The impact of quantum computing on the pharmaceutical industry relies on the development, manufacture, and delivery of molecules. This book explains the profound impact that quantum computing could have on the pharmaceutical industry and presents user examples of its application. I hope after reading this book, healthcare professionals and pharmaceutical companies will be well-advised to assess the quantum computing opportunity for themselves and begin laying the groundwork for securing their place in this new competitive and technological landscape.

Dr. Kamla Pathak
Dean and Professor
Faculty of Pharmacy
Uttar Pradesh University of Medical Sciences
Saifai, Etawah
Uttar Pradesh, India

Preface

The majority of drug candidates have low solubility, rapid blood clearance, poor targeting, and, usually, poor ability to penetrate cell membranes. Some of these obstacles can be addressed by drug delivery systems (DDSs) that improve drug delivery to the site of action. Drug delivery technology is advancing significantly, and controlling the precise level and/or location of a given drug in the body reduces adverse effects, lowers dosages, and makes new therapies possible.

Nevertheless, there are still significant obstacles to delivering certain medications to particular cells. Drug delivery methods change pharmacokinetic, pharmacodynamic, and drug release patterns to enhance product efficacy and safety, as well as patient convenience and compliance. Computational approaches in drug development enable quick screening of a vast chemical library and identification of possible binders by using modeling, simulation, and visualization tools. Quantum computing (QC) is a fundamentally new computing paradigm based on quantum mechanics rules that enables certain computations to be conducted significantly more rapidly and effectively than regular computing, and hence this has huge promise for the pharmaceutical sector.

Significant advances in computational simulation are making it easier to comprehend the process of drug delivery. This book explores an important biophysical component of DDSs, and how computer modelling may help with the logical design of DDSs with enhanced and optimized characteristics. The book concentrates on computational research for various important types of nanocarriers, including dendrimers and dendrons, polymers, peptides, nucleic acids, lipids, carbon-based DDSs, and gold nanoparticles.

This new technology's advancement promises a significant and positive effect on society, both independently and in conjunction with existing technologies. With ethical research, development, and implementation, the trickle-down consequences of the quantum revolution should enhance the lives of many people. Understanding the effects that quantum computing could have on the pharmaceutical business is therefore crucial.

This book contains 14 chapters, written by top researchers from many parts of the world. The book is profusely referenced and copiously illustrated, and all chapters were deliberately reviewed and revised to meet the highest standard of publication. We are deeply grateful to everyone who helped with this book and greatly appreciate the dedicated support and valuable assistance rendered by Martin Scrivener and the Scrivener Publishing team during its publication.

The Editors
April 2024

Acknowledgments

Writing a book seemed difficult, but working on this had been a delightful experience. It was more rewarding and motivating than I had imagined. I would like to begin by expressing my gratitude to the Almighty, whose unending blessings and supreme power enable us to accomplish all our goals.

Although this period of my life was filled with many ups and downs, I am eternally grateful to my family for their continuous support and encouragement that made it possible to complete this task.

I am deeply indebted to the management at Galgotias University. I want to thank you all for letting me serve, for being a part of our amazing organization, and for motivating me every day.

Many thanks to all contributors. Without their participation, this goal could not have been achieved. Last but not least, I would like to thank our publisher for their support, innovative suggestions, and guidance in bringing out this edition.

Acknowledgments

1

Quantum Computational Concepts and Approaches in Drug Discovery, Development and Delivery

Dhanalekshmi Unnikrishnan Meenakshi[1]*, Suresh Manic Kesavan[2], Arul Prakash Francis[3] and Shah Alam Khan[1†]

[1]*College of Pharmacy, National University of Science and Technology, Muscat, Oman*
[2]*College of Engineering, National University of Science and Technology, Muscat, Oman*
[3]*Centre of Molecular Medicine and Diagnostics (COMManD), Saveetha Dental College and Hospitals, Saveetha Institute of Medical and Technical Sciences, Saveetha University, Chennai, India*

Abstract

The use of quantum computing approaches in drug discovery, development, and delivery is on rise. Quantum computing (QC) is becoming more popular as a cost-saving measure in the research, production, and manufacturing of drugs. The use of molecular dynamics and computation tasks has greatly enhanced drug design and optimization of drug delivery systems. This QC method aids in investigating several issues that are difficult to examine in laboratory studies. This chapter discusses how QC simulations can assign a specific drug target site by anticipating the target, resulting in a successful drug development process. It also highlights the important applications of several algorithms in drug delivery and disease diagnosis. Additionally, computational methodologies for various drug delivery systems and carriers are discussed. This chapter also discusses ongoing challenges in the pharmaceutical industry and concerns regarding the chances of accomplishment of QC concepts and technology.

**Corresponding author*: dhanalekshmi@nu.edu.om
†*Corresponding author*: shahalam@nu.edu.om

Rishabha Malviya, Sonali Sundram and Dhanalekshmi Unnikrishnan Meenakshi (eds.)
Drug Delivery Systems Using Quantum Computing, (1–26) © 2024 Scrivener Publishing LLC

Keywords: Drug discovery, drug delivery, quantum computing, simulations, optimization, molecular dynamics, algorithms

1.1 Introduction

Drug development and discovery to identify new chemical entities to treat or cure human diseases is a very long, complex, expensive, and tedious process that often fails [1]. Over the years, the scientific community has tried its best to decrease expenditure, speed up the process of drug development, and progress the accomplishment frequency of identifying drugs or molecules of natural or synthetic origin [2, 3]. In the late 20th century, the paradigm slowly shifted from Ethnopharmacology, a traditional discovery method from medicinal plants, to a computational chemistry-based technique called computer-aided drug design (CADD) to identify a lead molecule capable of binding to a target protein to exhibit favorable therapeutic effects and pharmacokinetic profile [4]. Discovery and optimization of lead molecules following the high-throughput screening (HTS) over hit libraries using a CADD approach can be done either by structured-based or ligand-based drug design [5]. While the latter is favored in the absence of target architectural data and uses a quantitative structure–activity relation (QSAR) model, the former approach is used when knowledge of the 3D structural information of the biological target is available for molecular docking (pharmacophore modeling). Although most of the drugs in the pipeline have been discovered using CADD, the higher computational cost to study the molecular dynamics of protein–ligand interactions and the reliability of the current statistical techniques available to study pharmacokinetic profile limits its usefulness [6–8]. The challenges and limitations of CADD could be overcome with the help of an emerging quantum computing (QC) technology that can handle and stimulate larger and more complex chemical structures more efficiently. QC application will certainly benefit the pharmaceutical industry from innovation in drug discovery to the manufacture and development of promising therapeutic modalities [8].

Constructing a new drug for a chronic illness in the medical context was mainly based on new pills. Various drugs for identifying energetic facets in conventional therapies like penicillin have recently been developed. In natural compounds, organic molecules that aid in medical purposes to discover ingredients such as cells or unbroken life forms are used in pharmaceutical manufacturing. This is known as traditional pharmacology. As the DNA sequence has enabled massive cloning techniques and improved protein refinement, HTS with various libraries has become more common. The therapy of disease by checking for large molecules using genetic goals is known as reverse pharmacology. By screening activity to supply cells, countless collisions can be obtained, and animal tests for adequacy have also been performed. Pipelines for pharmaceutical research are lengthy, complex, and dependent on a variety of factors. Additionally, QC is crucial to the creation of medication delivery systems. There are still problems that need to be tackled despite the vital role that QC has played in many stages of drug development and technological advancements. Precision parameters related to these virtual screening methods are constrained by energy- or similarity-based scoring concerns. Before receiving market permission, the virtual screening technique still needs to validate the compounds through preclinical and clinical tests.

Quantum computing may carry out difficult tasks like interpreting sensor data or analyzing complex interrelationships and variables that affect the quantity or reliability of carriers in medication delivery systems through the use of machine learning (ML) algorithms and artificial intelligence (AI) [9]. High-precision imaging of the produced formulation, particularly morphology, may be provided by QCs with very sophisticated processing and computing capabilities [10]. QC's role in disease diagnosis is to create new interfaces between drug development, production, therapeutic effects, and diagnosis. Rapid cancer detection can be achieved using QCs and hence providing enhanced therapy. Besides, adverse effects can be reduced by devising precise radiation plans with an appropriate dose for targeting cancer cells [11]. Using Shor's algorithm for data collection and creating novel datasets develop a novel method for cancer detection and the stage identification of various diseases seems to be relaxed [12]. High-precision radiation beams can be directed at cancer cells using QCs, and they can also anticipate TP53 gene alterations, which are crucial to the pathophysiology of several cancer types [13, 14]. With a significant amount of high-quality data, ML approaches provide a variety of techniques that can enhance innovation and strategic planning for some well-posed topics. ML applications are possible across the entire research process. Different applications have taken different approaches to their research, and most of these methodologies have produced reliable predictions and findings. Comprehensive and integrated high-dimensional data must still be derived in all areas.

As mentioned, drug discovery, development, and delivery are increasingly using QC techniques in diverse ways. The use of QC is growing in popularity as a way to produce and manufacture pharmaceuticals more cheaply. The design of drugs and the optimization of drug delivery systems have both been considerably improved by the use of molecular dynamics and computational tasks. Investigation of several topics that are challenging to evaluate in laboratory investigations is aided by this QC strategy. To achieve a successful medication development process, this chapter explores how QC simulations might assign a specific drug target location by predicting the target. It also emphasizes the significant roles that various algorithms play in the administration of medications and the identification of diseases. There is also a discussion on computational approaches for different medication delivery systems and carriers. This chapter also addresses persistent difficulties and challenges faced by the pharmaceutical industry.

1.2 Algorithms and QC in Pharma

The pharmaceutical industry revolves around the creation of molecular formulations that are then turned into pharmaceuticals to treat or cure diseases. As mentioned, the use of QC is growing in popularity as a way to produce and manufacture pharmaceuticals more cheaply. Pharmaceutical companies spend a whopping investment across all industries. For decades, pharmaceutical companies are the early users of digital tools like CADD, to improve the R&D process. Recently, AI has been used in pharmaceutical research and development and the next technological edge to focus on is QC. It has been stated that leaders in the pharmaceutical sector should respond to and discuss a set of critical concerns to choose the best course of action in the computation/simulation era.

1.2.1 Algorithms

Businesses are increasingly utilizing ML algorithms to make life easier to keep up with consumer expectations as the world becomes "smarter." They can be found in consumer electronics (for example, facial acknowledgment for unlocking cellphones) or in the case of credit card fraud detection (such as setting up alerts for unusual purchases). Figure 1.1 illustrates the types of ML that are frequently used in research. AI and ML are divided into two categories: supervised learning and unsupervised learning.

1.2.2 Supervised Learning

It uses labeled datasets in supervised learning and this method is also a type of ML. The databases are employed to supervise processes so that they

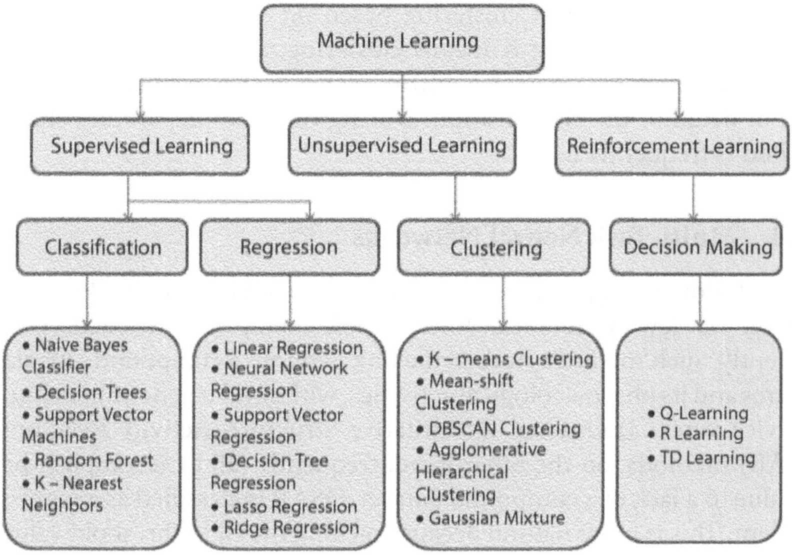

Figure 1.1 Types of machine learning (ML).

can correctly identify data or forecast results. The type can be tested for accuracy and learned over time by using labeled inputs and outputs.

There are two kinds of supervised learning problems in data mining: classification and regression.

An algorithm is used to accurately assign test data into specified groups, such as distinguishing apples from oranges. In this modern era, supervised learning algorithms can classify spam and place it in a different folder from your mail inbox. Linear classifiers, support vector machines, decision trees, and random forests are examples of classification algorithms. Regression is a class of supervised learning models that may predict numerical values using a range of corporate data sources. The three most common regression algorithms are logistic regression, polynomial regression, and linear regression.

1.2.3 Unsupervised Learning

In this method, ML algorithms are used to evaluate and cluster unlabeled datasets. Without any human interference, these algorithms can identify hidden patterns in data (thus the term "unsupervised"). For tasks such as clustering, association, and dimensionality reduction, unsupervised learning models are used. Unlabeled databases are grouped using the data mining approach of clustering based on similarities and contrasts. For instance, when using K-means clustering, associated data points are divided into different classes, with the K value representing the size and level of detail of the classification. During the data preparation step, this method is frequently used.

1.2.4 Multi-Task Neural Networks

Computational drug discovery methods, which have been widely studied over the last few decades, can find possible therapeutic interactions more efficiently. Such methods studied the link between a compound's insulative features and its pharmacological activities, with similar genetic benchmarks and vice versa. Traditional Quantitative Structure–Activity Relationship (QSAR) methods, on the other hand, frequently fail to accomplish precision due to a lack of a compound dataset. QSAR is classified as a regression problem. This is more natural because picking the right threshold values to find whether a particular drug is active or inactive is tricky and pervasively results in information loss. Researchers use RF as the foundation learner. To represent molecules, we use the functional-class fingerprints (FCFP) method is preferred. Researchers have empirically discovered that they are the most successful in prediction.

1.2.5 Graph Convolution

Each biblical SMILES is first converted into a binary coordinate with 75 measures per atom. This technique uses an initial vector as input and the vector is then converted into a single dense layer after this process was repeated. The dense layer's numerical vectors are added together in the graph collection layer to produce the compound's "neural fingerprint."

1.3 Potential of QC in Drug Discovery

Quantum computers act on quantum bits (*qubits*) and thus can perform computational tasks such as prediction of ADMET properties of ligands based on molecular descriptors, unfolding of protein structures, chemical structures behavior, reactivity, their interaction, and simulation with biological targets much faster than the present day's conventional computers. It will be able to help throughout the processes of drug discovery, production, and marketing but the major impact would be on the discovery phases. It will find application in finding a target protein and its structure, hit generation and identification, lead molecule identification, and optimization of molecular properties (ADMET, solubility, dosing, etc.) [8, 15, 16].

QC could be used to enhance the accuracy and efficiency of existing CADD techniques to match the ligands with the target proteins. QC techniques could be very helpful in obtaining information (3D chemical structure) on larger protein folding using amino acid sequence and to develop a hypothesis for the pathophysiology of a disease. Scientists may be able to screen computational libraries of ligands onto the target protein's various potential structures in tandem using QC. Due to computational limitations and time constraints, current methods typically limit the target molecule's structural flexibility. These limitations can make it harder to find the best therapeutic agents [8, 15]. By applying ML techniques to find novel structure–property correlations, QC may eventually enhance the production and validation of hypotheses. Once the QC approach reaches a certain degree of maturity, it might be able to create new kinds of candidate drug collections that are not only restricted to tiny molecules but may also comprise proteins and immunoglobulins [8, 15, 17]. QC may improve the *de novo* design process and thus will help in the optimized synthesis of new lead molecules, and active pharmaceutical ingredients (quantum

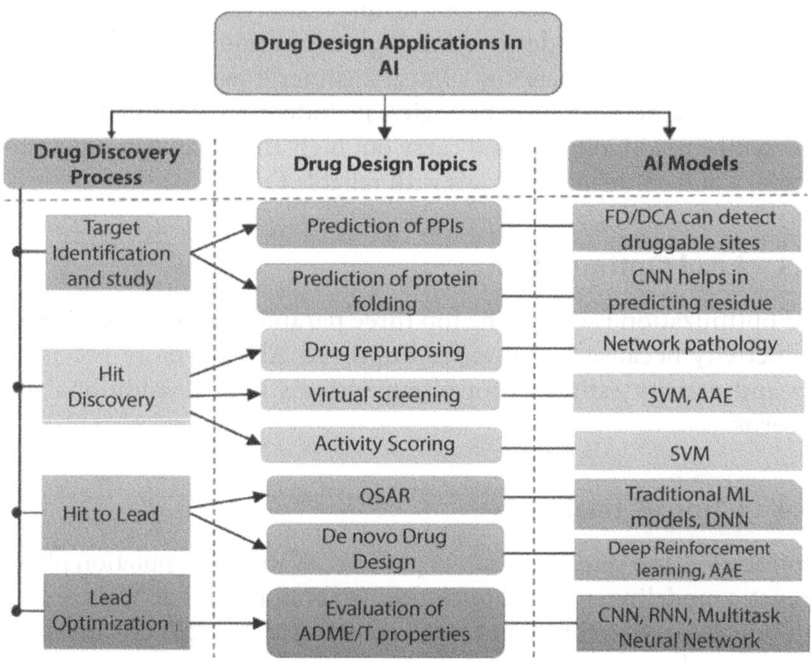

Figure 1.2 Different stages in the drug development process for QC application.

molecular design) as well as in the discovery of organic biodegradable compounds (quantum annealing) [8, 17, 18]. The following QC application cases will occur at different periods throughout the drug development process as shown in Figure 1.2.

1.3.1 Target Recognition and Validation

Protein 3-D architectures may be predicted with extreme precision using QC throughout target acquisition. Obtaining high-quality structural data is a time-consuming and risky process. Despite being a turning point in ML-driven protein folding, Google's DeepMind's Alpha Fold still fails to tackle each of the problems that conventional programming simulations confront, such as protein complex creation, molecular docking, and post-translational modifications. The ability of QC to explicitly tackle electrons, which makes correlations the most challenging problems to solve, may be helpful.

1.3.2 Production and Validation of Hits

The ability of QC to deal with multiple complex phenomena at the same time would be extremely useful during hit creation and validation. Pharma businesses utilizing modern technology may employ CADD on therapeutically active candidates in a linear manner. Pharma firms may apply all phenomena to particular biologics, including recombinant protein or semi-synthesized biologics, and carry out high-throughput computational discovery and confirmation trials with reliable QC.

1.3.3 Lead Optimization

Lead optimization is one of the top three parameters for increasing R&D productivity because it allows for improved ADME; more precise toxicity and activity estimates for organ systems besides additional safety concerns.

1.3.4 Clinical Trials

Identification and categorization of patients, as well as population pharmacogenetic modeling, may aid in clinical trial improvement. QC could help with venue selection throughout trial planning and execution and could be used to supplement causality analysis for adverse occurrences to improve active safety surveillance.

1.3.5 Linking and Generating Data

The meta-level of research and development is primarily concerned with linking relevant data—for example, by employing good (conceptual) management to establish reasonable data point relationships. QC could be used to generate bogus data using ML algorithms to replace missing data points throughout the investigation. This could be especially helpful in cases where data are scarce, such as rare diseases, which can be addressed by using fictitious datasets. QC will set a new standard for training ML models in terms of speed.

1.4 QC and Drug Delivery

Designing a drug or predicting the mode of delivery is mainly aimed at enhancing the efficiency of the drug by targeting the appropriate site of action and thus reducing toxicity to the healthy cells. The primary requisites for optimal drug carriers are to protect the medications against inactivation and deliver them to the target site in a controlled manner [19, 20]. On the other hand, unsuitable translocation, uncertain distribution, accumulation, or penetration of drugs, and lacking target specificity can limit the drug delivery to specific sites. Nanotechnology-based strategies evolved as an effective tool for early disease diagnosis and drug carrier by extending the retention of therapeutics at target sites, hence improving their efficiency as well as safety profiles [9]. Multi-functional nanoparticle (NP)-based carriers delivering the therapeutics in a controlled manner resulting in cancer cell death is a promising anticancer approach [21]. Despite the great success in various approaches to nanotechnology, different polymers like NPs have failed in the clinical trial stages as reported [19]. However, the challenges faced towards the clinical translation of nano-based formulations may greatly decrease the interest of investors in investments involving nanomedicine-based research and development. In recent years, unique opportunities have been created by computational models that helped in the development of new hypotheses or solutions to various challenging issues. Quantum technologies over the last decade have focused on highly sophisticated platform designs, enhanced drug design processes or therapeutic interventions, comparing large-scale compounds, predicting drug targets and drug–target interaction, and developing self-monitoring intelligent drug delivery systems [22, 23]. QC creates a significant breakthrough in current science and technologies. Apart from protecting sensitive data, telecommunication, code-breaking, and cyber security, chemical reactions can be predicted using quantum computers. Quantum computers are also used for optimization issues, developing new therapeutic agents, solving modeling effective therapeutics by computing protein folding sequences, and acquiring the transcription of DNA and various diseases' diagnosis. In addition, it also predicts the properties and structural features of the compounds and their interactions, thus finding novel drugs and indications [24, 25].

Computational methods developed for predicting drug–carrier interactions and associated mechanisms bring a breakthrough in drug delivery. Computational predictions before the laborious experimentation might reduce the huge investments. The traditional treatment approaches can be revolutionized using appropriate algorithms, software, or methods for obtaining suitable computational power and improved performance of models. Computational drug delivery prediction is a fast-growing sector of the pharmaceutical industry. The essential criteria of any drug delivery system are to deliver the optimal dosage of the active ingredient in a safe manner. As systematizing state-of-the-art drug delivery takes years, computer-aided drug delivery techniques can be exploited to design a sophisticated and novel drug delivery system as this reduces the period and cost of the process. The computational pharmaceutics-based drug design like QSAR, ligand docking, etc. has peaked in the last decade as it helps in screening thousands of molecules in a short time and at a low cost [26]. Mathematical-based computer models are used in computer-aided drug delivery to mimic the drug delivery system similar to the human body. The preliminary step in computer-aided drug delivery is to form a mathematical model and then it should be optimized using the governing equations of various phenomenon's diffusion, absorption, etc. The outline of the computer model used to simulate the effectiveness of a drug delivery process is based on the data obtained from mathematical models and equations. The flowchart of an optimized computational drug delivery system is given in Figure 1.3.

Computational techniques like Molecular Mechanics (MM), Monte Carlo (MC), and Molecular Dynamics (MD) use empirical models for interpreting the molecular or atomic interactions, especially in large molecular systems. The accuracy of classical types of computational approaches was decided based on the parameters illustrating the inter-atomic potentials between the atoms and molecules [22]. In recent years, multi-scale models have been designed and developed to evaluate various chemical reactions and adsorption or distribution of drugs. Multi-scale models reported are predicting adverse drug effects during the early steps of drug discovery, evaluating drug transportation process and delivery, and estimating drug

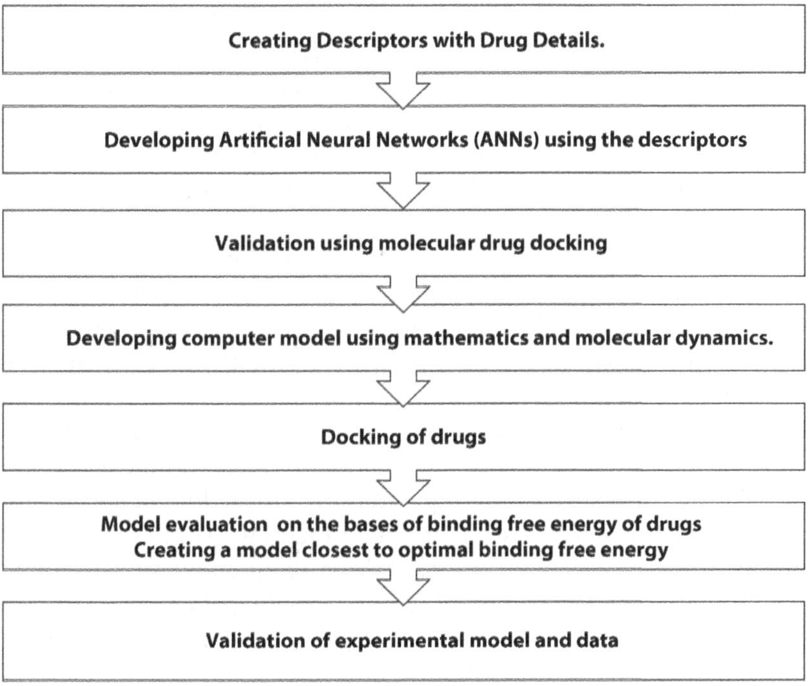

Figure 1.3 Flowchart of an optimized computational drug delivery system.

toxicity [27]. On the other hand, the tumor growth or therapy response was predicted using patient-specific multi-scale models [28]. Microfluidic models that enable restating the heterogeneous microenvironments of tumors and creating drug screening strategies in a patient-specific manner are reported [29].

Computational modeling is used often to design novel drug delivery systems like nanoparticles and better understanding of existing ones. In the field, modeled processes are becoming increasingly important. These processes include modeling the physical interaction between the nanoparticle drug delivery system and cancer cells as well as the release of medications at the tumor site. A powerful and essential set of computational techniques, including docking, molecular dynamics simulations, and QM/MM, is needed to explain the biological environment that becomes inert outside of the human cell. By offering targeted selectivity, the surface characteristic of functionalized drug delivery carriers would be significant in determining the effectiveness of the targeted drug delivery. The use of computational modeling tools can enable better design of drug delivery systems, particularly nanoparticles, by providing insights into the dynamic delivery process [30]. Research experts use multiscale computational methods to create a targeted medicine delivery. For reference, they built a few models of targeted drug administration within vascular settings and adopted a framework for multiscale modeling of the drug delivery system with formulas. Particulate modeling uses coupled Brownian adhesion dynamics, where it has also been shown that individual nanostructures move in and adhere to the blood [31].

The drug transportation and delivery mechanism in a dynamic biological system has been quantitatively described and in-depth examined using multi-scale modeling of the targeted drug delivery system. Such thorough data are highly helpful in estimating the dose quantity and toxicity as well as the effectiveness of a specific nanocarrier's delivery for a given vascular disease. The true physiological state of drug distribution must be accurately modeled computationally. The majority of recent studies concentrate on simulating nanoparticle binding and transport within an idealized vascular environment, such as straight and branching channels. The vascular geometry could be recreated from CT/MRI scanned pictures to replicate the genuine vascular environment. Under specific patient vascular geometry and hemodynamic parameters, a simulation technique can be used to estimate nanocarrier bio-distribution and the effectiveness of delivery. It can also be used to design and create nanoparticles for optimum targeting effectiveness and minimal drug dosage [30].

1.5 QC in Drug Delivery Modalities

A data-driven predictive system based on the ML technique has been utilized to evaluate the effectiveness of drug dose in drug delivery modalities by researchers. A quantitative structure–property relationship (QSPR) model was used to estimate the stability of the complexes produced by several cyclodextrins, including the well-known but poorly soluble antibiotic cefuroxime axetil. The recommended approach is relevant to structure-based drug design (SBDD), which generates leads suitable for clinical testing while being infrequent [32, 33]. By employing quantum computational modeling and simulations, it can study the principles of the relationships between medications and carriers as well as between drugs and biological molecules, and different drug delivery systems are displayed in Figure 1.4. The way that pharmaceutical companies administer pharmaceuticals may fundamentally change as a result of the computation of pharmaceutical science and computer-assisted medication delivery. To explore factors including dosage frequency, medication distribution, and drug release

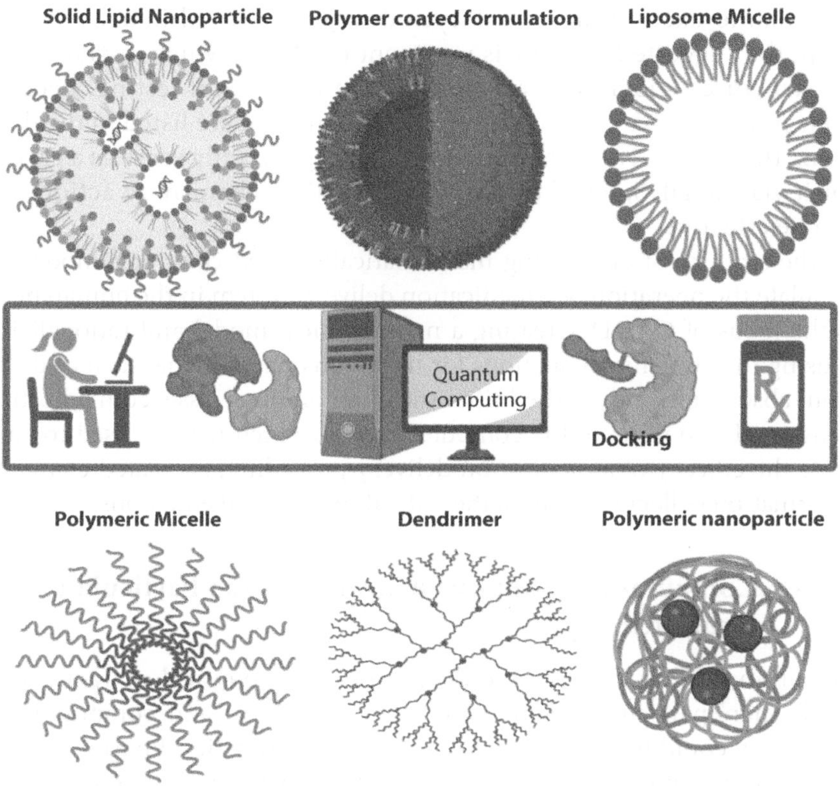

Figure 1.4 Drug delivery systems.

profile in patients, computational techniques can be utilized to generate models based on genomic data and cellular responses [34].

Repositioning or repurposing a medicine might reduce the effort and money involved in its advancement because various stages of that drug's development might be skipped [35]. Drug delivery methods have been researched recently and can be an important tool for repositioning [36]. However, as *in vitro* tests are laborious and time-consuming, experimental validation of DDS is not a simple process. Numerous computational strategies, including deep learning (DL) and ML techniques, have been proposed as solutions to these restrictions [9]. The efficiency of drug dosing in drug delivery modalities is determined by a data-driven predictive system built on ML techniques. Mizera *et al.* established a quantitative structure–property link in their study from 2020. Their group developed a QSPR to forecast the strength of complexed molecules generated by the widely used, insoluble antibiotic cefuroxime axetil and several cyclodextrins [32]. Although uncommon, the suggested strategy is pertinent to SBDD, which uses molecular structure to identify leads appropriate for clinical testing. Because their proposed material contains components that are listed in the drug bank, the micelle NPs that surround the drug not only serve as a carrier for it via the DDS, but they also have a dual mechanism of action that can be effective.

The strategy of developing mathematically based computer models to simulate the operation of a medication delivery system in the human body is the focus of CADD. Creating a mathematical model and rationalizing it using the governing equations for numerous phenomena, such as diffusion, absorption, etc., is the initial stage in computer-assisted medication delivery. The outline of the computer model, which may be used to simulate the efficacy of a medication delivery procedure, is created using the information collected from mathematical models and equations.

1.5.1 Computational Approach Towards Nano Particulate Drug Delivery

Quantum mechanics, MD, and drug loading are physical characteristics of the nanoparticulate medication delivery system and computational chemistry could be able to validate these characteristics. Pharmaceutical interactions and compatibility with solid dispersion could be predicted using mathematical modeling [37]. To examine the relationships between drugs and SLN, MD simulations are used to model the SLN first. According to Sun *et al.* (2014) [37] and Ramezanpour *et al.* (2016) [38], the developing concept of drug administration through nanoparticulate systems heavily relies on computer modeling to assess translational and distortion changes upon interaction with membranes, genetic material, etc. To promote gene targeting using nanocarriers, computational techniques can also be used to improve factors such as drug loading, drug release, distribution, drug–carrier interaction, permeability, and site specificity [39].

1.5.2 Computational Approach for Bone Drug Delivery

By using conventional methods, it is exceedingly challenging to deliver a medicine to the bone because of several restrictions, including drug dispersion and non-target specificity. Implants can be utilized to bypass these restrictions and induce localized effects, but it is highly challenging to conduct pharmacokinetics investigations using 2-D bone models that rely on hit-and-miss analyses [40]. The necessary information for creating an efficient drug delivery can be obtained through computer simulation: Zetos 3-D Bioreactor for drug delivery to trabecular bone, and Abaqus, a computer-aided engineering program, for mathematical simulation to obtain desired parameters in drug delivery system [41].

1.5.3 Computational Approach for Polymeric Drug Delivery

The topic of chemotherapeutics has tremendous potential for polymeric drug delivery methods. The main issue with use of nanoprecipitation to entrap drugs in nanocarriers is the polymers' efficiency. To achieve greater drug entrapment efficiency, one can make use of MD and other computational simulations based on mathematical equations. Drugs that would typically be very lipophilic are provided in preparation with certain other compounds known as additives rather than on their own. Blending screening experimentations with pattern classification and mathematical programming can be used to optimize this drug formulation, but MD simulation can also be an effective auxiliary tool in this context [42]. As an illustration, an analytical simulation model of cyclodextrin–drug combinations is frequently used as a method to enhance the delivery of medications that are not easily soluble [43, 44].

1.5.4 Computational Approach for Microsphere Drug Delivery

Advanced drug delivery systems featuring high precision over release kinetics are needed for next-generation therapies. Multi-particulate pharmaceutical delivery systems frequently include microparticles, microspheres, and microcapsules because they have therapeutic and technological benefits. In recent years, microspheres have been used for their site-specific features and controlled release. The microspheres are employed to deliver site-specific medications to the brain, liver, bone, etc., but it is exceedingly challenging to examine the *in vitro* release profile in 2D models. However, merging the existing mathematical models with the contemporary computational fluid dynamic models can aid in enhancing site specificity and controlled-release properties [45]. A simulated brain, liver, or bone model can be created using computational fluid dynamics in conjunction with mathematical models to conduct improved *in vitro* investigations. The model can assist in identifying the proper drug release profile as well as additional elements like local tissue binding and concentrated drug zones in the target organ. The architecture and other physicochemical characteristics of microspheres would be further reinforced by data methods, three-dimensional contouring graphics, and multiple linear regressions [46].

To choose the best manufacturing settings for the microspheres and produce the required drug concentration profile, it is helpful to have a reliable computational model of drug release from polymer microspheres. Using a mathematical modeling tool for the creation of paclitaxel microspheres, the primary principles influencing the pharmacokinetic profile throughout targeted paclitaxel administration were discovered. Intratumoral drug concentration was then quantitatively modeled and also discovered the medication release profile that enhances tumor cell killing. The main mechanisms governing drug transport inside a solid tumor are described by a reaction–diffusion model developed by researchers. Such processes include specific and nonspecific drug binding in the intracellular medium, drug diffusion and binding to drugs in the intercellular medium, drug clearing from the intercellular medium through ruptured microvessels, passive absorption of free interstitial drugs by the intracellular medium, and particular and nonspecific drug binding in the interstitial medium. Once extensively confirmed, a viable mechanistic model of this kind could be used to design pharmaceutical microsphere manufacturing as well as a tool for organizing research [47].

1.5.5 Computational Approach for Dendrimer-Based Drug Delivery

Dendrimers are immensely branched macromolecules wherein every branch is known as a generation. Molecule designing is more efficient and straightforward thanks to molecular mechanics and associated algorithmic techniques, which offer greater and wider molecular awareness. Developing innovative porters for both medicinal and diagnostic compounds can effectively take leverage of the polymerized dendritic polymer's repeating structure [48]. Computational designs support improved efficiency by enhancing vision and understanding of dendrimers and their interactions. Computational simulations based on MARTINI parameters can be used to design stable dendrimers [49]. The computational techniques can also be used to explore their interactions with other molecules, solvent and counter ions, and changes in pH [50]. Density functional theory was used to examine how Picoplatin, an anticancer medication, interacts with polyamidoamine dendrimers of the zero generation (G0). As per the dendrimer and medication structures, drug-loaded PAMAM dendrimers may produce two distinct kinds of polymer backbone that can engage with the medicines. AMD drug can be positioned within the PAMAM spaces using Cl and NH_3 ends. The outcomes showed that PAMAM dendrimer's association with picoplatin-based anticancer medications is physisorption-based. According to pertinent geometry, adsorption energy, molecular orbitals, and quantum molecular descriptor data, the dendrimer's core is perhaps the most stable place for drug content [51]. The creation of much more efficient drug delivery systems may be facilitated by the use of QC and ML for a faster examination of the actions of active compounds.

1.5.6 Computational Approach for Carbon Nanotube-Based Drug Delivery

The objective of the systemic delivery strategy is to provide a significant improvement for delivering medicine to a certain location purely based on mathematical parameters. Due to their extensive physicochemical properties and propensity to interact with biological membranes, carbon nanotubes (CNTs) have drawn a lot of interest during the past 10 years [52]. CNTs can be utilized in drug delivery for site-specific activity and to design a carrier based on CNT computer simulations, and these simulations can be used to determine its interaction with the lipid bilayer, drug, and other biomolecules [53]. The delivery systems process employing an SWCNT is simulated employing MD computations [54]. The use of drug carriers including single-walled carbon nanotubes (SWCNT), multi-walled carbon nanotubes (MWCNT), boron nitride nanotubes (BNNT), and graphene sheets—both modified and pure—could represent the next phase of therapeutic approaches. Utilizing these as biochemical carriers is made possible by both the characteristics associated with their specific configuration and the electrical characteristics. Numerous studies and experiments on the targeted release of therapeutics utilizing nanoparticles have been conducted over the years [55, 56]. The main goal of this research was to shed light into the relationship between carmustine and the GR enzyme by examining the persistence and bonding specificity of carmustine. This research can be expanded to consider the possible implications of computational methods in the redesign of upgraded drug carrier systems as well as the modification of existing medications to find additional targets. The two key components of the suggested delivery mechanism are targeting within the cell and ripping through the cell membrane. A simulation theorem and the appropriate magnetic flux composition both have been suggested for these two processes. A translocation profile of 1 to 100 and requirements for passing the cell membrane between 30 s and 5 min are provided by the program's architecture [57].

1.6 QC Applications in Pharma Industry

Finding and developing tiny entities and conglomerates to treat diseases is the primary job of pharmaceutical businesses. The industry is using QC as a natural candidate for molecular formations. Compared to traditional computers, QC can more accurately anticipate and simulate the characteristics, makeup, and interaction of such compounds. Presumably, QCs are capable of resolving all aspects of the issue, particularly quantum interactions. Pharmaceutical companies routinely process molecules with non-QC technologies like MD and DFT for computer-assisted drug discovery. To accurately forecast physicochemical structure, QC may create the most up-to-date CADD tools. The association of pharmacological compounds with important biological proteins can be predicted, which can be used to influence the design process in several different ways. When employing QC, investigators can simultaneously screen several potential target components against computational libraries. By employing ML techniques to find novel structure–property connections, QC may enhance the generation and validation of hypotheses. With regard to the size of the molecules, new kinds of drug-candidate libraries can be produced using QC technology. High-throughput methods are used to automatically check physiologically important domains versus drug-like compounds from a massive structural library [58].

The futuristic, hypothetical concept of quantum computing from science fiction is no longer valid. Although still in its infancy, this promising new technology is developing quickly and will soon play a vital role in changing drug discovery, development, and drug delivery [59, 60]. In fields of the pharma industry including target identification, drug delivery, disease diagnostics, toxicity analysis, logistics, literature research, and more, quantum computing might speed up any problem currently addressed with conventional computers while also opening up new applications. It is anticipated that a 1% to 5% improvement in the $1.5 trillion pharmaceutical research industry will increase the revenue by $15 to $75 billion and this attracts the focus toward QC [61]. Pharma industries started drafting their QC strategies that stand to benefit from the technology's early benefits.

Drug discovery and delivery were one of the main targets of using QC [61, 62]. A considerable expenditure of both resources, especially financial resources, is required to gain the quantum advantage. It is becoming clear that leaders in the pharmaceutical industry need to develop a quantum strategy to navigate what seems to be a more quantum future [15, 58].

It is prominent that the influence of every innovation is predicted to increase with the use of AI and QC together. The strength of AI-QC tends to make accessible a range of various pharmaceutical problems and their rationalization that have not been addressed previously due to a lack of suitable analytical tools, demonstrating the potential applications of these emerging multi-dimensional approaches [15]. Depending on the size of the quantum gear, algorithms will add different amounts of value. Small-molecule pharmaceuticals are typically described by less developed quantum computers, but biologicals will only be addressed as QC develops. Pharmaceutical businesses' primary activity is to find and create biological molecules and large molecules that may help treat illnesses and disorders. Pharma is a natural contender for QC given its emphasis on molecular structures. A large number of pharma startups are specialized in QC. Drug development is a primary concern for some, whereas QC is emphasized by others [15].

Pharma companies still need to conduct clinical trials after the medicine is discovered to confirm its effectiveness and determine how the human body reacts. Testing that takes place entirely virtually in a simulated environment, or *in silico* testing, is a novel sort of testing that QC can provide. With the use of *in silico* testing, it is simple to test medications on a variety of patients and big trial sets, and the tests can be accelerated to last even hundreds of years in a matter of days. For instance, *in silico* medicine created a new drug candidate using a simulated algorithm in just 46 days. QC is expected to shorten medication research and testing by thousands, if not millions.

Pharmaceutical companies are now able to compare hundreds of millions of complicated chemicals at once using QC. This allows researchers to look for matches between molecules and forecast both the beneficial and harmful impacts of a new therapeutic strategy. Results and design time are both enhanced. It will be interesting to watch which therapeutic sectors in the pharmaceutical industry will profit from QC technologies first. The effectiveness of the targeted drug delivery would be greatly influenced by the surface characteristics of functionalized nanoparticles, which would provide targeted selectivity. A computational modeling tool will provide insights into the dynamic delivery process, allowing for improved nanoparticle design [32, 33].

The data-driven study, artificial neural prognostication, and chemical monitoring will be used to link the life span or life cycle of drugs with AI and/or QC. This knowledge will be used to (1) effectively appreciate the implications of data processing, (2) streamline the experimental method, (3) find new drug targets as well as composites, and (4) strategize for or rethink upcoming pharmaceutical challenges. Due to a lack of suitable analytical tools, many pharmaceutical difficulties and their rationalization have not previously been addressed. The power of AI-QC makes these problems and their rationalization accessible, illustrating the breadth of potential applications of these emerging multidimensional approaches [62, 63].

1.7 Challenges or Prospects

To begin with, quantum computers are prohibitively expensive: A full QC system can cost several million dollars. While this is not a tremendous expense for many large pharmaceutical businesses, the fact that quantum computers are notoriously error-prone adds to the complexity. According to experts Scott Pakin and Patrick Coles in *Scientific American*, "quantum computers are extraordinarily difficult to engineer, build, and program." "As a result, they are hampered by mistakes like noise, flaws, and the loss of quantum coherence, which is critical to their functioning and effectiveness." The tendency to be excessively error-prone is, in fact, the biggest and most significant stumbling block in the way of QC. Quantum computers are extremely susceptible to environmental elements like vibrations and temperature variations, resulting in "quantum decoherence," which effectively renders the machine no more effective than a traditional computer. Decoherence becomes more frequent when quantum computers add more qubits, and faults within qubits cannot be fixed without destroying their quantum state, further complicating matters. We have gone a long way since Einstein dismissed quantum entanglement as "spooky action at a distance," as he put it. However, we still have a long way to go before we can claim a practical quantum advantage. QC and QML, on the other hand, contain the enormous disruptive potential for the pharmaceutical industry, with the potential to lead to the creation of novel, more powerful, and safer medications, as well as improved patient outcomes.

1.8 Conclusion

The use of quantum computers to complete a variety of tasks considerably more swiftly and efficiently than classical computers is one of the noteworthy discoveries brought about by enormous advancements in pharmaceutical technology. Rapid genome analysis and sequencing, early tumor recognition, the rapid discovery of new molecules, targets, or theranostic agents and evaluation of their behaviors, deeper comprehension of complicated data structures, mechanical failures, and the assimilation and analysis of massive amounts of data all benefited from QC. The use of QC and ML to accelerate the analysis of biological or medical data, identify the chemical processes that underlie the actions of drug candidates, and develop patient-specific treatment plans based on genomics information can lead to the creation of more efficient and less harmful medications. The significance of computational technologies in developing medications and formulations, as well as their drawbacks and potential remedies should be focused on by future research scientists. QC is particularly valuable in designing better drugs, drug carriers, and formulations and also to address issues that are difficult to be explored by humans. QC's potential to stimulate complex and more complicated molecules and the development of its drug-delivery technology could alter the pharmaceutical industrial landscape.

References

1. Drews, J., Drug discovery: A historical perspective. *Science*, *287*, 1960–1964, 2000.
2. Hughes, J.P., Rees, S.S., Kalindjian, S.B., Philpott, K.L., Principles of early drug discovery. *Br. J. Pharmacol.*, *162*, 1239–1249, 2011.
3. Harvey, A.L., Natural Products in Drug Discovery. *Drug Discov. Today*, *13*, 894–901, 2008.
4. Macalino, S.J.Y., Gosu, V., Hong, S., Choi, S., Role of Computer-Aided Drug Design in Modern Drug Discovery. *Arch. Pharm. Res.*, *38*, 1686–1701, 2015.
5. Zhang, S., Computer-aided drug discovery and development. *Methods Mol. Biol.*, *716*, 23–38, 2011.
6. Usha, T., Shanmugarajan, D., Goyal, A.K., Kumar, C.S., Middha, S.K., Recent updates on computer-aided drug discovery: time for a paradigm shift. *Curr. Top. Med. Chem.*, *17*, 3296–3307, 2017.
7. Frye, L., Bhat, S., Akinsanya, K., Abel, R., From computer-aided drug discovery to computer-driven drug discovery. *Drug Discov. Today Technol.*, *39*, 111–117, 2021.

8. Cao, Y., Romero, J., Aspuru-Guzik, A., Potential of quantum computing for drug discovery. *IBM J. Res. Dev.*, *62*, 6:1–6:20, 2018.
9. Hassanzadeh, P., Atyabi, F., Dinarvand, R., The significance of artificial intelligence in drug delivery system design. *Adv. Drug Deliv. Rev.*, *151–152*, 169–190, 2019.
10. Dilsizian, S.E. and Siegel, E.L., Artificial intelligence in medicine and cardiac imaging: harnessing big data and advanced computing to provide personalized medical diagnosis and treatment. *Curr. Cardiol. Rep.*, *16*, 441, 2014.
11. Parsons, D.F., Possible medical and biomedical uses of quantum computing. *NeuroQuantology*, *9*, 596–600, 2011.
12. Wang, L., Early diagnosis of breast cancer. *Sensors*, *17*, 1572, 2017.
13. Solenov, D., Brieler, J., Scherrer, J.F., The potential of quantum computing and machine learning to advance clinical research and change the practice of medicine. *Mo. Med.*, *115*, 463–467, 2018.
14. Perez, J.R., Kouroupis, D., Li, D.J., Best, T.M., Kaplan, L., Correa, D., Tissue engineering and cell-based therapies for fractures and bone defects. *Front. Bioeng. Biotechnol.*, *6*, 105, 2018.
15. Zinner, M., Dahlhausen, F., Boehme, P., Ehlers, J., Bieske, L., Fehring, L., Toward the institutionalization of quantum computing in pharmaceutical research. *Drug Discov. Today*, *27*, 378–383, 2022.
16. Blunt, N.S., Camps, J., Crawford, O., Izsák, R., Leontica, S., Mirani, A., Moylett, A.E., Scivier, S.A., Sünderhauf, C., Schopf, P. *et al.*, A perspective on the current state-of-the-art of quantum computing for drug discovery applications. *arXiv*, *2206*, 551, 2022.
17. Lam, Y.H., Abramov, Y., Ananthula, R.S., Elward, J.M., Hilden, L.R., Nilsson Lill, S.O., Norrby, P.O., Ramirez, A., Sherer, E.C., Mustakis, J. *et al.*, Applications of quantum chemistry in pharmaceutical process development: current state and opportunities. *Org. Process Res. Dev.*, *24*, 1496–1507, 2020.
18. Batra, K., Zorn, K.M., Foil, D.H., Minerali, E., Gawriljuk, V.O., Lane, T.R., Ekins, S., Quantum Machine Learning Algorithms for Drug Discovery Applications. *J. Chem. Inf. Model.*, *61*, 2641–2647, 2021.
19. Hassanzadeh, P., Atyabi, F., Dinarvand, R., Ignoring the Modeling Approaches: Towards the Shadowy Paths in Nanomedicine. *J. Control. Release*, *280*, 58–75, 2018.
20. Hassanzadeh, P., Atyabi, F., Dinarvand, R., Application of Carbon Nanotubes for Controlled Release of Growth Factors or Endocannabinoids: A Breakthrough in Biomedicine. *Biomed. Rev.*, *27*, 41–50, 2017.
21. Hedvat, M., Emdad, L., K. Das, S., Kim, K., Dasgupta, S., Thomas, S., Hu, B., Zhu, S., Dash, R., A. Quinn, B. *et al.*, Selected approaches for rational drug design and high throughput screening to identify anti-cancer molecules. *Anticancer Agents Med. Chem.*, *12*, 1143–1155, 2012.
22. Bravyi, S., Smith, G., Smolin, J., Trading Classical and Quantum Computational Resources. *Phys. Rev. X*, *6*, 021043, 2015.

23. Sutradhar, K.B. and Sumi, C.D., Implantable Microchip: The futuristic controlled drug delivery system. *Drug Deliv.*, *23*, 903579, 2016.
24. Mueck, L., Quantum reform. *Nat. Chem.*, *7*, 361–363, 2015.
25. Jain, M. and Kumar Chaturvedi, S., Quantum computing-based technique for cancer disease detection system. *J. Comput. Sci. Syst. Biol.*, *7*, 95–102, 2014.
26. Leelananda, S.P. and Lindert, S., Computational methods in drug discovery. *Beilstein J. Org. Chem.*, *12*, 2694–2718, 2016.
27. de Benedetti, P.G. and Fanelli, F., Multiscale quantum chemical approaches to qsar modeling and drug design. *Drug Discov. Today*, *19*, 1921–1927, 2014.
28. Li, Y., Stroberg, W., Lee, T.R., Kim, H.S., Man, H., Ho, D., Decuzzi, P., Liu, W.K., Multiscale modeling and uncertainty quantification in nanoparticle-mediated drug/gene delivery. *Comput. Mech.*, *53*, 511–537, 2014.
29. Cybulski, O., Garstecki, P., Grzybowski, B.A., Oscillating droplet trains in microfluidic networks and their suppression in blood flow. *Nat. Phys.*, *15*, 706–713, 2019.
30. Liu, Y., Shah, S., Tan, J., Computational modeling of nanoparticle targeted drug delivery. *Rev. Nanosci. Nanotech.*, *1*, 66–83, 2012.
31. Kaddi, C.D., Phan, J.H., Wang, M.D., Computational nanomedicine: Modeling of nanoparticle-mediated hyperthermal cancer therapy. *Nanomedicine*, *8*, 1323–1333, 2013.
32. Mizera, M., Muratov, E.N., Alves, V.M., Tropsha, A., Cielecka-Piontek, J., Computer-Aided Discovery of New Solubility-Enhancing Drug Delivery System. *Biomolecules*, *10*, 913, 2020.
33. Cho, T., Han, H.S., Jeong, J., Park, E.M., Shim, K.S., A novel computational approach for the discovery of drug delivery system candidates for COVID-19. *Int. J. Mol. Sci.*, *22*, 2815, 2021.
34. Herwig, R., Computational modeling of drug response with applications to neuroscience. *Dialogues Clin. Neurosci.*, *16*, 465–477, 2014.
35. Wishart, D.S., Feunang, Y.D., Guo, A.C., Lo, E.J., Marcu, A., Grant, J.R., Sajed, T., Johnson, D., Li, C., Sayeeda, Z. *et al.*, DrugBank 5.0: A major update to the drugbank database for 2018. *Nucleic Acids Res.*, *46*, D1074–D1082, 2018.
36. Lim, S., Park, J., Shim, M.K., Um, W., Yoon, H.Y., Ryu, J.H., Lim, D.K., Kim, K., Recent advances and challenges of repurposing nanoparticle-based drug delivery systems to enhance cancer immunotherapy. *Theranostics*, *9*, 7906–7923, 2019.
37. Sun, Y., Yu, B., Wang, G., Wu, Y., Zhang, X., Chen, Y., Tang, S., Yuan, Y., Lee, R.J., Teng, L. *et al.*, Enhanced antitumor efficacy of vitamin E TPGS-emulsified PLGA nanoparticles for delivery of paclitaxel. *Colloids Surf. B: Biointerfaces*, *123*, 716–723, 2014.
38. Ramezanpour, M., Leung, S.S.W., Delgado-Magnero, K.H., Bashe, B.Y.M., Thewalt, J., Tieleman, D.P., Computational and experimental approaches for investigating nanoparticle-based drug delivery systems. *Biochim. Biophys. Acta*, *1858*, 1688–1709, 2016.

39. Jämbeck, J.P.M., Eriksson, E.S.E., Laaksonen, A., Lyubartsev, A.P., Eriksson, L.A., Molecular dynamics studies of liposomes as carriers for photosensitizing drugs: development, validation, and simulations with a coarse-grained model. *J. Chem. Theory Comput.*, 10, 5–13, 2014.
40. Cardoso, L., Fritton, S.P., Gailani, G., Benalla, M., Cowin, S.C., Advances in assessment of bone porosity, permeability and interstitial fluid flow. *J. Biomech.*, *46*, 253–265, 2013.
41. Spencer, T.J., Hidalgo-Bastida, L.A., Cartmell, S.H., Halliday, I., Care, C.M., In silico multi-scale model of transport and dynamic seeding in a bone tissue engineering perfusion bioreactor. *Biotechnol. Bioeng.*, *110;* 1221–1230, 2013.
42. Mehta, C.H., Narayan, R., Nayak, U.Y., Computational modeling for formulation design. *Drug Discov. Today*, *24*, 781–788, 2019.
43. Huang, T., Zhao, Q., Su, Y., Ouyang, D., Investigation of Molecular Aggregation Mechanism of Glipizide/Cyclodextrin Complexation by Combined Experimental and Molecular Modeling Approaches. *Asian J. Pharm. Sci.*, *14*, 609–620, 2019.
44. Zhao, Q., Miriyala, N., Su, Y., Chen, W., Gao, X., Shao, L., Yan, R., Li, H., Yao, X., Cao, D. *et al.*, Computer-aided formulation design for a highly soluble lutein-cyclodextrin multiple-component delivery system. *Mol. Pharm.*, *15*, 1664–1673, 2018.
45. Heljak, M.K. and Swieszkowski, W., *In silico* model of bevacizumab sustained release from intravitreal administrated PLGA drug-loaded microspheres. *Mater. Lett.*, *307*, 131080, 2022.
46. Jabar, A., Madni, A., Bashir, S., Tahir, N., Usman, F., Rahim, M.A., Jan, N., Shah, H., Khan, A., Khan, S., Statistically optimized pentazocine loaded microsphere for the sustained delivery application: formulation and characterization. *PloS One*, *16*, e0250876, 2021.
47. Lengyel, M., Kállai-Szabó, N., Antal, V., Laki, A.J., Antal, I., Microparticles, microspheres, and microcapsules for advanced drug delivery. *Sci. Pharm.*, *87*, 20, 2019.
48. Chawla, M., Kaushik, R.D., Singh, J., Manila optimization and computational studies evaluating molecular dynamics of EDA cored polymeric dendrimer. *Sci. Rep.*, *10*, 21977, 2020.
49. Huynh, L., Neale, C., Pomès, R., Allen, C., Computational approaches to the rational design of nanoemulsions, polymeric micelles, and dendrimers for drug delivery. *Nanomedicine*, *8*, 20–36, 2012.
50. Martinho, N., Silva, L.C., Florindo, H.F., Brocchini, S., Barata, T., Zloh, M., Practical computational toolkits for dendrimers and dendrons structure design. *J. Comput. Aided Mol. Des.*, *31*, 817–827, 2017.
51. Farmanzadeh, D. and Ghaderi, M., A computational study of PAMAM dendrimer interaction with trans isomer of picoplatin anticancer drug. *J. Mol. Graph. Model.*, *80*, 1–6, 2018.

52. Tabari, S.H., Jamali, Y., Poursalehi, R., Multi-scale simulation of carbon nanotubes interactions with cell membrane: DFT calculations and molecular dynamic simulation. *Proc. Mater. Sci.*, *11*, 423–427, 2015.
53. He, X.C., Lin, M., Li, F., Sha, B.Y., Xu, F., Qu, Z.G., Wang, L., Advances in studies of nanoparticle-biomembrane interactions. *Nanomedicine*, *10*, 121–141, 2015.
54. Deb, A. and Vimala, R., Graphene mediated drug delivery - A boon to cancer therapy. *Res. J. Pharm. Tech.*, *10*, 1571–1576, 2017.
55. Li, Z., de Barros, A.L.B., Soares, D.C.F., Moss, S.N., Alisaraie, L., Functionalized single-walled carbon nanotubes: Cellular uptake, biodistribution and applications in drug delivery. *Int. J. Pharm.*, *524*, 41–54, 2017.
56. Begum, S.S., Das, D., Gour, N.K., Deka, R.C., Computational modelling of nanotube delivery of anti-cancer drug into glutathione reductase enzyme. *Sci. Rep.*, *11*, 4950, 2021.
57. Sohrabi, N., Alihosseini, A., Pirouzfar, V., Pedram, M.Z., Analysis of dynamics targeting CNT-based drug delivery through lung cancer cells: Design, simulation, and computational approach. *Membranes*, *10*, 283, 2020.
58. Evers, M., Heid, A., Ostojic, I., Pharma's digital Rx: Quantum computing in drug development. Available online: https://www.mckinsey.com/industries/life-sciences/our-insights/pharmas-digital-rx-quantum-computing-in-drug-research-and-development (accessed on 14 January 2024).
59. Lang, V., Quantum Computing, in: *Digital Fluency: Understanding the Basics of Artificial Intelligence, Blockchain Technology, Quantum Computing, and Their Applications for Digital Transformation*, V. Lang, (Ed.), pp. 51–111, APress Media, Berkeley, CA, 2021.
60. Thomas, L.B., Mastorides, S.M., Viswanadhan, N.A., Jakey, C.E., Borkowski, A.A., Artificial intelligence: Review of current and future applications in medicine. *Fed. Pract.*, *38*, 527–538, 2021.
61. Orwell, D., Quantum pharma, in: *Physiomics*, Oxford, United Kingdom, 2022.
62. Cova, T., Vitorino, C., Ferreira, M., Nunes, S., Rondon-Villarreal, P., Pais, A., Artificial intelligence and quantum computing as the next pharma disruptors. *Methods Mol. Biol.*, *2390*, 321–347, 2022.
63. Mishra, R., Mishra, P.S., Mazumder, R., *Computational Biology for Clinical Research*, in: *Bioinformatics Tools and Big Data Analytics for Patient Care* Edited by Malviya, R., Sharma, P.K., Sundram, S., Dhanaraj, R.K., & Balusamy, B. (1st ed.), p 240, Chapman and Hall/CRC., New York, 2022.

2

Quantum-Enabled Drug Discovery Process

Abhishek Rao[1], Deepika Kumari[1], Satyendra Singh[1], Ketan Kumar[1] and Vijay Kumar Prajapati[2*]

[1]Department of Biochemistry, School of Life Sciences, Central University of Rajasthan, Bandarsindri, Kishangarh, Ajmer, Rajasthan, India
[2]Department of Biochemistry, University of Delhi South Campus, Dhaula Kuan, New Delhi, India

Abstract
Quantum computational tools and their application in the field of drug discovery and design are new and advanced implementation of computer science. Old computational techniques are need to be improved to perform the task in less time and with precise accuracy on the lowest scale. A quantum computer uses a fundamentally advanced computational process to detect molecules and entities at the quantum level. Quantum computer applications are exact and accurate about their detection so that they can analyze minute differences and changes that classical computers cannot. Drug designing and discovery involves advanced application methods like the Fragment Molecular Orbital (FMO) method with the analysis of each fragment and all of the components, including bond energies, orientation, ligation energies, π–π stacking, halogen–π, halogen bonds, cation–π, hydrophobic interactions, and non-classical hydrogen bonds. The interacting molecular structure has different characteristic properties such as self-diffusion constant, root mean square displacement of atomic coordinates from beginning coordinates, radius of gyration, radial density profile (only for spherical boundary conditions), dielectric constant, and radial distribution function, which are some of the parametrical attributes that can be implemented for precise verification. Such characteristics can be sorted and are detected by specific quantum applications like MM and QM/MM (quantum mechanics/molecular mechanics) molecular dynamics-based free energy perturbation (FEP), QuanPol, QuanPol Vibrational Spectrum Program (QPVIB), Geometry optimization, CHARMM, AMBER or QM/MM/continuum solvent MD simulations. All of these, and many

**Corresponding author*: vkp@south.du.ac.in

Rishabha Malviya, Sonali Sundram and Dhanalekshmi Unnikrishnan Meenakshi (eds.)
Drug Delivery Systems Using Quantum Computing, (27–56) © 2024 Scrivener Publishing LLC

other quantum computing applications, can be used as a mainstay for drug design and discovery.

***Keywords*:** Quantum computing, drug discovery, quantum mechanics, molecular modeling, fragment molecular orbital, QuanPol, geometric optimization, molecular dynamics

2.1 Introduction

Drug designing and discovery is a challenging, big-budget, and high-risk venture in terms of finances. A new drug will take an average of 10-15 years to reach the market and will cost at least 2.5 billion dollars. The clinical phases are especially expensive, and end-moment failures, particularly in the second phase of clinical trials, are very common. As it was in the era of initial stages of development, pharmacokinetic (PK) qualities are a frequent reason for clinical drug failure, as well as exorbitant costs and last-time failures at the time of drug efficacy and safety testing phase in clinical trials [1]. Nowadays, late attrition is brought on by a lack of efficacy and safety data, as opposed to the 1990s when pharmacokinetic (PK) characteristics were frequent cause of clinical drug failure [2]. Both academic and commercial drug development can benefit from the use of bioinformatics-based chemistry, which enables researchers to study the molecular dynamics, structure, and energetics of protein–ligand interaction [3].

On the other hand, the accuracy and effectiveness of frequently used empirical methodologies are put to the test by the complexity of protein–ligand binding. An advanced level of research is being done on methodologies based on quantum computing techniques to accommodate more accurate description of molecular bonding. Understanding bonding including targeted proteins and their ligand is crucial for rationalizing potency and selectivity as well as identifying new drug targets [4]. The accessibility of molecular structural data for the targeted protein, as well as the reliability of computer tools for data analysis, can speed up the drug discovery process, increasing its efficiency and cost-effectiveness. Although some medications act as covalent binding inhibitors, present drug discovery and designing efforts are converging on weak interactions with ligands [2].

A molecule's ability to form non-covalent ligation compounds with protein or nucleic acid and then exert its biological activity can be determined by its ability to have non-covalent interaction at the binding site. In most cases, protein–ligand interactions are reported in the literature by well-known weak ionic interactions, such as classical hydrogen bonds, π–π stacking, salt bridges, hydrophobic interactions, and statistical bond testing of tiny molecule crystal structures. Protein–protein interaction through Cambridge Structural Database (CSD) deposited ligand in the Protein Data Bank (PDB) has revealed the presence of a number of previously unknown weak interactions [3]. With the help of QM techniques, one may accurately describe receptor–ligand interactions, estimate binding affinities, and predict bond formation and dissolution. For inducing charge transfer, electronic polarization, metal coordination, and halogen bonding, they expend a lot of computing time [5]. In the complex network of biological systems, hybrid QM/MM process approaches, which included precision of QM with way less time in the calculation of MM, are used to get the best out of both disciplines. QM is crucial for parameterizing force fields for novel target types, such as DNA and RNA, protein, and other ligands or targets.

Today, sophisticated algorithms and unparalleled processing capacity have made QM approaches an unavoidable element of the CADD (computer-aided drug designing) toolset. The quantum mechanical framework *de facto* accounts for a number of interactions or processes that are typically absent from force fields, such as halogen bonding, covalent bond formation, charge transfer, and electronic polarization. This also removes the need for system-dependent parameterizations, making it applicable across the chemical space [6]. In this chapter, some of the quantum structural molecular dynamics applications in the index of lead drug development bended toward those situations where QM is essentially utilized to compute minimally one of the energy factors. It is noteworthy that, over the past decade, there is an increase in interest in the development of QM-based drug design approaches due to developments in molecular modeling and the unceasing rise of computational capacity.

2.2 Usual Challenges in Drug Designing and Discovery

There are many difficulties and hurdles in drug discovery and designing, like-, finding a drug molecule for clinical trial and then granting approval for commercial production. There are several difficulties and obstacles along this winding, lengthy journey that can prevent a molecule from developing into a medication for therapeutic use. This fact can be comprehended by understanding how few molecules each year successfully navigate all of these obstacles to become a stable component of disease-fighting medication molecules. On an average rate, only 40–50 drug molecules are approved by the FDA according to a published magazine [7]. These molecules are very particular and significant to a single disease, which means the molecule is acting on a single type of disease and it cannot be used for any other disease. The majority of these recently identified drug molecules are related to cancer treatment and specific lifestyle diseases. Now, this is not the only problem in drug discovery and design, being time-consuming and costly, failure in the last stages of trials, and side effects are also some of the major hindrances.

The amount of time spent in this whole process is again a very serious issue in drug discovery and design. An approach using classical computational techniques may take several decades before finding a new druggable molecule. Using the *in silico* method, from several different libraries, several random and rational hits take hours or days to find a similarity in the active site and its target [8]. This is followed by the alignment and orientation of the target and receptor with their docking mechanism and then by the binding between the receptor and ligand. The nature of the bond and quantity of the different types of bonds are parameters for scoring molecular dynamics of a particular receptor ligand alignment and stability [9]. Here, sheer imagination and algorithms work together to predict the orientation and stability of a molecule with each molecule's area of limitation and effectiveness. This portion of the research study takes an immense

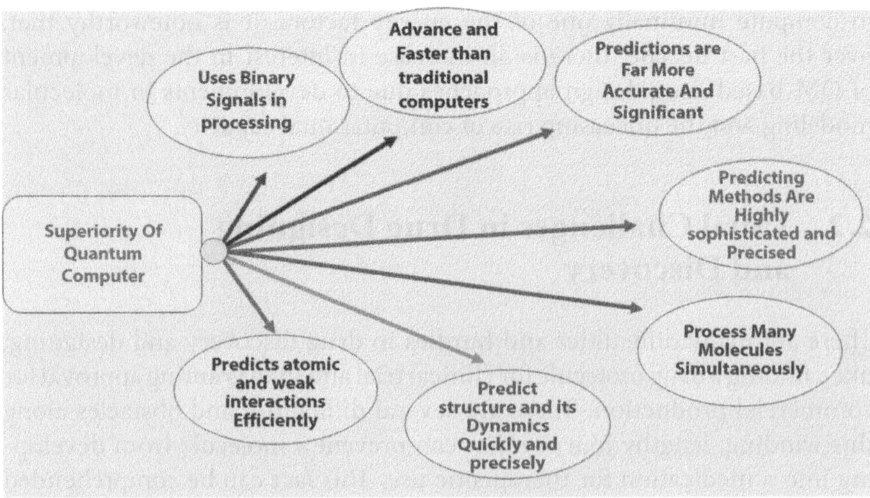

Figure 2.1 Superiority of quantum computational techniques.

amount of time because a classical computer cannot sort several libraries simultaneously and cannot accurately and precisely sort extensive and complex data sets (Figure 2.1).

2.2.1 Commercial Challenge

In recent years, there have been very few drugs, or none at all, that have been approved by the FDA for very generalized diseases like the common cold or flu or cholesterol-lowering agents like statin (mechanism like statin). Most of the drugs that have been approved are for diseases that affect a small population if we consider the entire global population as the set [10]. The majority of drugs recently developed are related to oncology and are therefore considered orphan drugs because of their particular target and therefore utilized by a small population [7]. The basic foundation of pharmaceutical industries is to collect significant revenue on the product, but when that consumable product is not meeting its mark and cannot generate the cost to which it is produced, it is very disappointing to the industry and makers [11]. The Pharma industry is also very determined to reducing the cost, especially because the product has to go through time patch, labor costing, clinical trials, and mass production and marketing. Now, all of these expenses should generate revenue, and an orphan drug molecule has a very small chance of generating that much revenue so the drug molecule [10] becomes very expensive and half of that small population is unable to buy that product.

2.2.2 Modification or Transitional Challenges

After investing a lot of money and time during *in silico*, *in vitro* and preclinical trials, the most crucial phase "clinical trials" has a high probability of the drug being rejected. This can be understood by realizing that only a very small number of molecules each year manage to get past all of these obstacles and join the permanent ranks of disease-fighting medication molecules. The reason behind the failure of drug molecules in this phase mostly is that the molecule acts sometimes differently in the human model than in the animal model [12]. Reasons may be undefined in this scenario like instigation and consequences of induced disease in the animal model, disease conditions are not exactly in the case of the human body, or maybe the pharmacokinetics can differ and the safety of the drug is compromised [13]. Thus, appropriate biomarkers, pharmacokinetics, and pharmacodynamics with ADMET and genetic evidence may be the reasons behind the failure of a drug molecule in clinical trials, which may cause a huge impact on the drug industry. The disease conditions in the animal model are not often induced satisfactorily. For example, the onset of depression condition in mice and humans is very different and behaviorally significant to both species. It means that the behavioral aspect of some conditions cannot validate its target [14]. *In silico* data analysis can only validate the target based on genetic evidence, which is the best-case scenario.

2.2.3 Unexplored Areas Under Classical Computational Techniques

Developing, designing, and repurposing drug molecules in pharma industries is a continuous process to fill the pipeline and meet medical requirements [15]. This vast field of research allows researchers to operate molecules on their limited terms. This propaganda requires sophisticated instruments and tools to identify and apply those molecules to become druggable molecules. With the help of classical computational techniques, this issue can be resolved up to a limit, but after that limit, researchers require more cutting-edge technology to go further and deeper into the world of molecules [16]. This picture can be understood by going a decade earlier in drug discovery and design; at that time, several easily detectable molecules were identified because they are evident and already known for their activities [17]. Thus, all of the low-hanging fruits have been discovered in the last decade or so, and now researchers require quantum tools and techniques to find better molecules, their binding, and molecular pathway more distinctively and prominently.

2.3 Medicinal Chemistry Through Quantum Mechanics

Understanding the interactions between ligand and receptor can be used to rationalize the potency and selectivity of small compounds. In this aspect, classical computational techniques and methods are unable to fully explain and theorize the structure of the target, visualization, and calculations based on the generated force field in molecular dynamics [18]. Quantum computational tools and techniques are very much able to address these shortcomings. For molecular recognition, noncovalent interactions are the primary targets. Several non-covalent interactions are present between molecules, but their concentration and energy field are low and dispersed, so a classical computer tool cannot recognize them as interactions, for example, π–π stacking, halogen–π, halogen bond, cation–π, hydrophobic interactions, non-classical hydrogen bond, etc. [19].

How active biomolecules interact and take actions on their target ligand can be understood by medicinal biochemistry and that can help in elucidating the structure–activity relationship, which may help in increasing the potency of the lead compound.

2.3.1 Underappreciation of Chemical Interactions in Protein–Ligand Complexes

Drug discovery and design is a multitasking process where a lead molecule must cross many barriers of potency, safety, pharmacokinetics, and toxicity profiles. In this regard, a 3D model of the target and receptors helps a lot and takes on the potency of the molecule. Based on the above statement, we may be able to say that computer-based programs of drug discovery and design are based on non-covalent interactions, whether small molecules may be a target or a recipient [19].

The molecules can form a number of non-covalent interactions and the majority of biological interactions are based on these non-covalent interactions. Scoring these interactions with algorithms and calculations based on bond energy provides us with enough information for validation of a particular target and receptor [20]. However, nowadays, we have run out of easily available drugs, so researchers like us require a more sophisticated and more parameters provide more vital information on the target for its activity on a particular target and random targets through the network analysis [21]. However, because the classical computer is not that precise and sensible to achieve that level of proximity with the molecule, quantum computational techniques, which are more reliable, have become useful [22]. In the next section of this chapter, we have tried to point out some of those interactions and how quantum computational techniques can help in this regard.

2.3.2 Non-Classical Hydrogen Bonding

In classical hydrogen bonding, the hydrogen atom is the bridge between the two electronegative atoms like nitrogen and oxygen, but in non-classical hydrogen bonds, CH acts as the bridge between the electronegative atoms, for example, (CH----O). The importance of this kind of interaction is that it is an integral part of stabilizing complexes related to ATP-competitive protein kinase, by forming a network of hydrogen bonds that provide a stabilized backbone hinge, and that stretch of hydrogen bond residues connects N and C lobe of kinase and anchors the adenine moiety of the ATP [23]. Classical computational techniques cannot identify this sort of interaction; it means the presence of carbon in the hydrogen bond [24]. Quantum computational techniques can point out the presence of carbon and the distance between electronegative elements of the hydrogen bond [25].

2.3.3 π–π Stacking

The electron system causes the attraction between the π-electron system of molecules and weak interactions. If this interaction is in the form of a stack, then we can identify it as π–π stacking. For example, benzene dimers are present in this form, which may be in a T-shape form, a sandwich form, or a parallel displaced form. The bond interaction energy lies between 1.8 and 2.8 kcal/mol [26]. Forming a stacking interaction between different moieties, benzene rings are arranged in a stacking manner, so to fulfill the gap, aromatic compounds are generally present in drug molecules [26]. As we already know, different aromatic rings have different properties, either physical or chemical, but in this context of solving the problem through a classical computer, these properties are usually underappreciated. Other than aromatic compounds, histidine with its imidazole ring engaged in classical hydrogen bonding and metallic cations, which coordinate cation–π interactions, hydrogen–π interactions, etc., is associated with π–π stacking. Visualization of these types of small interactions is not possible via classical computer tools, although they have an exciting role in scaffold replacement strategies based on their effect [27].

Interaction within a protein–ligand complex may form some π–π stacking interaction with the help of exposed amide group interactions, especially in trypsin-like proteins. However, this type of interaction is not between aromatic rings, so for them, we use π–π stacking involving a non-aromatic group. Another problem with π–π stacking is that it is also non-visualizable [28].

2.3.4 Hydrophobic Bond Interactions

Recognition of biomolecules depends heavily on interactions between hydrophobic bonds. With classical computational techniques, we can easily visualize and place hydrophobic interactions within the molecule but quantifying and evaluating its energy in the interactions of the molecule is not possible [7]. Quantum computational techniques, can evaluate and quantify hydrophobic interactions in molecular interactions, thus become valuable [29]. Hence, quantum computational techniques can help in the improvement of dispersion potential in the force field, and give a proper view of the overall stability and entropy of a molecule [30].

2.3.5 Coordination from Water

For defining and classifying purposes, the fragment molecular orbital (FMO) method detects cationic and anionic states and complex weak interactions within biomolecules in the mixture. Finding the orbital energies and inter-fragment interaction energies of water molecules exposed to biomolecular structure systems might be significant when utilizing FMO calculations [31]. Hence, we can see that adding the surrounding biomolecule is very much significant in calculations and future predictions. The above FMO calculation is an explicit approach for this purpose, and another one is the implicit approach in which solvent is taken as a continuum model [23].

2.3.6 Explicit Water

If there is a need to understand the role of water that surrounds biomolecules, and involves the formation of hydrogen bonds, then there is a transfer of charges, and now it is essential to involve explicit water. This causes electrostatic changes in charged solutes, which depend on the absence and presence of the water molecules. Water is also involved in the intramolecular interactions of the biomolecules [32]. The concept of shell solvent and crystal solvent appears in this aspect as how intriguing solvent is with biomolecules. Shell solvent is present from the surface of that molecule to its inner core up to almost 8 Å. In this 8 Å of solvent shell, there are two layers: the first layer of the solvent shell is up to 4 Å from the shell surface, which is involved in the solute's energy and charge of its surface, and the second or inner solvent shell is involved in the flexibility of the molecule related to its degree of freedom [33].

2.3.7 Implicit Water

Analyzing the molecule with x-ray crystallographic structure, the molecule is very important in predicting binding properties. As a component of the hydrogen bond network, the water molecule plays a role in ligand binding, according to FMO calculations [7].

2.4 Interaction Analysis

2.4.1 Fragment Interaction Energy

Inter-fragment interaction energy and decomposition analysis of inter-fragments are crucial ways to identify the mechanism leading to the main stability of the biomolecule. It can be explicitly said that solute–solvent interactions can be divided into FMO and decomposition energy [19]. Now, under the ideal condition, absolute electrostatic interaction between two-point charges can be screened out, which is called a solvent screening effect. This is essential regarding drug discovery and design, which quantum computational techniques can fully understand. The solvent screening effect can substantially reduce the excessive electrostatic interaction energy between changed fragments, which now hardly affects the surface charges [15]. If the two charged fragments are rarely exposed to the surface, solvent or neutral, then there will be a minimum effect on the IFIE or there will be no effect. In the case of protein target interaction, a fragment of a protein molecule cannot be uniformly exposed to solvent. Thus, simply put, we can say that the screening between two charged fragments is based on their charge, solvent exposure, and space between the molecules that are interacting [17]. In drug discovery and design, we have already focused on the interactions; the screening effect and decomposition energy of a molecule is a very promising concept through which a biomolecular interaction can be understood properly and applied in this field [34].

The idea of an energetic map is based on the screening effect and PIEDA (pair interaction energy decomposition analysis) [7]. This process will allow a researcher to identify the attractive and repulsive site of the binding site providing options for new binders to attach at different new sites.

2.4.2 Binding Free Energy

The fundamental goal of the drug discovery and design process is to determine the entire amount of intermolecular interactions that exist between the ligand and the target by accurately predicting the binding free energy of a relevant biomolecular analysis [5]. With the help of the quantum computational application FMO, it can be accurately determined.

There are several ways to find out this binding free energy with the help of FMO, such as an interaction-based approach in which solute is unpolarized and desolvation of the molecule is involved, and it is also cost-effective as well, whereas the supermolecule approach is moderately costly with the same feature [29].

2.4.3 Fragment Molecular Orbital (FMO) Process and Analysis

For rationalizing the potency and stability of a biomolecular interaction between a protein and a ligand, understanding their intramolecular interaction is very much necessary. The problem arises when with the already available structure of a molecule of interest, but the complexity of the interaction cannot be understood by classical computational techniques [35]. Hence, quantum computational techniques are exceptionally available to overcome these difficulties. Concerning quantum computing techniques, the ligand–target interaction occurs because the FMO is the primary application for combinational precision, speed, and the capacity to describe key interactions in determining the function of each small fragment in the combined molecular structure using FMO accessibility (Figure 2.2), which divides a big molecule into its smaller fragments [36]. The energy stabilization of the ligand and receptor fragment allows quantifying these interactions [29]. Now, we have the overall interaction energy as we sum up all the small fragment binding energies and predictions can now be made around the binding efficiency and stability of the molecular interaction.

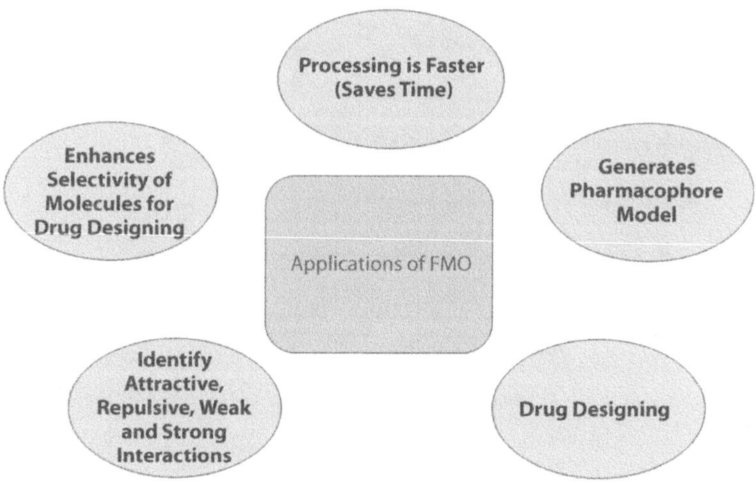

Figure 2.2 Applications of FMO.

FMO calculations are a universal algorithmic technique used in quantum mechanics that can be accessed on any atom, whether these atomic molecules are soluble or insoluble in form, like membrane-attached protein. For chemical and structural analysis and prediction, the FMO method is applied with the General Atomic Molecular Electronic Structure System (GAMESS) [37]. Here, FMO considers the self-consistent mutual polarization of the fragments considered and the charges that flow between the small fragments of the molecule. FMO calculations have many steps and very diverse forms of algorithms that have mathematical formulations, but some of the major steps in the progression of this process are described briefly, as the detailed process is challenging to describe in this chapter (Figure 2.3).

- **Fragmentation:** In this step, whole molecular atoms are assigned into a systemic fragment form.
- To fully understand intra-fragment charge transfer, all the fragments (fragment self-consistent field) are now calculated with other quantum effects.
- The third step accounts for all the inter-fragment charge transfer, i.e., fragment pair self-consistent field.
- In the last step of this process, all of the properties of that molecule are taken into account like energy, gradient, inter- and intramolecular forces, etc. [7].

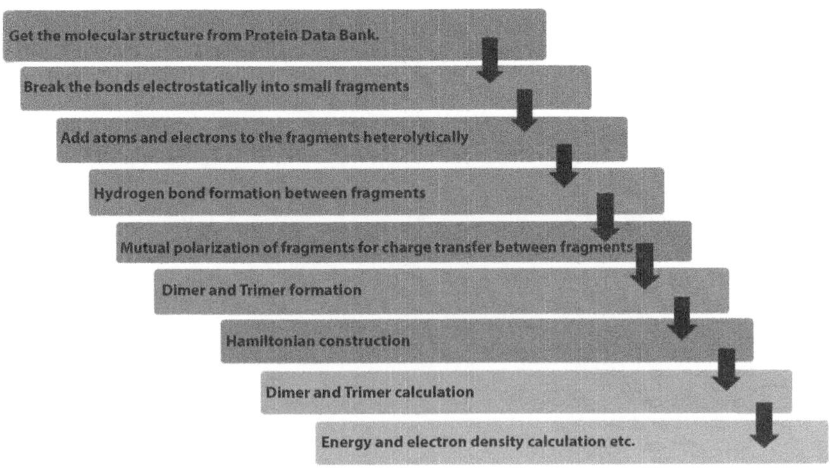

Figure 2.3 Workflow for FMO application.

2.5 Geometric Optimization

Most of the methods that can determine the molecular structure are based on energy derivatives for nuclear coordination, and that is energy gradient, with the possible lowest energy structure. In the past, molecular mechanics (MM) techniques have been used for structural reduction; more recently, however, developed versions of MM and quantum mechanics (QM) have gained their reputation in this field [38]. This is significantly approved for protein–ligand interactions since ligands may not have specific features or ligands, such as transition metals, which may not equate accurately inside force fields. A particle system is divided into small fragments in the FMO process, and each single fragment, known as monomer, and dual fragments, known as dimer, are subjected to QM calculations [28]. The overall properties are estimated using the many-body expansion from the monomer and dimer properties.

2.5.1 Analyzing Gradients

Constructing an estimated FMO gradient is possible by ignoring the response that arises in CP equations (specific heat equation). Numerous tests show that inaccuracies for this gradient are in an approximation of 103 hartree/Bohr. Geometry optimizations are necessary as a result of real life. A tight convergence criterion (about 103 hartree/Bohr) cannot be employed because of the residual error, if it is, the outcome will be erroneous [39]. Optimization frequently involves an oscillation of the geometry. The gradient and energy would never drop below the necessary level. This predicted gradient poses a severe problem for numerical techniques. The two-body level is where most FMO2 geometry optimizations are carried out. However, there are specific situations when the more accurate FMO3 technique should be taken into consideration. Due to an inaccurate valuation in mixture charge transfer, FMO2, for instance, accurately estimates the value of the coordination number for solvent cation or anion. Monomers, dimers, and trimers are computed in FMO3 [32].

Flexible systems do not convert their minimum energy stability structure in a single step. For simpler systems, an FMO optimization can be carried out using various techniques, such as RHF or DFT (including a few hundred atoms) [40]. Now, for a larger system, there are four methods (Table 2.1) [7]. Only a small portion of atoms may be expediently optimized. There are three ways to do this: standard FMO calculations, flexible domain calculation (which involves recalculating the entire system), and

Table 2.1 Summary of FMO methods used for a large system [7].

Method	Optimized area	Details
FMO-DFTB	Whole system	Require DFTB parameter
FMO/FD	Active site	Coordinates are frozen
FMO/MM	Whole system	Utilizes GAMESS/Tinker
FMO-HF3c	Whole system	Active sites may be optimized

Table 2.2 Geometry optimization of different processes [7].

Process	Description
The minimum energy crossing point	Crossing across surface
Saddle point location	Search for the transitional state
Intrinsic reaction coordinates	Mapping for reaction path
Minimum location	Energy minimization for process

non-flexible domain calculations (analysis of active site), but utilizing the latter is more effective (Table 2.2) [7]. Four geometry optimizations are distinguished: a conventional optimization and three strictly constrained optimizations.

2.6 GAMESS: A Computational Technique for Biochemical Simulations

2.6.1 Introduction to Biochemical Simulations

Modeling biochemical processes has been done using a variety of ways. Some of these strategies have been used over time. This chapter introduces the quantum chemistry package GAMESS and visualization tools that supplement GAMESS [41]. Calculation accuracy is provided via quantum mechanical (QM) techniques. The modeling of biological responses of the chemical system is appealing. There are many problems with conventional QM methods like Hartree–Fock (HF), density functional theory (DFT), and Miller–Plesset perturbation theory [42]. The computational time and memory requirements of a system expand super linearly with size, whereas they do not in a multiprocessor (MP) or connected cluster (CC). Therefore these are challenging to apply in biological models. Several approaches are quicker, like molecular dynamics force fields. To duplicate QM results at a high level, mechanics (MM) or more modern, cutting-edge approaches should be used [43]. In classical force fields, the potential energy of the system is roughly represented by the multiplication of parts, characterizing bound and non-bonded relations between atoms. It requires more processing to solve the Schrodinger equation in QM. The force field is used, and the parameter transferability significantly impacts MM accuracy [44]. For a process to work out, QM and MM can be combined in QM/MM procedures. The context behind this idea is to use the less expensive MM force field on the rest of the system while using a precise but expensive QM approach on a small active spot. Large active sites may be necessary for specific reactions, such as proton or electron movement in QM/MM. Additionally, accuracy suffers due to the QM/MM border [34]. The QM/MM methods include GAMESS, with QuanPol integrated GAMESS.

2.6.2 Parametrizing Quantum Mechanics Method for Simulations

Several parametrized QM techniques are included in GAMESS that can be used to duplicate either experimental data or sophisticated QM computations. Since there are no specific atomic types and covalent bonds can be formed and broken, parametrization is typically less constrained than force fields. Consideration is given to both polarization and charge transfer. Another method for accelerating QM computations is fragmentation [45]. Implementation of GAMESS for elongation method does not use fragments, whereas the system is gradually lengthened, yet a differently optimized bunch of orbitals are roughly equivalent with fragments (Table 2.3). This process is particularly suitable for DNA. Another method that works on the divide-and-conquer principle was used to study protein–ligand complexes [29]. Enzymatic reaction pathways have been mapped using EFMO.

2.6.3 Quantum Mechanics and Molecular Mechanics Associated with GAMESS

Integrated quantum and molecular mechanics (QM/MM) techniques use a precise but expensive quantum mechanics method to describe a small

Table 2.3 Parameters for processing in GAMESS for biochemical structural simulations [7].

S. no.	Operation	Accessibility	Functional attribute
1.	RMI	Accessible	Based on a simple semi-empirical method
2.	EFP1 (flexible)	Accessible	Based on the oldest generation application of significant fragment potential
3.	Density function tight bonding	Accessible	Density-based access to inter-fragmental bonding
4.	Harty-Fock with 3 corrections	Accessible	Latest algorithm-based program for fragmental application
5.	PM-6	Inaccessible	Also based on a semi-empirical method but is not accessible to allocate proper functioning
6.	Effective fragment potential 2	Accessible	The latest tool for evaluating effective fragment potential; advanced branch of EFP 1

crucial component of a molecule structural system that is an active site [36]. The remaining components of the arrangement are expressed by a quick molecular mechanics generated force field, striking a balance between process validation and speed. Now, the MM process of the major part of the system is sufficient, and the preciseness of QM/MM is collated to QM valuation for the complete arrangement. On the other hand, mounting up the QM/MM process is challenging and requires knowledge of the underlying chemical processes [28].

There are two different kinds of QM/MM approaches: addition and subtraction. The QM and MM components are disjoint in the additive process and the total energy of the system (E_{tot}) (Equation 2.1) is

$$E_{tot} = E_{QM} + E_{MM} + E_{QM/MM} \qquad (2.1)$$

The EQM/MM interaction occurs between the MM and QM regions, which have corresponding energies of EQM and EMM. The electrical embedding is described by the additive scheme [46].

The system is divided into several stacked layers using the subtractive method. The entire system is modeled using MM in the upper layer. The QM and MM approaches independently describe the active site in the deepest layer. The system's overall energy (Equation 2.2) is calculated using

$$E_{tot}^{OUTER} = E_{MM}^{OUTER} + (E_{QM}^{OUTER} - E_{QM}^{INNER})\ldots\ldots \qquad (2.2)$$

The labels "inner" and "outer" describe layers. By creating high- and low-level QM and MM zones (three layers), this technique can be simply extended to encompass more than two levels; hence, two to three layers are typically employed [19].

Implementations of subtractive QM/MM systems include the IMOMM, SIMOMM, and ONIOM strategies. The mechanical embedding is described by the subtractive scheme [12]. GAMESS, a QM tool with many uses, integrates MM techniques. Most QM/MM implementations use third-party MM software [22]. Packages are transported using interfaces, some of which are included in the software. GAMESS is available for use immediately [21] (Table 2.4). The CHARMM software was one of the first few that use MM to connect with GAMESS applied for additive QM/MM strategy [47]. One drawback of this interface is that it cannot be set up automatically and necessitates manual modifications to the GAMESS source code.

Table 2.4 Fragmentation processes in GAMESS for biochemical simulations [7].

Process	Functioning point	Accessibility
FMO	Fragment molecular orbital	Accessible
ELG	Elongation	Accessible
EFMO	Effective fragment molecular orbital	Accessible
SMF	Systematic molecular fragmentation	Inaccessible
MTA	Molecular tailoring approach	Inaccessible
DC	Divide and conquer	Accessible

2.6.4 Introduction to QuanPol

QuanPol can be merged with MMFF94 force fields, saddle point search, and harmonic frequency computation using HF, DFT, DFTB, GVB, geometry optimization, MCSCF, MP2, TDDFT, or TDDFTB wave functions (Hessian), as well as CHARMM, AMBER, MM, QM/MM, or QM/MM/continuum solvent MD simulations [6].

The complete energy of a molecule structural arrangement defined by QM/MMpol/continuum solvent (Equation 2.3) is estimated in QuanPol for the GVB, DFT, DFTB, HF, and MCSCF methodologies, and it can be shown as

$$\mathbf{E}_{\mathbf{QM/MMpol/C}}=\mathbf{E}_{\mathbf{QM}}+\mathbf{E}_{\mathbf{MM}}+\mathbf{E}_{\mathbf{chg}}+\mathbf{E}_{\mathbf{rep}}+\mathbf{E}_{\mathbf{disp}}+\mathbf{E}_{\mathbf{pol}}+\mathbf{E}_{\mathbf{sol}}\ldots\ldots\ldots\ldots\ldots \quad (2.3)$$

EQM calculates the QM region and represents the total QM energy (including nuclear repulsion) [48]. These methods are Hamiltonian, with MM atoms and a continuous solvent polarizing the wave function. The covalent potential terms and paired intermolecular interaction potential terms are both in EMM (non-polarizable charged interactions and Lennard-Jones/their derivatives) [49]. The symbol Echg represents the electrostatic interaction energy between QM electrons/nuclei and MM atomic point charges. Between QM/MM atoms, the interaction energies of repulsion and dispersion are Erep and Edisp, respectively. Epol is the induced dipole polarization energy of the MM atoms in QM/MMpol or QM/MMpol/continuum. In QM/MMpol/continuum, Esol is the induced surface charge-induced solvation free energy [50]. Due to the SCF provision of these QM methods, induced dipoles and surface charges are included in QuanPol's HF, DFT, DFTB, GVB, and MCSCF methods using another SCF methodology [2]. In SCF, calculations and electron density are answered repetitively and self-consistently.

After SCF combined with many quantum processes like QuanPol, wave function, induced dipoles, induced surface charges, induced dipole polarization energy, and induced surface charge solvation free energy are all obtained [51]. The variational technique is advantageous since it is simple to examine the analytical gradient. Among the various methods established using the QuanPol program is the QM/MMpol/continuum technique [7] (Table 2.5).

Table 2.5 Different QuanPol software and their functional attributes [7].

S. no.	Operation	Function in support
1.	Quantum mechanics	Applications related to quantum support like DFT, DFTB, MCSCF, etc.
2.	Molecular mechanics	Different applications related to force field and molecular structure like CHARMM, AMBER, etc.
3.	QM/MM	Application method for capping of H-atom
4.	Infrared spectrum	Applicable in the Hessian method for QM/MM
5.	Molecular dynamics	Functions can be applied as radial distribution and diffusion constant
6.	Geometric optimization	Methods like Hessian-guided, steepest design, and transition state search is applicable
7.	Solvation method	Applicable for continuum method
8.	Free energy simulation	For this method, sampling can be processed by free energy sampling and umbrella sampling
9.	Boundary condition	For this application on solvation conditions, one can use periodic, spheric, and open boundary method

2.6.5 QuanPol: Covalent Boundary Treatment

Using the hydrogen (H) atom capping method, covalent bond breaks between the QM and MM sections of QuanPol can be managed [10]. When a protein's C_α–C_β bond is broken, for example, an H atom is substituted for the C_α to saturate the side chain's open bond that is the QM region. In the QM/MM computation process, the C_α atom preserves its MM covalent properties, but not its force field charge, polarizability, Lennard Jones potential, and/or effective Gaussian potentials [51]. The total of the QM force on the H atom and the MM force on the C_α atom is the force on the C_α atom. The H and C_α atoms must have the same coordinate, carbon mass, force, and velocity in all calculations [46]. QuanPol automatically generates an H–C bond length of 1.5 Å using an effective Gaussian repulsive potential for capping the H atoms. All QM methods can be used interchangeably with the capping H atom method [11].

2.6.6 GO and Harmonic Vibration Frequency

Hessian vibrational analysis, pure MM and QM/MM geometry optimization (including transition state search), and QuanPol can attain frequencies and thermochemical properties. The MM process initiates with molecular structures that can be optimized and have their complete system's frequency calculated. However, when the molecular system is large, partial geometry optimization and the Hessian process can be carried out for a subset of the atom. For QM/MM systems, geometry optimization can be performed for 1000–2000 QM and MM atoms around the QM center [33], and local vibration frequency and thermochemistry can be determined using Hessian calculations for only 20–30 QM atoms [45].

2.6.7 Molecular Dynamics Simulation

QuanPol includes subroutines for evaluating various critical MD simulation features. In meaningful volume, periodic boundary conditions (PBC) or spherical boundary conditions (SBC) can be utilized for molecular dynamics study and prediction [47]. Properties of molecular structure like self-diffusion constant, root mean square displacement of atomic coordinates from beginning coordinates, radius of gyration, radial density profile (only for spherical boundary conditions), dielectric constant, and radial distribution function are some of the more parametrical attributes that can be implemented [52]. MD step checks out the dipole moment of user-selected MM atoms, the dipole moment of QM atoms, and the dipole moment of the overall QM/MM system. To create the temporal autocorrelation functions of these dipole moments and the related infrared spectra, an additional process that can be combined with this is the QuanPol Vibrational Spectrum Program (QPVIB) [44].

2.6.8 Free Energy Perturbation (Deviation of Energy) Simulation

QuanPol invariably supports the process with the help of MM and QM/MM molecular dynamics-based free energy perturbation (FEP). The Gibbs free energy change (ΔG) from state A to state B can be calculated from the internal potential energy change (ΔE) determined by MM or QM/MM processes (constant pressure and temperature) [37]. Two different states of any molecule can generally be distinguished from one another by molecular structure, electrical configuration, distinct types, and quantities of atoms. In quantum computational tools, FEP techniques can be used to determine the free energy profiles with its reaction mechanism, i.e., the PMF (potential of mean force), solvation and binding free energies, relative reduction potential, and acid–base ionization constant [40].

There are possibilities for single and dual topologies. MD simulation techniques are used to perform FEP computations. When the single topology technique is used, the MD simulation is carried out with just one set of atoms. Atoms are referred to as "perturbation atoms" sometimes and their force field properties are altered accordingly (for example, mass, charge [53], bond angle bending force constant, bond stretching force constant, Lennard–Jones potential, and others), and they may have the same cartesian product but the force field parameters are the same [14].

QuanPol is capable of calculating a wide range of distinct free energy changes. The difference between the charged interactions and Lennard-Jones interaction energies of the "perturbation atoms" and the other atoms can be explained as ΔE for a pure MM system [17]. The resulting free energy change is the one caused by the solution. Single and dual topological models are there to calculate the change in the solvent-free energy. Now, with this in addition to the solvation potential energy of the "perturbation atoms," the covalent and non-covalent potential energy can also be included in ΔE [52]. In this case, it is likely to evaluate the change in the free remaining energy caused by the perturbation atoms' geometrical configuration. The free energy that remains in the system is PMF [41].

2.6.8.1 QuanPol Process Umbrella Sampling

The process of finding relevant energies in the molecular structure starts with umbrella sampling, which QuanPol supports; the Weighted Histogram Analysis Method may be used to analyze 1D and 2D histograms and produce PMF curves using the "QuanPol Weighted Histogram Analysis Program" (QPWHA) tool (WHAM) [7].

2.6.8.2 QuanPol Process Thermodynamic Integration (TI)

By utilizing the TI sampling procedure in MD, QuanPol may determine average force, i.e., Boltzmann averaged interpretation among two atoms within the alignment of limited distances within a reaction range (e.g., 1.5 Å to 5.0 Å). The potential of mean force (sometimes referred to as the change in free energy) over the displacement is equal to the mean force times the displacement. It can be applied to MM and QM/MM MD simulations. Any combination of the two atoms might be MM or QM [7].

2.6.9 Setting Up Valuation and Calculation

Performing pure MM calculations before performing QM/MM calculations is necessary. Pure MM computations with QuanPol require a collection of molecular Cartesian coordinates. Using visualization tools, small molecules can be manually drawn or imported from a PDB file. QuanPol can create the force fields that the user requests using these Cartesian coordinates or PDB files. The generated MM force field can be applied to MD simulation and geometry optimization [17]. Proteins and nucleic acids can contain more than 10,000 atoms in their totality in an aqueous solution. The system should be brought into equilibrium via MD simulation [42]. MM/MD simulations combine free energy simulations, and geometrical changes are attained after equilibration. By converting a set of MM molecules into a QM molecule ($DATA group), it is possible to use QuanPol to calculate QM/MM [35]. $FFDATA and $DATA are both present, and QuanPol will automatically perform QM/MM molecular calculations. The QM computer analysis can be governed by the GAMESS algorithmic calculation formulas [49]. QuanPol automatically modifies original capping H atom properties for QM/MM bonded atoms (one or multiple) [15]. Normally, no human intervention is required, while the QM/MM settings can always be adjusted using the $QUANPO group keywords. For the QM/MM input file, several QM techniques are applied interchangeably with no modifications to the MM component or QM/MM setup [31]; through this formulation, the acquired results are very much equivalent.

2.6.10 Molecular Modeling and Visualization Software

There are many GAMESS input and output format files that can certainly visualize the molecular system, and some of them are presented in Table 2.6.

Three types of GAMESS visualization packages are defined. The first group of programs, Molden, MacMolPlt, ChemCraft, and Gabedit, are created specifically for use by quantum chemists, employing sophisticated quantum chemistry techniques to analyze small- to medium-sized chemical systems [2]. Among other things, these tools enable one to view molecule electrostatic potential, electron density, normal vibrational modes, and IR spectra [54]. This type of software is typically not made to run on

Table 2.6 Molecular modeling and visualization software with their details [7].

S. no.	Operation	Accessibility	Functional description
1.	Multibody FMO	Accessible	Utilizes a lower amount of time
2.	FMO 1	Accessible	Single-body expansion of FMO
3.	FMO 2	Accessible	Double-body expansion of FMO
4.	FMO 3	Accessible	Triple-body expansion of FMO
5.	HOP	Accessible	For accessibility to inter-fragmental boundary
6.	BDA and BAA	Accessible	For understanding, a bond-detached atom and bond-attached atom in a multi-molecular structure
7.	Adaptive frozen orbital	Accessible	For accessibility to inter fragmental boundary
8.	Polarizable continuum model	Accessible	Applicable through induced solvent charge
9.	Solvent model density	Accessible	Utilizes continuum solvent model
10.	FMO/effective fragment potential	Accessible	For explicit solvent model
11.	FMO/reference interaction site model	Inaccessible	Statistical solvent model
12.	FMO/MD	Accessible	Structural analysis for fragment-based model
13.	FMO/intrinsic reaction coordinate	Accessible	Analyses a reaction path

(*Continued*)

Table 2.6 Molecular modeling and visualization software with their details [7]. (*Continued*)

S. no.	Operation	Accessibility	Functional description
14.	FMO/frozen domain (dimer)	Accessible	Optimizes active sites based on dimer optimization
15.	Pair interaction energy	Accessible	Applicable with inter-fragment interaction energy
16.	Pair interaction energy decomposition analysis	Accessible	Analyses bounding energy of the disassembled fragments
17.	Hessian	Accessible	Analyzes second derivatives of the fragments to evaluate harmonic vibrational frequencies
18.	RAMAN spectra	Accessible	Evaluate fragment bonds and their types through normal RAMAN spectra
19.	NMR	Accessible	Analyzes shifts through nuclear magnetic resonance
20.	FMO/density optimizing state	Accessible	Analyzes the electronic structure of material through the density of states

massive computers. They are all used to generate GAMESS input files, except for Molden. Additionally, MacMolPlt supports FMO fragmentation, albeit to a lower extent [55, 56].

The next group of visualization programs includes Avogadro, VMD, PyMol, and WinMostar, which were all compensated with large molecule visualization. Domain-specific frameworks PyMol and VMD for advanced programming languages (TCL and Python) allow external software file formats through scripts or plugins. The benefit of this approach is that it enables the user to create new functionalities. These programs support GAMESS output files and are utilized to view some of the outcomes of quantum chemistry valuation. However, in order to use PyMol with GAMESS, programmed plug-ins are required [54].

Avogadro and WinMostar are examples of add-on software. Despite having a more limited capability than the second group, they are more valuable to quantum chemists [57]. Both Avogadro and WinMostar offer a visual GAMESS input file generator. In contrast with the previous software group, they are more suited to large-scale chemical systems and have powerful macromolecule creation and structural analysis capabilities [58].

The third class of visualization tools, which have improved fragmentation functionality and were created for FMO, includes Facio and FU. While FU has a well-organized biochemical system support, Facio also has a well-organized FMO input file creator that supports the majority of FMO extensions. Both programs allow for the visualization of FMO findings [59].

2.7 Conclusion

This chapter presents a demonstration of molecular docking applications based on quantum mechanics regarding drug designing and discovery, where at least one of the energy terms is explicitly calculated using QM. It is important to note that, in the last decade, there has been a spike in interest in developing QM-based drug design due to increased knowledge in molecular modeling and an inexorable rise in computing capacity. However, the hunt for small and simple molecules as new therapeutics continues because of their benefits of buccal cavity absorption (oral bioavailability), blood–brain barrier crossing and cell permeability, and comparably inexpensive production.

Artificial intelligence and the use of virtual assistants are currently transforming the drug discovery research process. These systems give a thorough overview of current SAR, evaluate trends, and might one day help with the development of novel compounds. Quantum computational techniques have some potential limitations because they rely on crystal structure accuracy. Uncertainty in measuring electron density maps and inaccurate model refining may cause a sizable portion of noncovalent interactions to be overlooked. Weaker contacts are anticipated to be impacted at least as much as relatively stronger connections, such as conventional hydrogen bonds. The recovery of previously undiscovered interactions can benefit from the application of QM/MM optimization. Although it has some limitations, a thorough investigation of molecular interactions provides its capability to be used in logical drug design for potential compounds. GAMESS was initially designed to model atoms and small molecules, but over time, it has expanded to incorporate techniques for managing huge systems. Fragmentation process like EFMO and FMO, as well as QM/MM, do not fall within this restriction, GAMESS has a hard-coded limit of 2000 QM atoms, which can be ascended by recompilation. Most FMO calculations allow for any size of atoms and pieces. There is undoubtedly a realistic limit in terms of memory and computing costs. The biggest FMO run to date (FMO/MD using DFTB) had over 1.2 million atoms. A program called GAMESS can be used to model biological processes. Using DFTB, the largest FMO run yet contained more than 1.2 million atoms. Modeling biological processes can be done using a tool called GAMESS. QM calculations can generate an accurate description of minute molecules and biomolecules, even if it increases complexity and processing time. This chapter investigated a variety of QM-based approaches to predict ADMET properties, certainly focusing on the reaction mechanism of drug metabolism, a challenging area for existing approaches.

Using QM in conjunction with MM (QM/MM methods) can also help improve knowledge of the mechanisms of cytochrome P450 substrate oxidation from the perspective of protein conjugation and recognition. It is also important to note the excellent outcomes attained by merging machine learning techniques with experimental data, not just QM-derived physiochemical descriptors. The electrostatic potential and solvation energies determined from QM is also important in *ab initio* logP calculations. As a result, new QM techniques for ADMET prediction will have a greater impact on the drug development process.

References

1. Chen, H. *et al.*, The rise of deep learning in drug discovery. *Drug Discov. Today*, 23, 6, 1241–1250, 2018.
2. Brogi, S. *et al.*, *In silico* methods for drug design and discovery. *Front. Media SA*, 8, 612, 2020.
3. Batool, M., Ahmad, B., Choi, S., A structure-based drug discovery paradigm. *I.J.M.S.*, 20, 11, 2783, 2019.
4. Bera, I. and Payghan, P.V., Use of molecular dynamics simulations in structure-based drug discovery. *C.P.D.*, 25, 31, 3, 2019.
5. Riccardi, L., Genna, V., De Vivo, M., Metal–ligand interactions in drug design. *N.R.C.*, 2, no. 7, 100–112, 2018.
6. Velmurugan, D., Pachaiappan, R., Ramakrishnan, C., Recent Trends in Drug Design and Discovery. *Curr. Top. Med. Chem.*, 20, 19, 1761–1770, 2020.

7. Heifetz, A., *Quantum mechanics in drug discovery*, Springer, New York: Humana Press, 2020.
8. Patel, L. *et al.*, Machine learning methods in drug discovery. *Molecules (Basel, Switzerland)*, 25, 22, 5277, 2020.
9. Mucs, D. and Bryce, R.A., The application of quantum mechanics in structure-based drug design. *E.D.D.*, 8, 3, 263–276, 2013.
10. Bova, F., Goldfarb, A., Melko, R.G., Commercial applications of quantum computing. *E.Q.T.*, 8, 1, 2, 2021.
11. Abramov, Y.A., Sun, G., Zeng, Q., Emerging Landscape of Computational Modeling in Pharmaceutical Development. *J. Chem. Inf. Model*, 62, 5, 1160–1171, 2022.
12. Sulimov, V.B. *et al.*, Docking Paradigm in Drug Design. *Curr. Top. Med. Chem.*, 21, 6, 507–546, 2021.
13. Ferreira, L.L. and Andricopulo, A.D., ADMET modeling approaches in drug discovery. *D.D.T.*, 24, 5, 1157–1165, 2019.
14. Stitou, M. *et al.*, Virtual screening of PEBP1 inhibitors by combining 2D/3D-QSAR analysis, hologram QSAR, homology modeling, molecular docking analysis, and molecular dynamic simulations. *J. Mol. Model*, 28, 6, 145, 2022.
15. Yilmazer, N.D. and Korth, M., Recent progress in treating protein–ligand interactions with quantum-mechanical methods. *I.J.M.S.*, 17, 5, 742, 2016.
16. Merz Jr, K.M., Using quantum mechanical approaches to study biological systems. *A.C.R.*, 47, 9, 2804–2811, 2014.
17. Kairys, V. *et al.*, Binding affinity in drug design: Experimental and computational techniques. *Exp. Opin Drug Discov.*, 14, 8, 755–768, 2019.
18. Adeniyi, A.A. and Soliman, M.E., Implementing QM in docking calculations: is it a waste of computational time? *D.D.T.*, 22, 8, 1216–1223, 2017.
19. Olsson, M.A. and Ryde, U., Comparison of QM/MM methods to obtain ligand-binding free energies. *C.T. C.*, 13, 5, 2245–2253, 2017.
20. Platania, C.B.M. and Bucolo, C., Molecular Dynamics Simulation Techniques as Tools in Drug Discovery and Pharmacology: A Focus on Allosteric Drugs. *Methods Mol. Biol.*, 2253, 245–254, 2021.
21. Saikia, S. and Bordoloi, M., Molecular docking: Challenges, advances and its use in drug discovery perspective. *Curr. Drug Targets*, 20, 5, 501–521, 2019.
22. Rifaioglu, A.S. *et al.*, Recent applications of deep learning and machine intelligence on *in silico* drug discovery: Methods, tools and databases. 20, 5, 1878–1912, 2019.
23. Ertl, P. *et al.*, Chemical Reactivity Prediction: Current Methods and Different Application Areas. *Mol. Inform.*, 41, 6, e2100277, 2022.
24. Christensen, A.S. *et al.*, Semiempirical quantum mechanical methods for noncovalent interactions for chemical and biochemical applications. *Chem. Rev.*, 116, 9, 5301–5337, 2016.
25. Cavasotto, C.N., Aucar, M.G., Adler, N.S., Computational chemistry in drug lead discovery and design. *I.J.Q.C.*, 119, 2, e25678, 2019.

26. Kulkarni, P.U., Shah, H., Vyas, V.K., Hybrid Quantum Mechanics/Molecular Mechanics (QM/MM) Simulation: A Tool for Structure-Based Drug Design and Discovery. *Mini- Rev. Med. Chem.*, 22, 8, 1096–1107, 2022.
27. Paton, R.S., Goodman, J.M., Hydrogen bonding and π-stacking: how reliable are force fields? A critical evaluation of force field descriptions of nonbonded interactions. *C.I.M.*, 49, 4, 944–955, 2009.
28. Ryde, U. and Soderhjelm, P., Ligand-binding affinity estimates supported by quantum-mechanical methods. *Chem. Rev.*, 116, 9, 5520–5566, 2016.
29. Liu, J. and He, X., Fragment-based quantum mechanical approach to biomolecules, molecular clusters, molecular crystals and liquids. *P.C.C.P.*, 22, 22, 12341–12367, 2020.
30. Rasool, R.U. *et al.*, Quantum Computing for Healthcare: A Review, *Future Internet*, 15, 3, 94, 2023.
31. Borbulevych, O.Y., Martin, R.I., Westerhoff, L.M., The critical role of QM/MM X-ray refinement and accurate tautomer/protomer determination in structure-based drug design. *J. Comput. Aided Mol. Des.*, 35, 4, 433–451, 2021.
32. Quinn, T.R. *et al.*, Automated fitting of transition state force fields for biomolecular simulations. *PloS One*, 17, 3, e0264960, 2022.
33. Cova, T. and Pais, A., Deep Learning for Deep Chemistry: Optimizing the Prediction of Chemical Patterns. *Front. Chem.*, 7, 809, 2019.
34. Muller, C., Rabal, O., Diaz Gonzalez, C., Artificial Intelligence, Machine Learning, and Deep Learning in Real-Life Drug Design Cases. *Methods Mol. Biol.*, 2390, 383–407, 2022.
35. Steinbrecher, T.B. *et al.*, Accurate Binding Free Energy Predictions in Fragment Optimization. *J. Chem. Inf. Model*, 55, 11, 2411–20, 2015.
36. Zhu, Z., Xu, Z., Zhu, W., Interaction Nature and Computational Methods for Halogen Bonding: A Perspective. *J. Chem. Inf. Model*, 60, 6, 2683–2696, 2020.
37. Lavecchia, A., Deep learning in drug discovery: Opportunities, challenges and future prospects. *D.D.T.*, 24, 10, 2017–2032, 2019.
38. Nero, T.L., Parker, M.W., Morton, C.J., Protein structure and computational drug discovery. *B.S.T.*, 46, 5, 1367–1379, 2018.
39. Lim, H. *et al.*, Drug discovery with fragment molecular orbital (FMO), in: *XXXVIII Symposium of Bioinformatics and Computer-Aided Drug Discovery*, 2022.
40. Maier, S. *et al.*, Comparative assessment of QM-based and MM-based models for prediction of protein-ligand binding affinity trends. *Phys. Chem. Chem. Phys.*, 24, 23, 14525–14537, 2022.
41. Chudyk, E.I. *et al.*, QM/MM simulations as an assay for carbapenemase activity in class A β-lactamases. *Chem. Commun. (Camb.)*, 50, 94, 14736–9, 2014.
42. Hornak, V. *et al.*, Comparison of multiple Amber force fields and development of improved protein backbone parameters. *Proteins*, 65, 3, 712–25, 2006.
43. Rezáč, J., Riley, K.E., Hobza, P., S66: A Well-balanced Database of Benchmark Interaction Energies Relevant to Biomolecular Structures. *J. Chem. Theory Comput.*, 7, 8, 2427–2438, 2011.

44. Singh, S. *et al.*, High-throughput virtual screening of small-molecule inhibitors targeting immune cell checkpoints to discover new immune therapeutics for human diseases. *Mol. Divers.*, 27, 2, 729–751, 2023.
45. Gorgulla, C. *et al.*, Emerging frontiers in virtual drug discovery: From quantum mechanical methods to deep learning approaches. *Curr. Opin. Chem. Biol.*, 69, 102156, 2022.
46. Cavasotto, C.N. and Aucar, M.G., High-Throughput Docking Using Quantum Mechanical Scoring. *Front. Chem.*, 8, 246, 2020.
47. Pandey, R.K. *et al.*, High-throughput virtual screening and quantum mechanics approach to develop imipramine analogues as leads against try- panothione reductase of leishmania. *BioMed. Pharmacother.*, 83, 141–152, 2016.
48. De Vivo, M. *et al.*, Role of Molecular Dynamics and Related Methods in Drug Discovery. *J. Med. Chem.*, 59, 9, 4035–61, 2016.
49. Thapa, B. *et al.*, Theoretical study of protein–ligand interactions using the molecules-in-molecules fragmentation-based method. *J. Chem. Theory Comput.,* 14, 10, 5143–5155, 2018.
50. da Silva, T.H. *et al.*, Using computers to ESKAPE the antibiotic resistance crisis. *Drug Discov. Today*, 27, 2, 456–470, 2022.
51. Watanabe, C. *et al.*, Development of an automated fragment molecular orbital (FMO) calculation protocol toward construction of quantum mechanical calculation database for large biomolecules. *Drug Discov. Today,* 19, 5–18, 2019.
52. Pandey, R.K. *et al.*, Febrifugine analogues as Leishmania donovani trypanothione reductase inhibitors: Binding energy analysis assisted by molecular docking, ADMET and molecular dynamics simulation. *J. Biomol. Struct. Dyn.*, 35, 1, 141–158, 2017.
53. Ma, Q. *et al.*, Scalable Electron Correlation Methods. 4. Parallel Explicitly Correlated Local Coupled Cluster with Pair Natural Orbitals (PNO-LCCSD-F12). *J. Chem. Theory Comput.*, 13, 10, 4871–4896, 2017.
54. Fedorov, D.G. *et al.*, Molecular Electrostatic Potential and Electron Density of Large Systems in Solution Computed with the Fragment Molecular Orbital Method. *J. Phys. Chem. A*, 123, 29, 6281–6290, 2019.
55. Senn, H.M. and Thiel, W., QM/MM methods for bio molecular systems. *Angew. Chem. Int. Ed Engl.*, 48, 7, 1198–229, 2009.
56. Mishra, R., Mishra, P.S., Mazumder, R., *Computational Biology for Clinical Research*, p. 240, Routledge Taylor & Francis Group, London, 2022.
57. Simoncini, D. *et al.*, Quality Assessment of Predicted Protein Models Using Energies Calculated by the Fragment Molecular Orbital Method. *Mol. Inform.*, 34, 2–3, 97–104, 2015.
58. Sladek, V. *et al.*, Protein Residue Networks from Energetic and Geometric Data: Are They Identical? *J. Chem. Theory Comput.*, 14, 12, 6623–6631, 2018.
59. Boniface, P.K., Sano, C.M., Elizabeth, F.I., Unveiling the Targets Involved in the Quest of Antileishmanial Leads Using *In silico* Methods. *Curr. Drug Targets*, 21, 7, 681–712, 2020.

3

Quantum Computing and Its Promise in Drug Discovery

Rakhi Mishra[1]*, Prem Shankar Mishra[2], Rupa Mazumder[1], Avijit Mazumder[1] and Shruti Varshney[1]

[1]Noida Institute of Engineering and Technology (Pharmacy Institute), Greater Noida, India
[2]Department of Pharmacy, School of Medical and Allied Sciences, Galgotias University, Greater Noida, India

Abstract
It is widely accepted that quantum computation and the science of quantum computation mechanics are excellent tools for drug design and discovery because of their high precision in estimating numerous relevant parameters. A scientist can use quantum computing to perform hundreds of millions of comparisons of complex molecules at the same time to examine matches between the molecules, and both the positive and negative effects of a new therapeutic approach can be predicted. These approaches make the most significant contribution to the development of drugs in clinical use or clinical trials. These methods emerge and develop in tandem with the experimental approaches used in drug design. Quantum Computational (QC) methods offer a highly precise description of protein–ligand interactions. This method defines nuclei and electron behavior and explains molecular interactions in terms of distribution and motion. Quantum computers have the potential to revolutionize computation by solving previously intractable problems. Quantum computers are extremely fast and efficient. They can perform calculations that would take decades or even millennia on today's supercomputers in a matter of seconds. Experts refer to this fact as quantum superiority over other traditional methods. Second, QCs can also be archetypal complicated structures that will allow researchers to understand the interaction of both small molecules and biologics before the preclinical and clinical stages of drug development. In practice, quantum computational methods will not be able to take over every task performed by classical computational methods—at least not anytime soon.

**Corresponding author*: rakhi.misra84@rediffmail.com

Rishabha Malviya, Sonali Sundram and Dhanalekshmi Unnikrishnan Meenakshi (eds.)
Drug Delivery Systems Using Quantum Computing, (57–92) © 2024 Scrivener Publishing LLC

However, in the future, these methods may be combined with traditional computational technologies such as machine learning (ML) and artificial intelligence (AI) to pave a new and groundbreaking path toward the goal of drug discovery and development.

Keywords: Quantum computing, computational techniques, quantum mechanics, drug discovery, drug design

3.1 Introduction

Quantum mechanics entails the use of quantum computing, which is employed to make significant advances in solving specific problems [1]. Quantum computing can be defined as the use of quantum mechanics to solve specific problems. Quantum computing (QC) is a fundamentally different approach to computing that is based on the laws of quantum mechanics [2]. Quantum computers are also able to sort issues in a matter of seconds that are so difficult to solve that even supercomputers struggle with them [3]. This accomplishment is referred to as "quantum supremacy." Physicist Richard Feynman noted in 1959 that as electronic components become smaller, quantum-mechanical effects that were predicted begin to manifest [4]. He proposed that these effects could be utilized in the creation of more effective computers, and he made this observation as early as 1959. Because of quantum processes such as "superposition" and "entanglement," these devices manage information in a manner that is fundamentally dissimilar to that of "classical" computers such as smartphones, tablets, or even the most powerful supercomputers available today [5].

The work that is done in pharmaceutical and chemical laboratories is undergoing a revolution as a result of the rise of computational technologies. Quantum chemistry serves as a useful tool in the investigation of the properties of active compounds as a result of the development of ever-faster and more accurate simulation algorithms [6].

Since the discovery of the first synthetic medication in 1869, the process of drug development has been considered a miracle, and computers have been shown to be of great assistance to the pharmaceutical industry throughout the process [7]. For this purpose, several different approaches including artificial intelligence and machine learning are utilized; nevertheless, these approaches are typically restricted to calculating molecules of a specific size or smaller. In order to circumvent this limitation and move forward with drug development, the area of biogenetics has begun incorporating quantum computing and quantum machine learning into its operations [8].

The phenomenon known as superposition is something that has piqued the curiosity of researchers using the tool of quantum mechanics, who are particularly interested in discovering ways to regulate it. The theory of quantum computers made enormous advances ahead during the 1980s and 1990s, well beyond Feynman's early expectations of what could be accomplished by such computers [9].

Quantum logic gates for a universal quantum computer were first proposed by David Deutsch of Oxford University in 1985. In 1994, Peter Shor of AT&T developed a quantum computer factoring algorithm that required no more than six qubits to solve a complex problem. Both of these achievements were monumental in the field of quantum computing. This approach enables certain computations to be performed significantly more quickly and efficiently than traditional computing does, and as a result, it holds a great deal of potential for use in the pharmaceutical industry. Leading pharmaceutical companies are forming quantum task forces and investing significant sums of money in order to investigate the potential applications of quantum computers and software in the fields of chemistry and biology [10]. These companies are aware of the potential game-changing nature of this approach. In order to provide a clearer explanation of what quantum computing is, we can say that it is the processing of information that is represented by unique quantum states [11]. The pharmaceutical industry devotes a staggering 15% of its sales to research and development, which is about 20% of the total R&D expenditure made by all companies operating in the world economy (Source). Such type of expenditure is for the concept of innovation. In their effort to improve the research and development process, pharmaceutical companies have been early adopters of the digital tools that computational chemistry has to offer, such as molecular dynamics (MD) simulations and density functional theory (DFT), for several decades [12].

Quantum mechanics, sometimes known as QM, is a helpful tool in the field of CADD study. The amount of time needed for chemical discovery and optimization can be significantly cut down with the use of high-throughput, *in silico* screening of ligand binding (also known as docking or QSAR). However, these quick approaches frequently lack the accuracy necessary for investigating the specifics of the binding mechanism. In the process of CADD, a rise in the accuracy of molecular simulations results in a harsh exponential growth in the amount of computation that must be performed [13]. The simulation of solids, molecules, atoms, nuclei, and subatomic particles might be rethought in a fundamentally different way if quantum computing becomes widespread. It might be able to help overcome the scaling limits of traditional computational approaches and make it possible to arrive at numerically precise solutions to the Schrodinger equation for big and complex molecular systems [14]. Quantum-mechanical calculations are used to describe the electronic behavior of atoms and molecules. Because of this, QM methods are suitable for the calculation of bond formation and dissociation energies, which is the site of failure for molecular mechanical calculations; however, QM methods are expensive to compute. Recent research and development in the pharmaceutical industry have included the application of artificial intelligence (AI). The next step in the evolution of computers is quantum computing [15].

Quantum chemistry has emerged as a useful technique for the investigation of the active compound's properties as a result of the development of ever-faster and more accurate simulation algorithms [16]. We anticipate that quantum computing will affect all stages of the drug development pipeline as quantum hardware continues to evolve in terms of both the quality of quantum bits (qubits) and the count and volume of quantum information [17].

3.1.1 Need for Quantum Computing

Up until this point, everywhere and every company has relied on supercomputers for the answers to difficult problems, although these machines are HUGE versions of traditional computers. They have thousands of huge CPU and GPU cores inside of them. The issue was that they were not effective in resolving a particular type of problem, which appears straightforward to the human mind. That is when the necessity for quantum computers started to become apparent [18]. What will we do, for instance, if there are 10 people at a dinner table and we need to figure out how to seat everyone so that everyone has a seat? At first glance, the problem appears to be a straightforward one involving permutations and combinations, which is precisely what it is. But once we receive the answer, we find that the number of possible permutations we have to check for just 10 persons is greater than 3 million. The majority of supercomputers are knocked out by this point in the game. The concept of quantum computing first emerged at this time [19].

The motivation for why we thought quantum computing was necessary is right in front of our eyes now:

1. The amount of working memory found in supercomputers is insufficient to store the innumerable real-world problem variants.

2. Supercomputers are required to perform a step-by-step analysis of each possible combination, which can be a time-consuming process [20].

3.2 Quantum Computing's Types, Applications, Generality, and Power

The Canadian business D-Wave has been successful in the development of the quantum annealer; nevertheless, it is difficult to determine whether or not it genuinely possesses any genuine "quantumness" at this point. In December of 2015, Google released test results demonstrating that its D-Wave quantum computer was 3,600 times quicker than a supercomputer at tackling specific, complex problems. These results lent legitimacy to the idea that a quantum computer could one day replace traditional computers [21].

These kinds of criticisms also shed light on the most significant limitation of quantum annealers, which is that they can only be designed to solve very specific optimization problems and have limited general applicability. This is a limitation that has been brought to light by the criticisms themselves [22]. The universal quantum is considered to be the holy grail of quantum computing since it has the potential to make calculations exponentially faster while also expanding their applicability. The construction of such a gadget, on the other hand, ends up presenting a number of significant technical obstacles. It was discovered that quantum particles are highly temperamental, and even the slightest interference from light or sound can cause errors in the computation process [23]. Calculating at an exponentially increasing rate is of limited utility if the results of those computations are inaccurate. There were 11 quantum computers in use worldwide as of June 2018. Some of these machines belonged to institutions such as IBM, Bristol University, and other universities in Australia [24]. There are 11 different quantum computers, but only three of them are necessary, and they are as follows:

3.2.1 Quantum Annealer

It is generally agreed that quantum annealing is the most effective method for solving optimization issues. The researchers are still looking for new approaches to investigate the various possible configurations of variables for the selection of the best technique. The use of quantum computing that is considered to be the least powerful and most specialized is known as quantum annealing. As a result, it is the quantum computer that is both the simplest and the least powerful to construct [25, 26]. When it comes to performing common computer operations like e-mail, video creation, gaming, and so on, the classical computer of today can surpass its quantum counterpart. On the other hand, the Quantum Annealer excels in its capacity to factor in extremely large numbers (thereby breaking encryption) and in its ability to find optimal solutions to different problems (like analyzing route data for mapping applications). Quantum annealing is the central processing method utilized by D-Wave Systems' quantum computer, which was one of the first commercially viable quantum computers to be announced by the company [27, 28]. Annealing is in its very early stages at D-Wave, and it is possible that in the future they will be able to harness this processing to give more qubits and, as a result, greater performance. A quantum annealer does not have any known advantages over traditional computing, according to the general opinion of the scientific community as a whole [29].

3.2.2 Analog Quantum

The ability of classical systems to solve the physics problems that analog quantum simulators help solve is orders of magnitude greater than what classical systems are capable of. Quantum simulators are capable of solving some of the most challenging problems in biology, such as modeling the folding of proteins. For example, proteins that fold incorrectly may cause the development of disorders like Parkinson's and Alzheimer's. Researchers will therefore be able to examine whether medications can be utilized for treatments with the assistance of a random computer model. Traditional computers of today cannot compare to their speed, and it offers great advantages in terms of computing. Using this technology paradigm, a number of firms are currently developing quantum computing systems [30]. These companies include Google, Microsoft, IBM Q, Rigetti, Google, Honeywell, and IonQ. This type of quantum computer is more challenging to fabricate, but it holds great promise for solving enormous problems in quantum chemistry, materials science, quantum dynamics, sampling challenges, and optimization issues. Using algorithms designed specifically for decrypting encrypted data, such as SHOR's or Grover's, the Analog Quantum computer is capable of breaching the currently accepted encryption standard. The Analog Quantum Computer will be the first type of computer to appear, and it will have performance levels that are exponentially higher than those of today's classical computers [31].

It will be possible for the analog quantum computer to simulate complicated quantum interactions that are intractable for any known conventional machine or combination of these machines. These interactions include quantum tunneling and quantum entanglement. It is hypothesized that the analog quantum computer will have anything from 50 to 100 qubits of storage space [32].

3.2.3 Universal Quantum

The Universal Quantum Computer possesses the maximum ability to solve problems but is also the most difficult to construct. For the performance of its function, a large amount of energy is utilized by this type and it consists of more than 100,000 qubits. In addition to the energy requirements, the Universal Quantum Computer requires cryogenic cooling to keep the temperature at an absolute minimum of -4 Kelvin so that it can function [33]. IBM is developing a chip for universal quantum computing, and the company is also conducting significant research and development in this area, focusing on both the hardware and software aspects of the problem. Additionally, businesses such as Google and Microsoft will be launching their own projects in this sector. Universal Quantum Computers have applications in areas such as safe computing, machine learning, quantum dynamics, challenges involving optimization, materials science, and quantum chemistry. It can decipher encrypted data even more quickly than any other type of quantum computer. In terms of power, performance, and capability, computing based on quantum mechanics is thus making quantum leaps forward [34–36]. The field is quickly developing into its mature state, although there are still many advances to come. Currently, businesses are developing the necessary hardware and software to lay the groundwork for the quantum computing revolution and its eventual implementation in our daily lives. The sector is through a very interesting period right now. This particular computer is thought to be the most powerful, the most general, and the most difficult to construct, posing a variety of significant technical obstacles. In a nutshell, this is the computer. According to the most recent projections, this device will have more than 100,000 qubits that are physically stored [37].

The universal quantum computer makes use of quantum mechanics to handle enormous amounts of data and carry out computations in innovative and potent new methods that are not currently feasible with the conventional computers that are available today. In the future, this kind of technological advancement could speed up the discovery of life-saving cancer drugs, unlock new capabilities in artificial intelligence through faster machine learning, or protect cloud computing systems from cyber-attacks entirely [38].

3.3 Drug Discovery and Quantum Computing

Throughout the whole of human history, the process of drug development has been one of the most difficult and labor-consuming processes in the field of pharmacology. Before a substance can even advance to the stage of undergoing safety testing and clinical trials, it must first pass a preclinical evaluation in which it is determined whether or not new chemical formulations have any biological activity and whether or not the intended therapeutic function is achieved by those formulations [39].

3.3.1 A Brief History of Drug Discovery

In the past, the process of drug discovery was much more reliant on serendipity as well as an older version of what we could now name Phenotypic Drug Discovery. Phenotypic evaluations of active components evaluate the effect of a drug from the outside in, but they do so without having an in-depth understanding of the substance's intrinsic mechanism of action. The plant world is the source of many medications, and the observation of the effects of these substances was the primary motivation for their discovery [40]. Knowing the mechanisms by which these substances worked was not even a consideration at the time. The anti-inflammatory salicylic acid, which is produced from willow bark, and digoxin, which is derived from foxglove, are just two examples of the many medications that are still used today that are derived from plants [41].

Even if phytochemical analysis is still considered a legitimate method of research in the present era, it should come as no surprise that this approach is not responsible for the vast majority of new drug discoveries. Just as salicylic acid was the precursor to an entire category of medications known as nonsteroidal anti-inflammatory drugs, so too has the process of drug discovery advanced to the point where it is no longer limited by the variety of plant species found around the globe. This procedure started with the development of chemical equivalents of existing medications that were already on the market [42–44]. This paved the way for mechanistically driven drug discovery, which entails the analysis of huge databases of chemical compounds that are known to interact with a biological process or pathway (target) of interest. Mechanistically driven drug discovery came about as a result of this. Although all three approaches are being utilized to a certain degree in the pharmaceutical industry today, the mechanistically driven approach to drug development is the one that requires the most financial investment and is also the most widely known [45].

3.3.2 Modern Drug Discovery

Combinatorial chemistry is a subfield of chemistry that emerged as a result of this approach to drug discovery, which profited from advancements in computing throughout the 20th and 21st centuries. In the field of pharmacological drug discovery, combinatorial chemistry works by first determining the targets on which substances act, and then categorizing the substances according to the mode of action they have on the target, which can be either positive (activators, potentiators, or agonists) or negative (inhibitors, blockers, or antagonists) [46]. Following this step, extensive chemical libraries of potential new medications are generated and then tested to see whether or not they are effective. When compared to real-time experimental investigations, doing analyses *in silico* through the use of computational chemistry is significantly more cost- and time-efficient [47]. This is because it is possible to perform studies with more precision. This does, however, come with some obvious drawbacks, such as the need to determine how difficult it is to synthesize compounds that may be found to interact with a target and the fact that actual screening must still be done for any candidates that are identified. Both of these things must be done [48, 49].

3.4 Role of Quantum Computing in Drug Discovery

Some of the ways how quantum computing helps in drug discovery are given below:

1. The primary mission of pharmaceutical companies is to discover and manufacture microscopic molecules and macromolecules that have the potential to treat a variety of illnesses and conditions [50]. Quantum systems, or systems operating on the principles of quantum physics, are molecules, including any that could be used in the production of medications [51].
2. It is anticipated that quantum computing will be more successful than conventional computers in the prediction of the

behavior or reactivity of drug molecules by understanding their structure [52]. When atomic-level interactions are complex, as they are in many compounds, exact procedures are computationally difficult for conventional computers, yet approximation methods are frequently lacking in adequate precision [53].

3. Quantum computers can effectively stimulate the entire issue, including interactions on the atomic level. As their capabilities increase, these quantum computers are going to become increasingly valuable [54]. Although quantum computing technology is difficult to comprehend intuitively, its significance could not be overstated: it will do certain sorts of computational tasks several folds faster when compared to traditional computers of today [55].

After it has fully developed, quantum computing may therefore be able to provide value throughout the entirety of the therapeutic value chain, from the stage of discovery through that of development, registration, and postmarketing [56].

3.5 Quantum Computing Methodology in Drug Discovery

Compared to the traditional method, the quantum method helps provide additional contextual information about shared characteristics between molecules that are being compared. Because of quantum, the scientists and researchers at Biogen were able to observe exactly how, where, and why chemical bonds matched. This offered superior insights and had the potential to assist and hasten the discovery of new drugs [57]. Quantum computing has the potential to revolutionize the healthcare business by enabling a wide variety of disruptive use cases for healthcare providers and health plans. These use cases include speeding diagnosis, personalizing treatment, and optimizing pricing. The application of quantum-enhanced machine learning algorithms is very pertinent to the industry [58].

The applications of the idea of quantum mechanics in the process of discovering new drugs and developing new drugs can be seen at various phases as tabulated in Table 3.1.

Table 3.1 Applications of quantum mechanism in drug discovery [59].

 **Research**	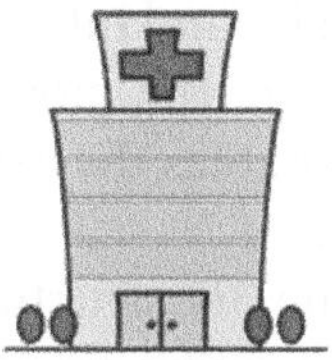 **Development**	 **Production**	 **Logistics and supply**	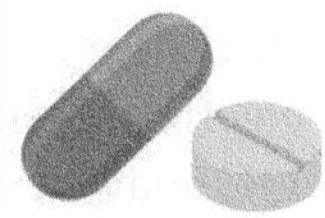 **Marketing**
Understanding the disease and developing a hypothesis	Identify patients and stratification	Calculate reaction rates	Network optimization	Advanced forecasting
Target finding	Patient pharmacogenetic modeling	Optimization of catalytic processes	Dynamic inventory/ warehouse/ procurement optimization	Patient understanding
Hit generation and identification	Site selection optimization	Product formulation		Tailored healthcare provider–patient engagement
Lead generation	Causality analysis for side effects	Quality monitoring		Automatic drug recommendations
Optimization of candidate properties		Predictive maintenance		
ADME activity and toxicity prediction for organ systems and other safety issues Dosing optimization				
Solubility optimization				

3.5.1 Target Identification and Validation

The selection of the computational quantum technique can be utilized to consistently anticipate the three-dimensional shape of proteins. The collection of good and sufficient information is a time-consuming process that, most of the time, ends in unsatisfactory output. AlphaFold and DeepMind were important steps forward in AI-driven protein folding, but it is still not possible to solve all of the issues that are present in traditional computing-based simulations [60]. These issues include the creation of protein complexes, interactions between protein and protein, and interactions between protein and ligand. Because the interactions are the most challenging to solve using traditional methods, QC, which enables the explicit treatment of electrons, could be of some assistance [61].

QC may be used to consistently predict the 3D structures of proteins, which can be useful during the target identification process. The collection of high-quality structural data is a time-consuming operation that frequently yields findings of poor quality [62]. Crystallization of many biologically significant proteins has not been achieved despite the best efforts of the research community. This could be because of the proteins' size, their solubility (particularly membrane proteins), or their inability to be expressed and purified in adequate quantities. When the business case for a given drug has the potential to be so strong, pharmaceutical companies will focus on the development of some drugs without even knowing the structure of a protein [63]. In subsequent phases of drug development, this requires the companies to accept the risk associated with a trial-and-error strategy.

3.5.2 Hit Generation and Validation

During both the hit generating process and the hit validation process, having QC that can handle complex phenomena in parallel would be quite valuable. Using the technology that is available today, pharmaceutical companies may only use CADD on drug candidates that are small to medium in size and must do so in a sequential manner [64]. Pharmaceutical companies would be able to extend the applicability of selected cases and undertake *in silico* search and validation studies at greater throughputs with sufficiently powerful QC. This use case would go beyond simply identifying the protein and would eventually embrace practically all of the biological world as it is currently understood. This stage would already offer prospective lead compounds that are more simple and faster to optimize if the hit-generation and validation approach used was robust enough [65].

3.5.3 Lead Optimization

QC may make it possible to improve absorption, distribution, metabolism, and excretion (ADME); obtain more accurate activity and toxicity estimates for organ systems; optimize dosage and solubility; and address other safety issues during lead optimization, which is one of the top three parameters that can be used to increase R&D productivity [66, 67].

3.5.4 Data Linkage and Generation

When it comes to research and development (R&D), the metalevel is primarily concerned with connecting relevant data, such as through employing efficient (semantic) management to establish logical linkages between data points [68]. When the readings and records during the drug research process are larger, the biological information that may be analyzed also becomes more complex [69]. QC could be utilized throughout the process of the study to generate bogus data by employing machine learning algorithms to make it appear as though there are missing data points [70]. This could be especially helpful in situations where there is a dearth of data, such as in the case of rare diseases; this dearth of data could then be alleviated by the deployment of manufactured datasets [71]. QC will establish a new standard in this regard with regard to the speed at which ML models are trained, the quantity of starting data that is required, and the level of accuracy [72].

3.5.5 Clinical Trials

It is possible that patient identification and categorization, in addition to population pharmacogenetic modeling, will be helpful in the improvement of clinical trials. When it comes to the planning and carrying out of trials, QC might make the venue selection process more effective. It is possible that QC may be employed as a supplement to causality analysis for adverse occurrences, which would help improve active safety surveillance [73, 74].

3.5.6 Molecular Formations

The primary focus of activities carried out by pharmaceutical businesses is the investigation and development of both small molecules and macromolecules with the potential to assist in the treatment of illnesses and diseases. Because of the industry's emphasis on the production of molecules, the pharmaceutical sector is a good choice for quantum computing [75]. Quantum systems, such as those that could be used in pharmaceuticals, include all molecules. This simply means that quantum physics principles underlie quantum systems. It is anticipated that QC will be able to more accurately analyze and simulate the structure, characteristics, and chemical nature of these molecules than traditional computing can [76]. For many compounds, atomic-level interactions are critical, but conventional computers cannot handle exact approaches, and approximate methods are often not accurate enough to be useful. Quantum computers have the potential, at least in theory, to effectively model the entire problem, including the interactions on the atomic level. The immense value that will be at stake will continue to increase along with the capabilities of these quantum computers [77].

Even though the technology that underpins quantum computing is rather challenging to comprehend straightforwardly (for an explanation of the fundamentals of quantum computing, it will be able to complete specific kinds of computational tasks at a rate that is exponentially higher than that of today's conventional computers) [78]. Therefore, once it has been perfected, QC has the potential to add value throughout the entirety of the drug value chain, from the stage of discovery through registration and postmarketing [79].

Computer-assisted drug development is a way that pharmaceutical companies use to process compounds. Some of these tools, such as MD and DFT, are not considered quality control tools (CADD) [80]. However, the traditional computers that they use have severe limitations, and the fundamental calculations that are required to precisely anticipate the behavior of medium-sized medicine molecules could take an entire lifetime to perform. CADD on quantum computers has the potential to expand the range of biological mechanisms that are amenable to CADD, cut down on the amount of time needed for screening, and cut down the frequency of whole phases of the development process which are required to repeat [81]. It could do this by removing some of the research-related "dead ends" that add a significant amount of time and cost to the discovery phase. Through its contribution to the accurate prediction of molecular properties, QC has the potential to make the existing CADD tools more efficient. This can have numerous effects on the research process, including the ability to predict how proteins fold and determine how potential drug candidates interact with biologically relevant proteins [82]. In this instance, QC may enable researchers to screen computational libraries against multiple conceivable target structures in parallel. Due to a lack of computational power and a lack of time, the approaches that are currently employed typically restrict the structural flexibility of the targeted molecule. Due to these restrictions, it is possible that the number of viable therapeutic candidates will decrease. Using machine learning (ML) algorithms to discover novel structure–property connections [83] may allow quality control to improve the generation and validation of hypotheses in the future. Once it reaches the required level of maturity, QC technology may be capable of developing new types of drug-candidate libraries that are not limited to small compounds but also include peptides and antibodies [84]. In these databases, small molecule library collections are not the only ones available. A large structural library of biologically relevant targets can be automatically screened against drug-like molecules using high-throughput methods. This could facilitate a more automated drug discovery process. Targets are selected based on their relevance to the drug in this method. QC may induce a paradigm shift in pharmaceutical R&D, moving beyond the digitally enabled R&D of today toward simulation-based or *in silico* drug discovery. This is a pattern that has also been observed in other industries. One could even envision QC causing such a shift. The following quality control use cases apply to many areas of drug discovery and are expected to occur at various points over a prolonged period. All of them, on the other hand, have the potential to make the development of targeted chemicals more precise and effective [85].

3.6 Examples of Companies Using Quantum Theory to Accelerate Drug Discovery

When the concept of modeling is considered for the chemistry of synthesis, organic, and medicinal nature as well as for different segments of the drug design process, the time-honored computational methodology known as molecular mechanics, abbreviated as MM, is the method of choice. On the other hand, MM approaches have a number of significant drawbacks, such as when they are applied to the investigation of electron-based features that are found in the interaction of drugs and receptors [86]. Quantum-mechanical (QM) procedures allow for a large increase in the accuracy of predictions and provide models of chemical and biological items and their interactions that are considerably more relevant; yet, QM methods are exceedingly (and frequently excessively) expensive to compute. Nevertheless, many developments that have taken place throughout the past few years have made it possible to expand horizons in this direction [87]. These developments include the introduction of the density functional theory (DFT), and also gain in computational power, leading to the introduction of distributed cloud-based computational infrastructures.

Given below is a brief overview of a few noteworthy computational companies, the majority of which are startups. These companies use quantum calculation methods for drug discovery, along with the application of the other techniques, and they promise to increase the positive graph of pharmaceutical research [88].

3.6.1 Aqemia

Aqemia is a spin-off company from École Normale Supérieure that uses innovative algorithms developed over the course of eight years of research directed by Dr. Maximilien Levesque, the company's CEO and co-founder. *De novo*, the structure-based design of lead-like compounds is the primary emphasis of the company, which utilizes both quantum and artificial intelligence in its work (AI). A one-of-a-kind algorithm based on quantum-inspired statistical mechanics can reliably predict the affinity between a drug and a therapeutic target 10,000 times more quickly than the other algorithms on the market. By receiving feedback from Aqemia's affinity predictor, the AI's ability to produce compounds with increasing accuracy can be improved. Aqemia was able to secure a total of $11.4 million in funding from several investors. A pilot study utilizing Aqemia's quantum physics-driven drug discovery technology was announced by Janssen and Aqemia in February 2022. The purpose of the pilot study was for Aqemia to predict the potency of small molecules for a specific target based on physics-based calculations using datasets chosen by Janssen [89].

3.6.2 Hafnium Labs

Hafnium Labs is a fledgling early research company. It was established in the Danish capital of Copenhagen in 2018. Engineers at a company are currently working on two high-precision simulation software packages for the physical properties of pure components and mixtures, and also for modeling electrolytes. Both Q-props and Epsilon, two different software programs, achieve a high level of accuracy in their predictions by combining the most recent discoveries and developments in quantum chemistry, artificial intelligence (AI), and cloud computing. Hafnium Labs can precisely anticipate chemistry through the utilization of its cloud-based computational capabilities, which speeds up the discovery of novel drugs, the development of new materials, and the execution of procedures. In addition to this, it employs a pay-per-use pricing strategy, which, in contrast to the typical license-based models used for goods of a similar nature, the company asserts offers more competitive pricing [90]. To date, Hafnium Labs has been successful in acquiring a total of $1.8 million in funding in the form of grants from the Innovation Fund Denmark, IBM, and the Climate-KIC Acceleration Programme (EIT).

3.6.3 Kuano

Beginning in the year 2020, Kuano was established as a business in the United Kingdom. This startup company is working on developing new artificial intelligence and quantum solutions for building enzymes. These solutions will address critical liabilities such as specificity, potency, and resistance. The research platform that Kuano uses incorporates quantum simulation, quantum-inspired artificial intelligence, and chemistry, and employs structural data on the enzyme or catalytic site that is the focus of the investigation. In addition to working with partners on developing next-generation inhibitors of clinically and commercially validated enzyme targets, the company has its internal research initiatives in areas such as epigenetics, protein degradation, immuno-metabolism, infectious illness, and others [91].

Kuano was able to secure grants and seed funding amounting to a total of $1.4 million from a variety of investors, including Innovate UK and O2h Ventures.

3.6.4 Menten AI

A software platform for the design of proteins is being developed by the Canadian startup Menten AI, which was established in 2018. The platform is powered by machine learning and quantum computing. The company makes use of its own patented algorithms for quantum optimization, which it says have the potential to dramatically increase the accuracy of medication discovery while simultaneously reducing both costs and the amount of time needed for its development. The researchers at Menten AI were the first to use a quantum computer to design a protein molecule, and they were also the ones who developed the first algorithm for constructing protein molecules that can be used on present and near-future quantum computers. According to the business, one design cycle can be completed in less than half a year, beginning with target selection and ending with *in vivo* efficacy testing. The Menten AI drug pipeline is now concentrating its efforts on developing peptide therapies for applications that have significant unmet medical requirements. Menten AI has established agreements with the D-Wave supercomputer, IBM-Q, and most recently, a newly announced collaboration with Xanadu. The company has secured a total of $4 million from a variety of investors [92].

3.6.5 Pharmacelera

Through their two core software packages, PharmScreen and PharmQSAR, the Barcelona-based computational startup Pharmacelera is implementing quantum theory in order to improve the design of pharmaceuticals. The first tool enables an accurate ligand-based virtual screening by making use of a high-precision 3D ligand-alignment method that is based on the interaction fields. In comparison to more traditional procedures and technologies, it has the potential to produce a greater variety of leads at a higher rate [93].

PharmQSAR is a 3D quantitative structure–activity relationship (QSAR) tool that makes it possible to do CoMFA and CoMSIA investigations by combining several fields of interaction.

Since its inception in 2015, the company has amassed a total of around $2.2 million in funding via grants and investments from pharmaceutical business angels. Here is an interview from the year 2020 with Dr. Enric Gibert, the co-founder and CEO of Pharmacelera. In it, he discusses the current state and potential future applications of quantum physics in the design of modern drugs [94].

3.6.6 PharmCADD

PharmCADD is a Southern Korea-based firm formed in 2019. The major technology platform of PharmCADD termed "Pharmulator" runs using deep neural network algorithms, molecular dynamic simulation, and quantum mechanics computation as fundamental technologies. Pharmulator can recreate 3D protein structures from amino acid sequences within just several seconds. The company cooperated with another Korean startup Genesen to explore medications for Multiple Sclerosis, and also secured an agreement with US biotech to discover new anti-coagulant treatments. Recently, the corporation made steps to grow into India, US, and France. The company raised a total of $22 million from SD Group and other investors [95].

3.6.7 Polaris Quantum Biotech

In Durham, there is a company called POLARISqb that has developed the world's first drug discovery software that was constructed specifically for quantum computers. This program combines artificial intelligence with a quantum approach. The TachyonTM drug design platform is at the core of the POLARISqb technology. This platform is utilized for the execution of distributed molecular design work across the cloud. This work is managed by an automated process that enables searching large chemical libraries while simultaneously running multiple projects in parallel. The company asserts that it can significantly quicken the process of designing drugs by creating proprietary software for quantum systems and that this would result in leads of a higher quality. It is possible for the TachyonTM system to be effective in treating a wide variety of diseases and indications because of its innate "agnosticism" [96].

POLARISqb and Fujitsu entered into a partnership in the year 2020 to jointly build a research platform with the goal of accelerating the development of medicines for dengue fever and other illnesses. In 2021, POLARISqb began working with Auransa on a joint project to address diseases that disproportionately affect women. A partnership between POLARISqb and PhoreMost was established in 2022 with the goal of utilizing the capabilities of quantum computing in the hunt for novel cancer treatments. To further the development of its platform, the company successfully raised a total of $2.1 million [97].

3.6.8 ProteinQure

ProteinQure is a startup company based in Toronto that was established in 2017. The company utilizes quantum computing, reinforcement learning, and atomistic simulations in the process of developing innovative protein therapeutics. They model fundamental processes, like the folding of proteins, as well as the underlying physics of the interactions between biomolecules by utilizing this diverse set of technological tools. ProteinQure is able to design small or other protein-based therapeutics including cyclic peptides, by utilizing its proprietary algorithms and the external supercomputing resources that are available to them. Additionally, they are able to investigate protein structures that do not have known crystal structures. The startup was able to secure funding from a variety of investors amounting to a total of $4.6 million, including Creative Destruction Lab and Felicis Ventures [98].

3.6.9 Riverlane

Riverlane, which has its headquarters in the United Kingdom, bills itself as "the world's first quantum engineering company" on its website. The company was established in 2016 in Cambridge, and has since raised a total of $24.1 million in order to improve its quantum hardware and software solutions for a large number of uses, including chemical and pharmaceutical research. In July of 2021, Riverlane collaborated with Rigetti Computing and Astex Pharmaceuticals, two companies that were already working together, to enhance quantum computing for the sake of drug discovery.

One of the challenges that currently face quantum computing is known as quantum error correction. Riverlane's operating system for quantum computers is known as Deltaflow.OS® and is a noteworthy project since it aims to solve this challenge [99].

3.6.10 Roivant Discovery

Vivek Ramaswamy established the publicly traded umbrella business Roivant Sciences in 2014. Roivant Discovery is the scientific arm of Roivant Sciences. Roivant Sciences' ticker symbol on the stock market is ROIV. Roivant Discovery was established through the acquisition of Silicon Therapeutics, a computational drug discovery company based in the city of Boston. Silicon Therapeutics was a developer of a research platform that combined physics-based molecular simulations, quantum physics, statistical thermodynamics, and molecular dynamics for the purpose of enhancing traditional drug discovery [100]. This led to the formation of Roivant Discovery.

Silicon Therapeutics has created its discovery pipeline of early small-molecule therapeutic candidates, and the company's primary focus was on innate immunity in the context of cancer and inflammation (4 in the discovery phase and 2 in the pre-clinical phase, as of June 2019). Additionally, the business was involved in the subject of conformational genetics, which seeks to establish connections between genetic mutations and biological activities.

Roivant Sciences completed the acquisition of Silicon Therapeutics in February 2021 for a total of $450 million in Roivant shares, in addition to potential further payments for regulatory and commercial milestones [101].

3.6.11 XtalPi

XtalPi is an artificial intelligence (AI) powered drug research and development company that was established in 2014 by a group of quantum physicists at MIT. The company's mission is to revolutionize the process of drug discovery and development by increasing the speed, scale, novelty, and success rate of the process. Although it is headquartered in Boston, the corporation has a significant presence in China both in terms of its operations and its research.

As a result of its Intelligent Digital Drug Discovery and Development (ID4) platform, it is now capable of high-precision prediction of small-molecule drug candidate physiochemical and pharmaceutical properties and crystal structures. These are essential components in successful drug research and development.

It was at this point that XtalPi had collected a stunning amount of $786.4 million from a variety of investors, some of which included Sequoia China, Tencent, and Google. As a result, it is now one of the most well-funded computational drug discovery firms available on the market. Pfizer is only one of the many pharmaceutical corporations that work together with this company on various research projects [102].

3.6.12 Zapata Computing

Boston-based Computing solutions for industrial and commercial usage are provided by Zapata Computing, which is a quantum software firm. These solutions can be used for a variety of applications, including pharmaceutical research. Zapata was established in 2017 and had its beginnings in the laboratory of Alán Aspuru-Guzik at Harvard, where he had been working on creating quantum computing approaches for chemical simulations. Zapata's origins may be traced back to that laboratory.

In total, Zapata was able to secure $67.4 million in funding from a variety of investors, some of which include Prelude Ventures and Comcast Ventures. Orquestra, the business's research platform, combines a robust software environment with quantum method libraries for use in fields such as chemistry, biopharma, machine learning, and optimization. This was accomplished through the construction of Orquestra by the company.

You can read or listen to our interview with Dr. Christopher Savoie, Co-founder and CEO of Zapata Computing, in which he discusses the current state of quantum computing, existing opportunities, and challenges, and provides his vision about what is coming next in the exciting field of quantum computing and quantum theory applications in drug discovery. Read or listen to the interview here [103].

3.7 Advantages of Quantum Computing

As opposed to conventional computers, which use bits that are either on or off, quantum computers use qubits, which can either be on or off, or both at the same time. This is a phenomenon known as superposition [104]. Because of this superposition, quantum computers can perform several calculations at the same time, making them significantly more efficient than computers that use classical technology as shown in Figure 3.1. Traditional computers are limited in the amount of data they can process in comparison to qubits. A computation that is done using a quantum approach can sample from a probability distribution of correctness because qubits use quantum-mechanical features to solve difficult equations in a probabilistic way [105]. Due to its combination of higher speed and probabilistic replies, quantum computing is best suited to a subset of computer needs and applications. Some examples of these applications and needs include optimization, chemical modeling, and artificial intelligence [106].

When compared to traditional computer methods, quantum computing offers the pharmaceutical industry a number of important benefits. In spite of the fact that the application of quantum computers is still in its infant stages, the pharmaceutical sector is through a period of revolutionary change as a direct result of the capability of quantum computers to

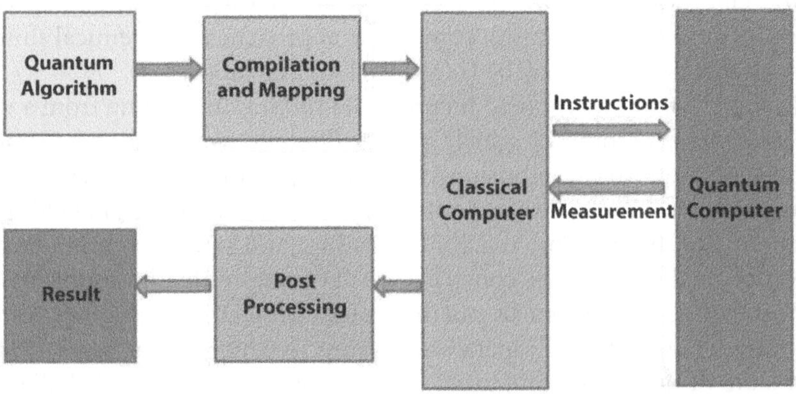

Figure 3.1 Working of quantum mechanics and quantum computer in compilation and processing of data.

significantly accelerate testing and enhance forecasting [107]. According to an article that was published in *Pharmaceutical Technology* in the year 2020, prominent pharmaceutical companies came together to form QuPharm to hasten the adoption of quantum computing in the industry [108].

One company that is working toward the objective of using quantum mechanics to speed up the discovery of new medications is called Quantum Brilliance. This company is based in Australia and Germany. Marcus Doherty, one of the company's co-founders, was quoted as saying, "It is possible to simulate a molecule with a quantum computer considerably more effectively than with a conventional computer because quantum computers follow the same laws that govern the behavior of molecules [109]. This means that quantum computers behave in the same way that molecules do. In order to run a simulation of two atoms on a quantum computer, you just need a certain number of qubits—say let's two—to carry out that simulation. On the other hand, in order to emulate a traditional computer, you would require at least four bits [110]. When compared to a quantum computer, this is an illustration of how the number of resources required by classical computer skyrockets whenever there is a need to simulate something quantum mechanical, such as a molecule" [111].

3.8 Applications of Quantum Computing in Drug Discovery and Development

The method of creating a new medication is one that is fraught with difficulty, fraught with risk, fraught with expense, and fraught with time. It is estimated that the traditional method of drug development, which culminates in the production of a new medicine that is suitable for sale, can take up to 15 years and cost more than one billion dollars [112]. Thankfully, this predicament has seen recent transformations because of the implementation of several novel technologies. The efficiency of the process of medicine development has been the focus of a significant number of recent innovations in both technology and methodology; among these, computational approaches have emerged as an essential component of a significant number of drug development programs [113]. In many discovery initiatives, methods such as ligand- or structure-based virtual screening are employed extensively at many different stages, from the detection of hits to the optimization of leads. To date, computational approaches have had a significant impact on the development of potential anticancer medications and drug candidates [114, 115] and provided new insights into the disease.

Designing potential anticancer treatments and drug candidates follows this pattern. In this article, we explore the concept of rational design and show some of the most representative examples of molecules discovered using it [116]. Research advancements and *in silico* drug design are used to demonstrate that new anticancer medications can be generated, and this is done through case studies that include successful results in the field of anticancer drug design [117].

3.8.1 A Future View of QC and Drug Discovery

Because of this, the development of quantum computing (QC) is such an exciting prospect for the sector. In contrast to classical computing, which can only store and process data in the form of bits that are either "on" or "off," quantum computing makes use of qubits, which can be either "on" or "off" or both at the same time [118]. In addition, the utilization of the benefits provided by quantum entanglement can be of great assistance to quantum computing. As a consequence of this, quantum computers are capable of giving orders of magnitude more computational power and speed than any classical machine is able to achieve.

The amount of computational power needed to simulate interactions between drug candidates and targets increases exponentially with the number of electrons in the compound in order to provide accurate simulations [119]. However, trade-offs can be made to boost simulation speed at the expense of accuracy. Even while this is a time-consuming process, it is nothing compared to the mammoth chore that this becomes when dealing with more complicated substances.

The use of quantum computing is another strategy that might be implemented to cut down on the amount of computer power that is necessary. In the field of drug development, artificial intelligence has already experienced enormous progress in both the realm of chemical simulations and compound synthesis, notably through the application of machine learning. This growth can be attributed to both of these domains. The amount of time that could be saved by the combination of machine learning and quantum computing would be phenomenal. Machine learning on its own can cut the amount of computer power needed for these sophisticated simulations by an exponential amount [120].

3.8.2 Current Developments in QC and Drug Discovery

Despite the fact that error-tolerant quantum processors are still some way off in the future, the field of quantum computing is already creating ripples in today's society. These processors would be able to repair errors quicker than they were being generated, and as a result, they would be able to endure the noise. On the other hand, pharmacology can anticipate enjoying the benefits of new technologies for a significant amount of time before that [121]. NISQ algorithms, which stand for noisy-intermediate scale quantum, are an aim that is significantly more attainable [122]. Roche formed a partnership with the British company Cambridge Quantum Computing (CQC) to develop and utilize NISQs in order to speed up Roche's research activities. Additionally, the CQC has built its very own platform for quantum computational chemistry; this platform is called as EUMEN [123].

ProteinQure, another firm that has achieved advancements in the field, has partnered with AstraZeneca and is one of the companies that has contributed to these advancements [124]. To find peptide-based pharmaceuticals and antibodies, which are both types of therapies that require extensive chemical simulations, the primary focus of ProteinQure's research and development efforts is directed toward the development of machine learning and quantum computing technologies [125]. The promise of quantum computing is that it will make it possible for pharmaceutical companies to develop unique and complicated structures that may be able to achieve goals that are not generally within the realm of possibility for conventional pharmaceutics. It is not anticipated that the cost of quantum computing would be particularly low, particularly while the technology is still in its infancy [126]. On the other hand, quantum computing and cloud computing can be coupled, which would make it possible for a greater variety of businesses to benefit from the capabilities of quantum computing and would also strengthen its adoption rates [127].

Similarly, the current cost of drug finding is around hundreds of millions to billions of dollars, with gargantuan investments required to bring a new drug to market, as well as a development time of ~10 years [128]. The average development cost of drugs (for which data were publicly accessible) approved by the FDA from 2009 to 2018 was $985 million. This figure increases exponentially for particularly complex treatments, such as those for cancer. As such, the benefits that quantum computing could bring to such a large market, in terms of both time and money, should not be discounted [129, 130].

While quantum computing may not yet be a mature technology, development in the sector is happening at a rapid pace. Pharmaceutical companies should take note to keep on top of these developments, as it is already making an impact on drug discovery and preclinical drug development [131–135]. The cycle of lead identification and optimization is shown in Figure 3.2.

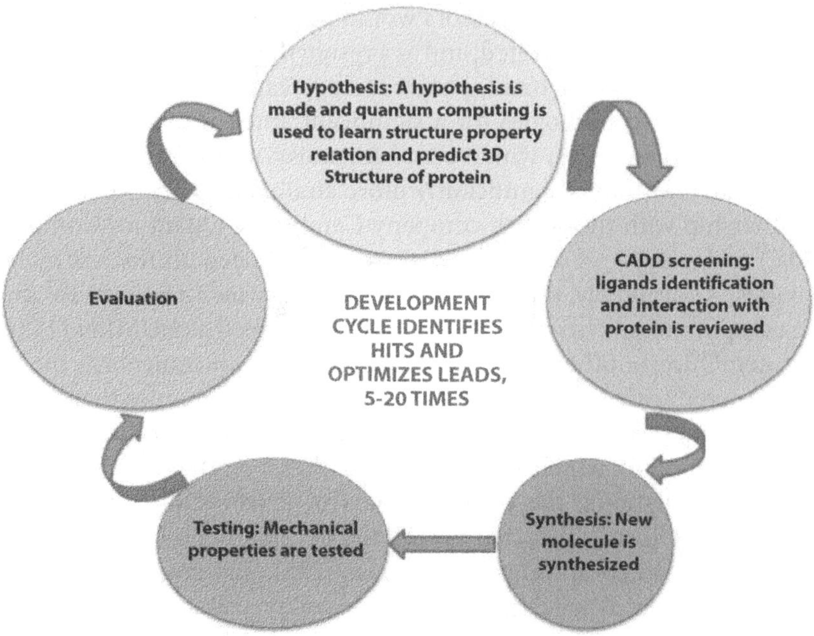

Figure 3.2 Whole cycle of lead identification and optimization [134].

3.9 Conclusion

It is likely that quantum computing will make existing CADD technologies more efficient by assisting in the more accurate prediction of molecular properties. This could be a result of the fact that quantum computing is more accurate. This can affect the development process in a number of different ways, including simulating the folding of proteins and the interactions of drug candidates with proteins that are physiologically relevant. Other ways in which this can affect the process include the following: In this particular case, researchers may be able to take advantage of quantum computing to test computational libraries in parallel against a variety of different conceivable target architectures. Because there is not enough processing power and time, the approaches that are currently available frequently limit the structural flexibility of the target molecule. This is because there is not enough processing power. Because of these limitations, it may be more challenging to determine which treatment options are the most effective. By utilizing machine learning (ML) strategies to discover new structure–property correlations, quantum computing may, in the long term, be able to facilitate an improvement in the process of hypothesis formulation as well as its validation. In the long run, this is something that could be accomplished. If the field of quantum computing reaches a particular degree of maturity, it is now a potent tool for the development of novel lead molecules that can play an eminent role in drug discovery and development. This capability, however, is contingent upon the field reaching the maturity level. In addition to very small molecules, the contents of these libraries may also comprise peptides and antibodies. It also imparts an advanced approach for a more automated approach to drug development, in which a huge structural library of physiologically relevant targets is automatically screened against drug-like molecules utilizing high-throughput technology. This would be done in the process of developing new drugs. This may speed up the procedure considerably. One could even imagine that quality control could be the impetus for a paradigm shift in pharmaceutical research and development, away from the technologically enabled R&D that exists today and toward simulation-based or *in silico* drug discovery. This would be a significant departure from the current state of the industry. This pattern has previously been spotted in a variety of different fields and industries.

References

1. Leelananda, S.P. and Lindert, S., Computational methods in drug discovery. *Beilstein J. Org. Chem.*, 12, 1, 2694–718, 2016 Dec 12.
2. Glassner, A.S., Quantum computing. *Comput. Graphics Appl.*, 21, 72–82, 2001 Nov-Dec.
3. Hughes RJ, Williams CP. Quantum computing: The final frontier?. *Intelligent Systems and their Applications*, vol. 15, pp. 10–18, IEEE, 2000 Sep.
4. Caraiman, S. and Manta, V.I., New applications of quantum algorithms to computer graphics: the quantum random sample consensus algorithm, in: *Proceedings of the 6th ACM Conference on Computing Frontiers*, pp. 81–88, 2009 May 18.
5. Suter, H.U., Marič, D.M., Weber, J., Thomson, C., Quantum Chemistry and Drug Design. *Chimia*, 49, 5, 125, 1995 May 31, Available from: https://chimia.ch/chimia/article/view/1995_125.
6. Yu, W. and MacKerell, A.D., Jr, Computer-Aided Drug Design Methods. *Methods Mol. Biol.*, 1520, 85–106, 2017, doi: 10.1007/978-1-4939-6634-9_5.
7. Narayanan, A., Quantum computing for beginners. *Evolutionary Computation, 1999. CEC 99. Proceedings of the 1999 Congress on IEEE*, vol. 3, 1999.
8. Svensson, F., *Computational Methods in Medicinal Chemistry: Mechanistic Investigations and Virtual Screening Development*, Acta Universitatis

Upsaliensis, Uppsala, 2015, ISSN 1651-6192 ISBN 978-91-554-9293-9 urn:nbn:se:uu:diva-259443. Available from: http://urn.kb.se/resolve?urn=urn:nbn:se:uu:diva-259443.

9. Svensson, F., Karlén, A., Sköld, C., Virtual screening data fusion using both structure- and ligand-based methods. *J. Chem. Inf. Model.*, 52, 1, 225–32, 2012 Jan 23, doi: 10.1021/ci2004835. - Dec 22. PMID: 22148635.
10. Lindh, M., Svensson, F., Schaal, W., Zhang, J., Sköld, C., Brandt, P., Karlén, A., Toward a Benchmarking Data Set Able to Evaluate Ligand- and Structure-based Virtual Screening Using Public HTS Data. *J. Chem. Inf. Model.*, 55, 343–353, 2015.
11. Svensson, F., Engen, K., Lundbäck, T., Larhed, M., Sköld, C., Virtual screening for transition state analogue inhibitors of IRAP based on quantum mechanically derived reaction coordinates. *J. Chem. Inf. Model.*, 55, 9, 1984–93, 2015 Sep 28.
12. Baldi, A., Computational Approaches for Drug Design and Discovery: An Overview. *Sys. Rev. Pharm.*, 1, 1, 332–337, January, 2010.
13. Nielsen, M.A. and Chuang, I.L., *Quantum Computation and Quantum Information*, Cambridge University Press, 2010.
14. Reggio, P.H., Computational methods in drug design: modeling G protein-coupled receptor monomers, dimers, and oligomers. *AAPS J.*, 8, 2, E322–36, 2006 May 12, doi: 10.1007/BF02854903, PMID: 16796383; PMCID: PMC3231557.\.
15. Izaac, J. and Wang, J., *Computational Methods in Quantum Mechenics*, vol. 13, pp. 2192–4791, Springer Cham, https://pages.vassar.edu/magnes/2019/05/10/computational-methods-in-quantum-mechanics/, doi: https://doi. org/10.1007/978-3-319-99930-2.
16. Szabo, A. and Ostlund, N., *Modern Quantum Chemistry: Introduction to Advanced Electronic Structure Theory, 2nd ed.*, Dover Publications, Courier Corporation, 1996.
17. Karton, A., A Computational chemist's guide to accurate thermochemistry for Organic Molecules. *Wiley Interdiscip. Reviews: Comput. Mol. Sci.*, 6, 3, 292–310, 2016. 10.1002/Wcms.1249.
18. http://www.physics.udel.edu.bnikolic/teaching/phys660/lectures/quantum_numeric.pdf
19. Hassanein, A.A. and Evans, M.W., *Computational Methods in Quantum Chemistry World Scientific Series in Contemporary Chemical Physics*, vol. 2, pp. 79–188, 1996.
20. Khandelwal, M. and Chatterjee, S.S., Quantum Computing: An Introduction. *Int. J. Sci. Eng. Res.*, 4, 5, 2293–2298, May-2013.
21. Gavrilov, Z., *Quantum Computation*, quant Comp.pdf (mit.edu).
22. Arora, S. and Barak, B., *Complexity Theory: A Modern Approach*, vol. 22, pp. 197–234, 2009.
23. Bais, A.F. and Farmer, D.J., Quantum Computation: The Physics of Information. *Handb. Philosophy Science, Philosophy Inf.*, 609–683, 2008,

ISSN 18789846, ISBN 9780444517265. https://doi.org/10.1016/ B978-0-444-51726-5.50020-0. https://www.sciencedirect.com/science/article/pii/B9780444517265500200).

24. Nielsen, A. and Chuang, I.L., *Quantum Computation and Quantum Information:10th Anniversary Edition*, 10th ed, Cambridge University Press, USA, 2011.
25. Zhou, Y., Tang, Z., Nikmehr, N., Babahajiani, P., Feng, F., Wei, T.C., Zheng, H., Zhang, P., Quantum computing in power systems. *IEnergy*, 12, 170-87, 2022 Jun.
26. Batra, K., Zorn, K.M., Foil, D.H., Minerali, E., Gawriljuk, V.O., Lane, T.R., Ekins, S., Quantum Machine Learning Algorithms for Drug Discovery Applications. *J. Chem. Inf. Model.*, 61, 6, 2641–2647, 2021 Jun 28, doi: 10.1021/acs.jcim.1c00166, Epub 2021 May 25. PMID: 34032436; PMCID: PMC8254374.
27. Zhou, T., Huang, D., Caflisch, A., Quantum mechanical methods for drug design. *Curr. Top. Med. Chem.*, 10, 1, 33–45, 2010, doi: 10.2174/156802610790232242.
28. Sliwoski, G., Kothiwale, S., Meiler, J., Lowe, E.W., Jr, Computational methods in drug discovery. *Pharmacol. Rev.*, 66, 1, 334–95, 2013 Dec 31, doi: 10.1124/pr.112.007336, PMID: 24381236; PMCID: PMC3880464.
29. Stenta, M. and Dal Peraro, M., An introduction to quantum chemical methods applied to drug design. *Front. Biosci. (Elite Ed)*, 3, 3, 1061–78, 2011 Jun 1, doi: 10.2741/e311, PMID: 21622114.
30. Ferrer, S., Ruiz-Pernía, J., Martí, S., Moliner, V., Tuñón, I., Bertrán, J., Andrés, J., Hybrid schemes based on quantum mechanics/molecular mechanics simulations goals to success, problems, and perspectives. *Adv. Protein Chem. Struct. Biol.*, 85, 81–142, 2011.
31. van der Kamp, M.W. and Mulholland, A.J., Combined quantum mechanics/molecular mechanics (QM/MM) methods in computational enzymology. *Biochemistry*, 52, 16, 2708–28, 2013 Apr 23, doi: 10.1021/bi400215w. Apr 12, PMID: 23557014.
32. Rogers, D.W., *Computational Chemistry Using the PC*, 3rd edition, Wiley, Hoboken (NJ, 2003.
33. Tannor, D.J., Introduction to Quantum Mechanics: A Time-Dependent Perspective. *J. Chem. Educ.*, 85, 7, 2006, doi: 10.1021/ed085p919.
34. Zhu, H., Tropsha, A., Fourches, D., Varnek, A., Papa, E., Gramatica, P., Oberg, T., Dao, P., Cherkasov, A., Tetko IV. Combinatorial QSAR modeling of chemical toxicants tested against Tetrahymena pyriformis. *J. Chem. Inf. Model.* 2, 48, 4, 766-84, 2008 Apr 28.
35. Jensen, F., *Introduction to Computational Chemistry*, 2nd ed., Wiley, Hoboken NJ, 1999.
36. Hehre, W.J., *A Guide to Molecular Mechanics and Quantum Chemical Calculations. Irvine (CA): Wavefunction*, 2003.

37. Dewar, M.J., Zoebisch, E.G., Healy, E.F., Stewart, J.J., AM1: A New General Purpose Quantum Mechanical Molecular Model. *J. Am. Chem. Soc.*, 107, 3902–3909, 1985.
38. Lipkowitz, B.K., *Reviews in Computational Chemistry*, vol. 22, pp. 167–217, John Wiley & Sons, 2015. ISBN: 978-1-119-62589-6.
39. Car, R., Introduction to density-functional theory and *ab initio* molecular dynamics. *QSAR Comb. Sci.*, 21, 97–104, 2002.
40. Gómez A, Leitao Á, Manzano A, Musso D, Nogueiras MR, Ordóñez G, Vázquez C. A survey on quantum computational finance for derivatives pricing and VaR. *Arch. Computat. Methods Eng.*, 29, 6, 4137–63, 2022 Oct.
41. Atkins, P.W. and Friedman, R.S., *Molecular Quantum Mechanics*, 4th ed, Oxford University Press, 2005.
42. Atkins, P.W. and Friedman, R.S., *Molecular Quantum Mechanics*, 3rd ed., Oxford University, 2005.
43. Zhu, H., Tropsha, A., Fourches, D., Varnek, A., Papa, E., Gramatica, P., Oberg, T., Dao, P., Cherkasov, A., Tetko, I.V., Combinatorial QSAR modeling of chemical toxicants tested against Tetrahymena pyriformis. *J. Chem. Inform. Model.*, 48, 4, 766-84, 2008 Apr 28.
44. Lewars, E.G., *Computational Chemistry*, Springer, Heidelberg, 2011.
45. Bao, G. and Suresh, S., Cell and molecular mechanics of biological materials. *Nat. Mater.*, 2, 715–725, 2003.
46. Lipkowitz, K.B., Abuses of molecular mechanics: pitfalls to avoid. *J. Chem. Educ.*, 72, 1070–1075, 1995.
47. Burkert, U. and Allinger, N.L., *Molecular Mechanics*, American Chemical Society, Washington, 1982.
48. Rappé, A.K. and Casewit, C.J., *Molecular Mechanics Across Chemistry*, University Science Books, Sausalito, CA, 1999.
49. Sauer, J. and Sierka, M., Combining Quantum Mechanics and Interatomic Potential Functions in *Ab Initio* Studies of Extended Systems. *J. Comput. Chem.*, 21, 1470–1493, 2000.
50. Menikarachchi, L.C. and Gascon, J.A., QM/MM approaches in medicinal chemistry research. *Curr. Top. Med. Chem.*, 10, 46–54, 2010.
51. Cho, A.E., Guallar, V., Berne, B.J., Friesner, R., Importance of accurate charges in molecular docking: quantum mechanical/molecular mechanical (QM/MM) approach. *J. Comput. Chem.*, 26, 915–931, 2005.
52. Marrone, T., Briggs, J.M., Mccammon, J.A., Structure-based drug design: computational advances. *Annu. Rev. Pharmacol. Toxicol.*, *37*, 71–90, 1997.
53. Raha, K., Peters, M.B., Wang, B., Yu, N., Wollacott, A.M., Westerhoff, L.M., Merz, K.M., Jr, The role of quantum mechanics in structure-based drug design. *Drug Discov. Today*, 12, 17-18, 725–31, 2007 Sep, doi: 10.1016/j.drudis.2007.07.006, Epub 2007 Aug 31. PMID: 17826685.
54. Peters, M.B., Raha, K., Merz, K.M., Jr, Quantum mechanics in structure-based drug design. *Curr. Opin. Drug Discov. Devel.*, 9, 3, 370–9, 2006 May, PMID:16729734.

55. Alves, C.N., Martí, S., Castillo, R., Andrés, J., Moliner, V., Tuñón, I., Silla, E., A quantum mechanic/molecular mechanic study of the wild-type and N155S mutant HIV-1 integrase complexed with diketo acid. *Biophys. J.*, 94, 7, 2443–51, 2008 Apr 1, doi: 10.1529/biophysj.107.107623, Epub 2007 Nov 2. PMID:17981909; PMCID: PMC2267120.
56. Zhang, X., Zhao, Y., Lu, G., Recent development in quantum mechanics/ molecular mechanics modeling for materials. *Int. J. Multiscale Comput. Eng.*, 10, 65–82, 2012.
57. Marx, D. and Hutter, J., *Ab initio* molecular dynamics: Theory and implementation. *Modern Methods Algo. Quan. Chem.*, 1, 301-449, 141, 2000.
58. Karelson, M., Lobanov, V.S., Katritzky, A.R., Quantum-Chemical Descriptors in QSAR/QSPR Studies. *Chem. Rev.*, 96, 3, 1027–1044, 1996 May 9, doi: 10.1021/cr950202r, PMID: 11848779.
59. Woods, C.J., Manby, F.R., Mulholland, A.J., An efficient method for the calculation of quantum mechanics/molecular mechanics free energies. *J. Chem. Phys.*, 128, 1, 014109, 2008 Jan 7, doi: 10.1063/1.2805379, PMID: 18190187.
60. Park, K., Götz, A.W., Walker, R.C., Paesani, F., Application of Adaptive QM/ MM Methods to Molecular Dynamics Simulations of Aqueous Systems. *J. Chem. Theory Comput.*, 8, 8, 2868–77, 2012 Aug 14, doi: 10.1021/ct300331f, Epub 2012 Jul 11. PMID: 26592126.
61. Cho, A.E., Chung, J.Y., Kim, M., Park, K., Quantum mechanical scoring for protein docking. *J. Chem. Phys.*, 131, 13, 134108, 2009 Oct 7, doi: 10.1063/1.3239504, PMID: 19814544.
62. Nantasenamat, C., Isarankura-Na-Ayudhya, C., Prachayasittikul, V., Advances in computational methods to predict the biological activity of compounds. *Expert Opin. Drug Discov.*, 5, 7, 633–54, 2010 Jul, doi: 10. 1517/17460441.2010.492827, Epub 2010 May 22. PMID: 22823204.
63. Quantum Mechanics (utexas.edu)
64. Dawson, J.F., *Quantum Mechanics: Fundamental Principles and Applications.* Lecture Notes, University of New Hampshire, Durham. 2009 Oct 14.
65. Abagyan, R., Lee, W.H., Raush, E., Budagyan, L., Totrov, M., Sundstrom, M., Marsden, B.D., Disseminating structural genomics data to the public: from a data dump to an animated story. *Trends Biochem. Sci.*, 31, 2, 76–8, 2006 Feb, doi: 10.1016/j.tibs.2005.12.006, Epub 2006 Jan 6. PMID: 16406633.
66. Dirac, P.A., Quantum Mechanics of Many-Electron Systems. *Proceedings of The Royal Society A: Mathematical, Physical and Engineering Sciences,* vol. 123, pp. 714–733.
67. Acharya, C., Coop, A., Polli, J.E., Mackerell Jr., A.D., Recent advances in ligand- based drug design: relevance and utility of the conformationally sampled pharmacophore approach. *Curr. Comput. Aided Drug Des.*, 7, 10–22, 2011.
68. Altschul, S.F., Gish, W., Miller, W., Myers, E.W., Lipman, D.J., Basic local alignment search tool. *J. Mol. Biol.*, 215, 3, 403–10, 1990 Oct 5, doi: 10.1016/ S0022-2836(05)80360-2, PMID: 2231712.

69. Anderson, A.C., Structure-based functional design of drugs: from target to lead compound. *Methods Mol. Biol.*, 823, 359–66, 2012, doi: 10.1007/978-1-60327-216-2_23, PMID: 22081357; PMCID: PMC3644978.
70. Baroni, M., Costantino, G., Cruciani, G., Riganelli, D., Valigi, R., Clementi, S., Generating Optimal Linear PLS Estimations (GOLPE); An Advanced Chemometric Tool for Handling 3D-QSAR Problems. *Quantitative Structure-Activity Relat.*, 12, 9–20, 1993, http://dx.doi.org/10.1002/qsar.19930120103.
71. Blunt, N.S., Camps, J., Crawford, O., Izsák, R., Leontica, S., Mirani, A., Moylett, A.E., Scivier, S.A., Sünderhauf, C., Schopf, P., Taylor, J.M., Holzmann, N., Perspective on the Current State-of-the-Art of Quantum Computing for Drug Discovery Applications. *J. Chem. Theory Comput.*, 18, 12, 7001–7023, 2022, doi: 10.1021/acs.jctc.2c00574.
72. Dirac, P.A.M., The Lagrangian in quantum mechanics. *Physikalische Z. der Sowjetunion*, 3, 1, 64–72, 1933.
73. Feit, A., Fleck, J., Steiger, A., Solution of the Schrödinger equation by a spectral method. *J. Comput. Phys.*, 47, 3, 412–433, 1982.
74. Feynman, P.R., Space-Time Approach To Non-Relativistic Quantum Mechanics. *Rev. Modern Phys.*, 20, 2, 367, 1948.
75. Foulkes, W., Mitas, L., Needs, R., Rajagopal, G., Quantum monte carlo simulations of solids. *Rev. Modern Phys.*, 73, 1, 33, 2001.
76. Troyer, M. and Wiese, U.J., Computational Complexity and Fundamental Limitations To Fermionic Quantum Monte Carlo Simulations. *Phys. Rev. Lett.*, 94, 170201, 2005 May.
77. Lipinski, C.A., Lombardo, F., Dominy, B.W., Feeney, P.J., Experimental and computational approaches to estimate solubility and permeability in drug discovery and development settings. *Adv. Drug Deliv. Rev.*, 46, 1–3, 3–26, 2001 Mar 1, doi: 10.1016/s0169-409x(00)00129-0. PMID: 11259830.
78. Luscombe, N.M., Greenbaum, D., Gerstein, M., What is bioinformatics? A proposed definition and overview of the field. *Methods Inf. Med.*, 40, 4, 346–58, 2001, PMID: 11552348.
79. Foloppe, N. and Hubbard, R., Towards Predictive Ligand Design With Free-Energy Based Computational Methods? *Curr. Med. Chem.*, 13, 29, 3583–3608, 2006, https://dx.doi.org/10.2174/092986706779026165.
80. de Beer, S., Vermeulen, N.P., Oostenbrink, C., The role of water molecules in computational drug design. *Curr. Topics Med. Chem.*, 10, 1, 55–66, 2010 Jan 1.
81. Ohlstein, E.H., Ruffolo, R.R., Jr, Elliott, J.D., Drug discovery in the next millennium. *Annu. Rev. Pharmacol. Toxicol.*, 40, 177–91, 2000.
82. Ooms, F., Molecular modeling and computer aided drug design. Examples of their applications in medicinal chemistry. *Curr. Med. Chem.*, 7, 2, 141–58, 2000 Feb, doi: 10.2174/0929867003375317, PMID: 10637360.
83. Crespo, A., Rodriguez-Granillo, A., Lim, T.V., Quantum-mechanics methodologies in drug discovery: applications of docking and scoring in lead optimization. *Curr. Topics Med. Chem.*, 17, 23, 2663–80, 2017 Sep 1.

84. Propst, C.L. and Perun, T.J., Introduction to computer-aided drug design. *Comput.-Aided Drug Des.*, 1–6, 1989 Feb 22.
85. Podlogar, B.L. and Muegge, I., Brice, L.J. Computational methods to estimate drug development parameters. *Curr. Opin. Drug Discov. Dev.*, *4*, 1, 102–9, 2001.
86. Maggio, R., Barbier, P., Fornai, F., Corsini, G., Functional Role of the Third Cytoplasmic Loop in Muscarinic Receptor Dimerization. *J. Biol. Chem.*, *271*, 31055–31060, 1996.
87. Filizola, M. and Weinstein, H., The study of G-protein coupled receptor oligomerization with computational modeling and bioinformatics. *FEBS J.*, 272, 12, 2926–38, 2005 Jun, doi: 10.1111/j.1742-4658.2005.04730.x, PMID:15955053.
88. Pazos, F., Helmer-Citterich, M., Ausiello, G., Valencia, A., Correlated mutations contain information about protein-protein interaction. *J. Mol. Biol.*, 271, 4, 511–23, 1997 Aug 29, doi: 10.1006/jmbi.1997.1198, PMID: 9281423.
89. Debroise, T., Hoste, R., Chamayou, Q., Minoux, H., Filoche-Rommé, B., Bianciotto, M., Rameau, J.P., Schio, L., Levesque, M., In silico drug repositioning for COVID-19 using absolute binding free energy calculations, in: *arXiv preprint*, 2021 Sep 8, arXiv:2109.03565.
90. Zohuri, B., *Molten Salt Reactors and Integrated Molten Salt Reactors: Integrated Power Conversion*, Academic Press, 2021 Jun 29.
91. Gouldson, P.R., Dean, M.K., Snell, C.R., Bywater, R.P., Gkoutos, G., Reynolds, C.A., Lipid-Facing Correlated Mutations And Dimerization In G-Protein Coupled Receptors. *Protein Eng.*, 14, 759–767, 2001.
92. Wiewiora, R.P., *Rigorous Construction of Markov State Models for Conformationally Selective Drug Design*, (Doctoral dissertation, Weill Medical College of Cornell University.
93. Vazquez, J., Deplano, A., Herrero, A., Gibert, E., Herrero, E., Luque, F.J., Assessing the performance of mixed strategies to combine lipophilic molecular similarity and docking in virtual screening. *J. Chem. Inf. Model.*, 60, 9, 4231–45, 2020 May 4.
94. Laine, E., Eismann, S., Elofsson, A., Grudinin, S., Protein sequence-to-structure learning: Is this the end-to-end revolution)? *Proteins: Structure, Function, Bioinf.*, 89, 12, 1770–86, 2021 Dec.
95. Cano, J., Kumar, R., Brankovic, A., Pavlou, D., Stavrouz, K., Gibert, E., Martınez, A., González, A., Quantitative characterization of the software layer of a HW/SW co-designed processor, in: *2016 IEEE International Symposium on Workload Characterization (IISWC)*, pp. 1–10, IEEE, 2016 Sep 25.
96. Labant, M., Fully Automated Luxury Drug Discovery: Lacking the molecular assemblers of science fiction, drug discovery is making do with AI-driven lead generation, robot-executed experiments, and advanced analytical technologies. *Genet. Eng. Biotechnol. News*, 40, 8, 18–20, 2020 Aug 1.

97. Cova, T., Vitorino, C., Ferreira, M., Nunes, S., Rondon-Villarreal, P., Pais, A., Artificial Intelligence and Quantum Computing as the Next Pharma Disruptors. *Artif. Intell. Drug Design*, 321–347, 2022.
98. Fingerhuth, M., Babej, T., Wittek, P., Open source software in quantum computing. *PloS One*, 13, 12, e0208561, 2018 Dec 20.
99. Fouad, H. and Moskowitz, I.S., Meta-agents: using multi-agent networks to manage dynamic changes in the internet of things, in: *Artificial Intelligence for the Internet of Everything*, pp. 271–281, Academic Press, 2019 Jan 1.
100. Ben-Shalom, I.Y., Lin, C., Radak, B.K., Sherman, W., Gilson, M.K., Fast Equilibration of Water between Buried Sites and the Bulk by Molecular Dynamics with Parallel Monte Carlo Water Moves on Graphical Processing Units. *J. Chem. Theory Comput.*, 17, 12, 7366–72, 2021 Nov 11.
101. Satyanarayana, M., Drug Discovery Roivant will buy Silicon Therapeutics. *Chem. Eng. News*, 99, 8, 13, 2021.
102. Sharma, K. and Manchikanti, P., Regulation of artificial intelligence in drug discovery and health care. *Biotechnol. Law Rep.*, 39, 5, 371–80, 2020 Oct 1.
103. Mauri, M., Gili, K., Perdomo-Ortiz, A., Evaluating generalization in classical and quantum generative machine learning models: Part I, in: *APS March Meeting Abstracts*, vol. 2022, pp. T37-012, 2022.
104. Filizola, M., Olmea, O., Weinstein, H., Prediction of heterodimerization interfaces of G-protein coupled receptors with a new subtractive correlated mutation method. *Protein Eng.*, 15, 11, 881–5, 2002 Nov, doi: 10.1093/protein/15.11.881, PMID: 12538907.
105. Filizola, M. and Weinstein, H., Structural models for dimerization of G-protein coupled receptors: the opioid receptor homodimers. *Biopolymers*, 66, 5, 317–25, 2002, doi: 10.1002/bip.10311, PMID: 12539260.
106. Soyer, O., Dimmic, M.W., Neubig, R.R., Goldstein, R.A., Using evolutionary methods to study G-protein coupled receptors. in: *Biocomputing*, 2002, pp. 625–636, 2001.
107. Banères, J.L. and Parello, J., Structure-based analysis of GPCR function: evidence for a novel pentameric assembly between the dimeric leukotriene B4 receptor BLT1 and the G-protein. *J. Mol. Biol.*, 329, 4, 815–29, 2003 Jun 13, doi: 10.1016/s0022-2836(03)00439-x, PMID: 12787680.
108. Rocheville, M., Lange, D.C., Kumar, U., Sasi, R., Patel, R.C., Patel, Y.C., Subtypes of the somatostatin receptor assemble as functional homo- and heterodimers. *J. Biol. Chem.*, 275, 11, 7862–9, 2000 Mar 17, doi: 10.1074/jbc.275.11.7862, PMID: 10713101.
109. Xu, J., He, J., Castleberry, A.M., Balasubramanian, S., Lau, A.G., Hall, R.A., Heterodimerization of alpha 2A- and beta 1-adrenergic receptors. *J. Biol. Chem.*, 278, 12, 10770–7, 2003 Mar 21, doi: 10.1074/jbc.M207968200, Epub 2003 Jan 15. PMID: 12529373.
110. Stanasila, L., Perez, J., Vogel, H., Cotecchia, S., Oligomerization of the alpha 1a- and alpha 1b-adrenergic receptor subtypes. Potential implications in receptor internalization. *J. Biol. Chem., 278*, 41, 40239–51, 2003.

111. Canutescu, A.A., Dunbrack RL Jr Cyclic coordinate descent: A robotics algorithm for protein loop closure. *Protein Sci.*, 12, 963–972, 2003.
112. PowerPoint Presentation (rajasthan.gov.in)
113. Drug Discovery and Quantum Computing - PharmaFeatures
114. Frush, E.H., Sekharan, S., Keinan, S., *In Silico* Prediction of Ligand Binding Energies in Multiple Therapeutic Targets and Diverse Ligand Sets-A Case Study on BACE1, TYK2, HSP90, and PERK Proteins. *J. Phys. Chem.*, *121*, 34, 8142–8148, 2017.
115. Yokokawa, F., Nilar, S., Noble, C.G., Lim, S.P., Rao, R., Tania, S., Wang, G., Lee, G., Hunziker, J., Karuna, R., Manjunatha, U., Shi, P.Y., Smith, P.W., Discovery of Potent Non-Nucleoside Inhibitors of Dengue Viral RNA-Dependent RNA Polymerase from a Fragment Hit Using Structure-Based Drug Design. *J. Med. Chem.*, 59, 8, 3935–52, 2016 Apr 28, doi: 10.1021/acs.jmedchem.6b00143, Epub 2016 Apr 8. PMID: 26984786.
116. Keinan, S., Hatcher, F.E., Shipman, W.J., Leveraging Cloud Computing for *In-Silico* Drug Design Using the Quantum Molecular Design (QMD) Framework. *Computing Sci. Eng.*, 20, 4, 66–73, 2018.
117. Schneider, G.M., Mind and Machine in Drug Design. *Nat. Mach. Intell.*, 1, 3, 128–30, 2019.
118. Gao J. and Truhlar D.G., Quantum mechanical methods for enzyme kinetics. *Annu. Rev. Phys. Chem.*, 53, 1, 467–505, 2002 Oct.
119. Brown, N., Fiscato, M., Segler, M.H.S., Vaucher, A.C., GuacaMol: Benchmarking Models for *de Novo* Molecular Design. *J. Chem. Inf. Model.*, 59, 3, 1096–1108, 2019 Mar 25, doi: 10.1021/acs.jcim.8b00839, Epub 2019 Mar 19. PMID: 30887799.
120. Mucs, D. and Bryce, R.A., The application of quantum mechanics in structure-based drug design. *Expert Opin. Drug Discov.*, 8, 3, 263–76, 2013 Mar, doi: 10.1517/17460441.2013.752812, Epub 2013 Jan 7. PMID: 23289945.
121. Preuer, K., Renz, P., Unterthiner, T., Hochreiter, S., Klambauer, G., Fréchet ChemNet Distance: A Metric for Generative Models for Molecules in Drug Discovery. *J. Chem. Inf. Model.*, 58, 9, 1736–1741, 2018 Sep 24, doi: 10.1021/acs. jcim.8b00234, Epub 2018 Aug 28. PMID: 30118593.
122. Glover, F., Kochenberger, G., Du, Y., Quantum Bridge Analytics I: a tutorial on formulating and using QUBO models. 4or. 17, 4, 335–71, 2019 Dec.
123. Ertl, P. and Schuffenhauer, A., Estimation of synthetic accessibility score of drug-like molecules based on molecular complexity and fragment contributions. *J. Cheminform.*, 1, 1, 8, 2009 Jun 10, doi: 10.1186/1758-2946-1-8, PMID:20298526; PMCID: PMC3225829.
124. Alejandro, C. and Lim, V.T., Quantum Mechanics Methodologies in Drug Discovery: Application of Docking and Scoring in Lead Optimization. *Curr. Top. Med. Chem.*, 17, 23, 2663–2680, 2017.
125. Alexeev, Y., Mazanetz, M.P., Ichihara, O., Fedorov, D.G., GAMESS as a free quantum-mechanical platform for drug research. *Curr. Top. Med. Chem.*, 12, 18, 2013–33, 2012, doi: 10.2174/156802612804910269, PMID: 23110536.

126. Balasubramanian, K., Mathematical and Computational Techniques for Drug Discovery: Promises and Developments. *Curr. Top. Med. Chem.*, 18, 32, 2774–2799, 2018, doi: 10.2174/1568026619666190208164005, PMID: 30747069.
127. Schneider, P. and Schneider, G., *De Novo* Design at the Edge of Chaos. *J. Med. Chem.*, 59, 9, 4077–86, 2016.
128. Markovic, M., Ben-Shabat, S., Dahan, A., Computational Simulations to Guide Enzyme-Mediated Prodrug Activation. *Int. J. Mol. Sci.*, 21, 10, 3621, 2020 May 20, doi: 10.3390/ijms21103621, PMID: 32443905; PMCID: PMC7279318.
129. Barbault, F. and Maurel, F., Simulation with quantum mechanics/molecular mechanics for drug discovery. *Expert Opin. Drug Discov.*, 10, 10, 1047–57, 2015 Oct, doi: 10.1517/17460441.2015.1076389. Epub 2015 Aug 8. PMID: 26289577.
130. Kollar, J. and Frecer, V., How accurate is the description of ligand-protein interactions by a hybrid QM/MM approach? *J. Mol. Model.*, 24, 1, 11, 2017 Dec 12, doi: 10.1007/s00894-017-3537-z. PMID: 29234892.
131. Dahan, A., Markovic, M., Keinan, S., Kurnikov, I., Aponick, A., Zimmermann, E.M., Ben-Shabat, S., Computational modeling and *in-vitro/in-silico* correlation of phospholipid-based prodrugs for targeted drug delivery in inflammatory bowel disease. *J. Comput. Aided Mol. Des.*, 31, 11, 1021–1028, 2017 Nov, doi: 10.1007/s10822-017-0079-5, Epub 2017 Nov 3. PMID: 29101519.
132. Söderhjelm, P., Kongsted, J., Genheden, S., Ryde, U., Estimates of ligand-binding affinities supported by quantum mechanical methods. *Interdiscip Sci.*, 2, 1, 21–37, 2010 Mar, doi: 10.1007/s12539-010-0083-0, Epub 2010 Jan 28. PMID:20640794.
133. Mishra, R., Mishra, P.S., Mazumder, R., Computational Biology for Clinical Research. Routledge Taylor & Franchis Group, 31, 240, 2022 Aug.
134. Kulkarni, P.U., Shah, H., Vyas, V.K., Hybrid quantum mechanics/molecular mechanics (QM/MM) simulation: a tool for structure-based drug design and discovery. *Mini Rev. Med. Chem.*, 22, 8, 1096–107, 2022 May 1.
135. Schneider, G., Mind and machine in drug design. *Nat. Mach. Int.*, 1, 3, 128–30, 2019 Mar.

4

Exploring Nano-Based Therapeutics by Quantum Computational Modeling

Ponduri Teja Kumar[1], Roja Rani Budha[2], G. Raghavendra Kumar[3], B. Nagamani[4] and G.S.N. Koteswara Rao[5*]

[1]Wuxi DEC Pharma, Wuxi City, Jiangsu Province, China
[2]Department of Pharmacology, H.K. College of Pharmacy, Oshiwara, Jogeshwari (W), Mumbai, India
[3]Department of Pharmaceutics, Narasaraopeta Institute of Pharmaceutical Sciences, Narasaraopeta, Pradesh, India
[4]Viswanadha Institute of Pharmaceutical Sciences, Mindivanipalem, Visakhapatnam, Andhra Pradesh, India
[5]Shobhaben Pratapbhai Patel School of Pharmacy & Technology Management, SVKM's NMIMS, Vile Parle (W), Mumbai, India

Abstract

Nano-based therapeutics are the outcome of applying nanotechnology in therapy that resulted in a combined advantage of enhanced drug delivery efficiency and targeted delivery. Along with the advantages, there are challenges to targeted delivery due to the size of the drug delivery system and the complex *in vivo* vascular system. To efficiently develop nanotherapeutics and deliver its benefits, it is imperative to understand the complex processing of developing nanotherapeutics.

Designing the optimal process is necessary to understand the nano-based therapeutic's targeted delivery mechanism. With the quantum computational simulation models, application in the complex processes of nano-based therapeutic research and development was believed to have the potential to revolutionize the nano-based pharma industry. Quantum computational modeling was estimated to have a drastic impact on the understanding of complex biological processes, solubility enhancement of drugs, drug-excipient compatibility, drug encapsulation, drug release, nanoparticle targeting, drug delivery and uptake, pharmacodynamics, and pharmacokinetics. The quantum computational models for the determination and distribution of nanocarriers in the body followed by systemic

**Corresponding author*: drgsnkrao@gmail.com

Rishabha Malviya, Sonali Sundram and Dhanalekshmi Unnikrishnan Meenakshi (eds.)
Drug Delivery Systems Using Quantum Computing, (93–140) © 2024 Scrivener Publishing LLC

administration in various routes, models for visualizing the drug release, optimizing the drug carrier design, drug distribution in the body over time, transportation of the contents of drug product across the compartments in the human body, and how they negotiate with the body's defense system are the priorities for the design of nanotherapeutics. The final critical step requires models to predict the risk factors and analyze the benefits of nano-based therapeutics in the hope of reducing the number of extensive and expensive clinical trials, if not totally avoiding them. Thus, quantum computational simulation is envisaged to be able to shorten the drug development timelines and lower the risk of failure. The present chapter discusses the role of quantum computational models (coarse-grained, molecular dynamics, particle-based, and physiologically based pharmacokinetic), providing deep insight into the design of nano-based therapeutics.

Keywords: Drug delivery, nano therapeutics, quantum computational model, molecular dynamic simulation, QSPR tools, PBPK models, protein corona, blood brain barrier

4.1 Introduction to Nano-Based Therapeutics

In the current world of making small, compact, targeted, and constantly changing objects, there is a revolutionary term called 'nano'. For rapidly evolving science community, 'nano' is a prefix in the international system of units to indicate the billionth of a measure as per the General Conference on Weights and Measures, 1960. The word 'nano' was derived from the Greek word 'nanos', which means 'dwarf'. It is a common practice to use 'nano' to describe any phenomena involving nanoscale (less than 100 nm) objects or processes occurring on a nanoscale particularly at the atomic or molecular level. The science of dealing with designing, using, creating, and producing structures and devices having a size of 100 nm or less has evolved into a branch called nanotechnology. The utilization of nanotechnological approaches thereby employing nanosized objects for the efficient delivery of therapeutic applications and diagnostic purposes belongs to a broad category of nano-based therapeutics. The growing importance of nanosized therapeutics is due to the fact that the nanosized objects will possess adjustable physicochemical, biological, optical, and mechanical properties that are different from the same materials in bigger sizes due to modified quantum mechanical properties that can be exploited by researchers for therapeutic applications. The well-known fact about nanosized objects is that the size matches the biological macromolecules, which enables the use of nanosized objects in both *in vitro* and *in vivo* applications. This has led to the rapid rise of nano-based therapeutics, which additionally consists of newly rising ideas and programs of molecular nanotechnology for designing nano-machines known as nano-robots [1].

The shape of nanosized objects can be hollow, solid, or porous depending on the type of material (gold, silica, iron oxide, lipid, and polymer) and requirement. These attributes in accord with high interactions and transfer potentialities made nanosized objects an alluring tool for the design of therapeutic carriers, biosensors, biological devices, and diagnostic procedures. The nanosized objects, using their inherently smaller size, can have the potential to penetrate tissues, can transit through biological barriers, and move freely inside the human body compared to bigger materials. Thus, they are able to interact with biologically relevant molecules such as proteins, antibodies, mammalian intracellular structures, mammalian cells, and viruses, which are typically in the nm to μm range.

Nano-based therapeutics is a constantly expanding field and is about the use of nanosized objects for the diagnosis, treatment, and cure of diseases and includes relevant areas of drug delivery, *in vivo* and *in vitro* diagnostics, sensory purposes, and active implantable medical devices. However, irrespective of the terminology, all are related to the design, creation, utilization, and regulation of therapeutic agents or devices that help to overcome the challenges posed by various sources to human health. The nanosized objects are the building blocks for the nano-based therapeutics [2–4].

With the conventional therapeutics, there are drawbacks of lower efficiency, non-specific activity, unwanted side effects, and poor bioavailability. The combination of the therapeutic agents with the nanosized objects showing unique physicochemical and biological properties and further designing them for targeting specific locations of the organs where the therapeutic agents need to be activated at the desired release rate are achievable with the advancements in nanotechnology and medical and biological sciences. This approach helps to deliver the desired concentrations of the therapeutic agents to the desired cells, and it may reduce the required doses when there is a narrow therapeutic index. It also augments the efficiency and tolerability in the biological systems. Thus, the unification of nanosized objects with therapeutic agents marks the new era of nano-based therapeutics for the benefit of the health care system [3]. The advantage of nano-based therapeutics is that it is a proven science and has commercially approved products that are increasing yearly. From the first nanotherapeutic product approved in 1995 by the U.S. Food and Drug Administration (USFDA), the number of products approved by USFDA and European Medicines Agency (EMA) for marketing has reached more than 80 products and even more products are under clinical evaluation. The class of nanotherapeutics that were approved belong to lipid-based nanoparticles, nanocrystals, polymer-based nanoparticles, dendrimer-based nanoparticles, protein-based nanoparticles, and inorganic nanoparticles [5–7]. There are many types of nano-based therapeutics based on the nanosized carrying objects that were classified in multiple ways by various publications provided in Table 4.1 [3, 8–11].

Table 4.1 Classifications of nanosized carrying objects.

Classification	Category	Examples
Based on dimension (number of dimensions of nanosized carrying objects outside the nanoscale, i.e., 100 nm)	Zero-dimensional nanosized carrying objects (all dimensions are well within nanoscale)	Graphene quantum dots, carbon quantum dots, fullerenes, magnetic nanoparticles, polymer dots [12]
	One-dimensional nanosized carrying objects (only one dimension is outside the nanoscale)	Nanotube, nanofiber, nanowire, nanorod [13]
	Two-dimensional nanosized carrying objects (only two dimensions are outside the nanoscale)	Graphene, metal nanosheets, nanoribbons, nanoplates, nanowalls [14]
	Three-dimensional nanosized carrying objects (only three dimensions are outside the nanoscale)	Nanocubes, nanocages, dendrimers [10, 15]
Based on application	Polymeric based	Dendrimers, polymeric micelles, drug conjugates, nanogels [3, 8]
	Lipid based	Liposomes, exosomes, lipid nanoparticles [3, 8]
	Carbon based	Carbon nanotubes, fullerenes, graphene, graphene oxide, nanodiamond [3, 8]
	Protein and nucleic acid based	Peptide-based nanoparticles, cell-penetrating peptides [8]
	Metal based	Gold nanoparticles, silver nanoparticles, iron nanoparticles [8]

These nanotherapeutics employ nanocarrier technologies of micelles, liposomes, nanorods, nano lipid carriers, metal-coated nanoparticles, single-walled carbon nanotubes, nanorods, solid lipid nanoparticles, nanocrystals, nanofibers, dendrimers, quantum dots, polymeric nanoparticles, nanoribbons, fullerenes, nanowires, nanodiscs, and nanospheres, to name a few; representative illustrations are shown in Figure 4.1 [6, 8, 16].

The rise in demand for nano-based therapeutics was due to the reduced toxicity of existing molecules and the improved efficiency of therapeutic agents. This is made possible by targeting the specific site of the infected organ. Considering the advantages of the nano-based therapeutics, scientists have worked on a wide range of therapeutic areas such as cancer, pulmonary diseases, viral infectious diseases caused by resistant strains, autoimmune diseases, cardiovascular diseases, neuro-degenerative diseases, ocular diseases, regenerative therapy, kidney diseases, central nervous system disorders, psoriasis, and blood purification. In recent times, successfully developed mRNA vaccines for COVID-19 were lipid nanoparticle-based nanotherapeutics, which received emergency use authorization

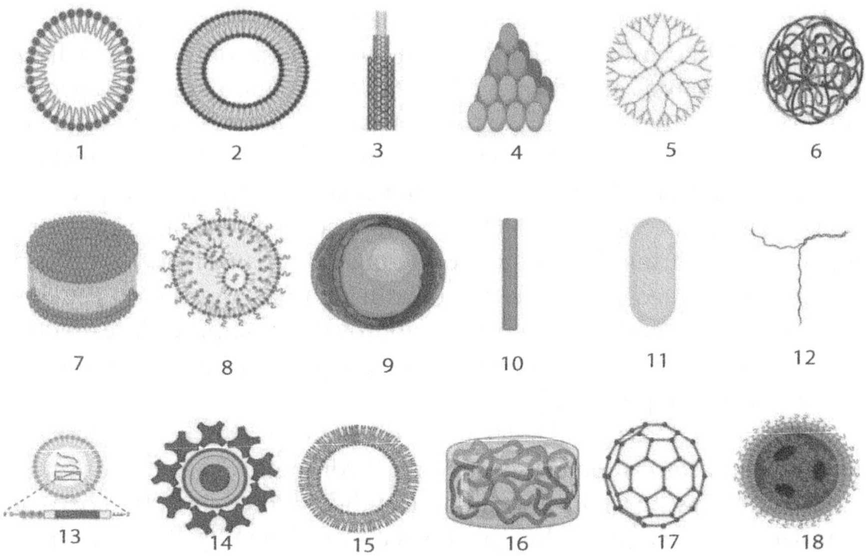

Figure 4.1 Nano-sized carrying objects used in nano-based therapeutics. (1) Micelle, (2) liposome, (3) carbon nanotube, (4) nanocrystal, (5) dendrimer, (6) polymeric nanoparticle, (7) nanodisc, (8) solid lipid nanoparticle, (9) nanosphere, (10) nanowire, (11) nanorod, (12) nanopore, (13) lipid nanoparticle, (14) quantum dot nanocrystal, (15) polymersome, (16) nanoribbon, (17) fullerene, (18) lipid nanoparticle (not drawn to the scale).

from the USFDA. The key ingredient in the fast-track development of these vaccines was nanotechnology. This evidence of success in an unforeseen health crisis such as COVID-19 with the aid of nanotherapeutics has not only undermined the skepticism on the use of nanotherapeutics during emergencies but gave hope to humankind to overcome such life-threatening challenges [17].

When compared with conventional therapeutics, the design and development of nano-based therapeutics was the most complex and challenging as it involves designing at the nanoscale using nanosized objects. However, it is worth studying this most advanced technology due to its vast potential advantages over conventional therapeutics with respect to lower toxicity, higher efficiency, and the possibility of personalized medication. Even though a significant number of nano-based therapeutics have been approved for clinical use, there remain plenty of unexplored areas that pose stiff challenges in the pathway of design and development at a rapid pace. As the field of nano-based therapeutics is still in its early stages, with the exception of limited success stories, ample effort is required to overcome these challenges for a healthier society.

The categories of challenges are broadly divided into the following: (i) delivering the drug across biological barriers, (ii) physicochemical and biological characterization of nanosized components of the drug delivery system, (iii) analysis and characterization of the nano-based therapeutics, (iv) formulation development and optimization of the nano-based therapeutics, and (v) scale-up and manufacturing of nano-based therapeutics. Another significant challenge that is not related to technical aspects but is related to the economic aspect is the high cost and the large amount of time needed to bring the product to the market. It is a big risk to develop and transfer the technology to manufacturing units and then meet clinical requirements and gain regulatory approval. These procedures require analytical instrumentation, manufacturing equipment and facilities, and highly skilled manpower [5].

Overcoming these challenges in a conventional way of experimentation and characterization using trial-and-error methods may take years or even decades owing to the poor understanding of the interaction between nanosized objects and biological molecules in bringing the nano-based therapeutic into clinical use with limited assurance of success. These risk factors made drug developers think about risk mitigation strategies and improvement approaches. It has led to a paradigmatic shift in the design and development approaches currently employed by drug developers. That shift is utilization of theoretical modeling set up in the design and development of nano-based therapeutics. The sophisticated modeling tools (both mathematical models and computational models) help not only to understand the physicochemical and biological characterization of nano-sized components of the drug delivery system but also to predict the *in vivo* characteristics of nano-based therapeutics [8, 18]. Of the two modeling tools, computational modeling is the most exciting tool for exploring the dynamics of nano-based therapeutics at both the molecular and quantum levels. This attribute made computational modeling as the most widely used simulation tool for the rational design and development of nano-based therapeutics [19]. The design and optimization of a chemical or biochemical process requires the understanding of the dynamics of reactions, which can be as simple as fragmentation and formation of chemical bonds, exchange of protons and electrons, and interactions between inductive forces and molecular systems [20]. The scope of computational models is broad and includes an integrated approach of artificial intelligence and multiscale modeling methods. In the context of computational modeling approaches with respect to nano-based therapeutics, researchers aim to provide deep insights into physicochemical, biochemical, and biological details across chemical molecule screening, polymorphism, chemical stability, formulation screening, and targeted delivery [21].

4.2 Introduction to Quantum Computational Modeling with Respect to Nano-Based Therapeutics

The scope of computational modeling techniques in nano-based therapeutics is broad and it involves but is not limited to quantum mechanics, molecular dynamics simulation, mathematical modeling, *in silico* modeling tools, physiologically based pharmacokinetic modeling, process simulation, artificial intelligence, and machine learning algorithms at different time scales and length scales [21]. The computational modeling techniques are categorized in multiple ways by various publications [18, 19, 21–25]. However, for clarity and to introduce as many models as possible, the following categorization was followed in the present chapter. The computational modeling approaches were broadly categorized under particle-based models and continuum-based models [18].

4.2.1 Particle-Based Models

The particle-based models employ details at the finer space–time scales for the characterization of nano-based therapeutics. They utilize length scale on the order of nm to μm and time scales of ns to μs for predicting the following interactions: therapeutic agent–nanosized objects, nanosized object–biological medium, and therapeutic agent–biological medium. Particle-based models utilize computational methods of quantum mechanics, molecular dynamics, Monte Carlo, and coarse-grained materials for the quantification of interactions. The particle-based models are based on quantum mechanics and classical mechanics. The quantum mechanics models are further categorized into *ab initio* quantum mechanics models, dynamic functional theory models, and semi-empirical quantum mechanics models [18].

The *ab initio* models at the highest level of detail provide the context to converge the exact solution of the Schrodinger equation with a number of estimations. These models utilize an electronic wave function model to calculate the Schrödinger equation. The significance of *ab initio* methods was in simulation of large biological molecules like carbohydrates, lipids, proteins, and nucleic acids for functional mechanisms such as photophysics and photochemistry of nucleic acids and nucleobases. The *ab initio* methods are further utilized as tools to verify the accuracy of other computational models or parameterization of material models. However, the *ab initio* methods are time-consuming and very expensive for large systems or macromolecules [18].

a. Dynamic Functional Theory Models

Dynamic functional theory models use total electron density and consider the effect of electron correlation. It makes this model suitable for large molecules with small computational demands. This is the most widely used quantum computational model and the operational cost is lower than the *ab initio* models while simulating medium-sized and large molecules. These have applications that investigate specific metal organic framework–drug molecule interactions and identify suitable conformations and adsorption sites between them. However, they lack a systematic development path and have reduced accuracy compared to *ab initio* models [19, 26, 27].

b. Semi-Empirical Quantum Mechanical Model

Semi-empirical quantum mechanical models are based on the simplified Hartree–Fock (HF) Hamiltonian operator using many approximations. They are introduced to simulate larger systems of atoms. The accuracy of the models and two electron interactions is taken from reliable empirical data or theoretical data. They are mostly used for simulation of larger systems than single atoms and for describing experimentally obtained drug release kinetics but not for the prediction of impact of formulation parameters. They can maintain high levels of accuracy with reduced computational cost in comparison with *ab initio* models [18, 19, 25].

All the quantum computational models discussed above are able to investigate the structure-based drug design and estimate the properties of various nanosized objects used as carriers such as carbon nanotubes, dendrimers, and metal organic frameworks. They have the capability to deal with small systems with few hundreds of atoms and limited degrees of freedom but are expensive for larger systems with thousands of atoms and even larger in size.

c. Molecular Dynamics Models

Molecular dynamics simulation models deal with the structural, dynamic, and thermodynamic properties of molecular systems and they start with the Protein Data Bank, which consists of experimental protein structures. Molecular dynamics models are the best suited for investigating the movement of atoms and molecules. The movement of molecules is nothing but movement of atoms, which is further the movement of atomic constituents. These models are based on the assumption that the movement of atoms is estimated to be the movement of their nuclei, which are considered as particles that act in accordance with Newton's laws of motion. Thus, the movement of atomic constituents, atoms, and molecules in a system is interconnected. The interacting particles in a system determines the trajectory of particles by solving Newton's equations of motion. The interacting particle's potential energies are calculated using molecular mechanics methods with different interatomic forces. This is because atoms, which are the fundamental entities of molecules, move due to the interatomic forces or force fields. The force fields are the most commonly used in simulating particle–particle interactions that can imitate structural and conformational changes. Molecular dynamics simulations display the structural changes with changes in environment such as pH and temperature. They can also reveal the dynamic process of protein aggregation and protein misfolding [16, 28–31]. Molecular dynamics models were used to design controlled-release anti-cancer nano-based therapeutics and simulate drug loading, prediction of ligand binding, nanosized carrying object's stability, interactions between therapeutic and cell membrane, and permeation properties [16].

Few publications describe the variations of atomistic all-atom molecular dynamics models, and coarse-grained molecular dynamics simulation models basically work on the principle of movement because particle interactions in a system were described by Newton's law of motion.

The common software packages used for the study of molecular dynamics simulations are GROMACS, AMBER, NAMD, and CHARMM [31].

d. Monte Carlo Simulation Models

Monte Carlo models are faster than molecular dynamic models. These simulation models sample the configuration of a system of molecules according to the probability of finding those configurations given by their Boltzmann factor. Their applications are to study drug release from the nano-based therapeutics and the design of nano-based therapeutics for targeting purposes, thus simulating the binding of nanosized objects to the specific targeted cells. However, Monte Carlo models are not capable of simulating complex macro molecules with many internal degrees of freedom [29, 30].

e. Coarse-Grained Models

All the models discussed above are related to the investigation of atoms and are linked to higher computational costs. The costs were reduced by removing the atomic interactions, and a set of new models called coarse-grained models were gaining significance. These models assume that each particle represents a group of atoms and this assumption facilitates simulations in larger time and length scales, which leads to enhanced efficiency and also acts as a link between molecular models and continuum models. The coarse-grained models have the objective of performing affordable simulations of complex systems mainly focusing on the critical characteristics of the systems. For simulation at the molecular scale, the force field MARTINI was most used by the researchers for its flexible operation and reliable prediction capabilities. Coarse-grained models are of two types, coarse-grained molecular dynamics and coarse-grained stochastic dynamics [8, 18, 24, 29, 32]. The application of coarse-grained molecular dynamics simulation models includes estimation of drug diffusion, drug release, interactions between drug and protein during protein corona formation, drug–lipid bilayer interactions, and design of targeted nano-based therapeutics. The application of coarse-grained stochastic dynamic simulation models includes estimation of drug loading, drug release, drug–lipid bilayer interactions, and distribution of nano-based carrying objects [18, 33, 34]. However, these methods cannot give accurate simulation when used at resolution levels coarser than one amino acid residue per coarse grain particle in proteins.

4.2.2 Continuum-Based Models

The continuum-based models consider materials as continuous masses, whereas the materials were considered as particles in particle-based models. There are two platforms of simulation in continuum-based models, non-spatial and spatial platforms [18, 25]. Non-spatial models are based on differential equations. These have found applications in modeling the effects of the distribution of drugs inside body compartments and predicting the drug concentration in body compartments. These can simulate the pharmacokinetic and pharmacodynamic responses of nano-based therapeutics. By these applications, non-spatial models have found a regular utilization in the development, design, and optimization of nano-based therapeutics. Even though non-spatial models are implicitly simulating the drug concentrations across body compartments, the research requires details such as transport of nanotherapeutics within interstitial space, and the adhesion of drug molecules to the tissues are beyond their potential. However, these details can be simulated by the application of spatial models that are based on partial differential equations. The finite element method can predict the interaction between therapeutic agent and cells, as well as distribution attributes of therapeutic agents in the blood vessel walls. The finite volume method can predict the drug transportation through the blood vessels [25, 35–39]. The overview of the computational models is provided in Table 4.2.

Table 4.2 Overview of computational models with their applications and limitations.

Approach	Category	Technique	Applications	Limitations
Particle-based models	Quantum mechanics models	*Ab initio* quantum mechanical models [22, 23].	Simulation of functional mechanisms of small molecule involving electronic interactions such as chemical bond fragmentation and formation.	Complex and tedious to simulate systems
		Dynamic functional theory [19, 25–27].	Therapeutic design and prediction of interaction between host and guest	No systematic path for development

(*Continued*)

Table 4.2 Overview of computational models with their applications and limitations. (*Continued*)

Approach	Category	Technique	Applications	Limitations
		Semi-empirical quantum mechanics [18, 25, 26, 40, 41].	Prediction of protein–ligand interactions, vaccine design, biologic therapeutic design	Not suitable for predicting atomic scale interactions.
	Classical mechanics models	Molecular dynamics [16, 28–31].	Design of efficient anti-cancer nano-based therapeutics, prediction of protein–ligand interactions, drug loading, stability of formulation, design of antibodies	Time for simulation increases with molecular size
		Monte Carlo [29, 30].	Evaluating the performance of metal organic frameworks in nano-based therapeutics, design of nano-based therapeutics for targeting purposes, simulating the binding of nanosized objects to the specific targeted cells.	Cannot be applied when higher degrees of freedom and complex systems.
	Coarse-grained models [8, 18, 24, 29, 32].		Structural characterization of nanocarriers, prediction of interactions between therapeutic agents and nanocarriers, simulation of permeability of nano-based therapeutics through cell membrane, prediction of therapeutic agent loading, and release profiles.	May not give accurate simulation when used at resolution levels coarser than one amino acid residue per coarse grain particle in proteins [42].

(*Continued*)

Table 4.2 Overview of computational models with their applications and limitations. (*Continued*)

Approach	Category	Technique	Applications	Limitations
Continuum-based models	Non spatial [18, 39].	Pharmacokinetic/ pharmacodynamic, physiological-based pharmacokinetic	Correlation of absorption, distribution, metabolism, and extraction properties of nano-based therapeutics for design, development, and optimization.	Cannot provide details beyond drug concentration in the specific site.
	Spatial models [18, 37–39].	Finite element method, finite volume methods	Simulation of movement of drug and drug delivery system inside the interstitial space.	Molecular details and PKPD details cannot be investigated together.

4.3 Exploration of Nano-Based Therapeutics

Nano-based therapeutics (NBT) are basically medicinal products where the therapeutic agent is wrapped or encapsulated or covered by a nanosized carrying object intended to be delivered to a specific targeted location in the body. The nanosized carrying object or nanocarrier that sometimes may act as a shield to the therapeutic agent modifies the release and carries the therapeutic agent from the site of administration to the site of action specifically. The world of nano-based therapeutics is finer in terms of space and full of mysterious mechanisms. Right from the administration of nano-based therapeutic into the body, the delivery of the therapeutic agent to the target cells involves complex mechanisms depending on the route of administration. The researchers must be aware of the challenges posed by the physiological barriers to nano-based therapeutics or be able to predict the fate of nano-based therapeutics in *in vivo* conditions [8, 43, 44].

When nano-based therapeutics are administered into the blood vessels, the formulation should remain intact only if stable enough, and if the nano-based therapeutics are unstable, there is the possibility of leakage of the therapeutic agent in the blood vessel itself, which causes an insufficient amount reaching the desired location. The intact nano-based therapeutics might interact with ions and biological molecules and fluids such as blood cells, plasma proteins, and other molecules that are encountered during its voyage through the blood vessel. Based on the nature of ions or biomolecules, plasma proteins are adsorbed on the nano-based therapeutic, and the properties of the nano-based therapeutics are modified *in vivo*. When biological molecules in the blood vessels interact with NBT, they form a complex described as protein corona by many authors [45, 46]. The nature of the interacting molecules and the physicochemical properties including morphology of nano-based therapeutics decide the composition and nature of the protein corona. This newly formed protein corona forms a new interphase between the nanosized objects within the nano-based therapeutics and biological medium. It will have a significant impact on the uptake of nano-based therapeutics, distribution, and therapeutic response. The protein corona may change the path of the nano-based therapeutics and it may also mask the targeting capabilities of the nano-based therapeutic's surface. Nano-based therapeutics conjugated with transferrin (a blood plasma glycol protein) has shown decreased target specificity due to interaction with proteins in the serum. Also, in blood circulation, there is the possibility of nano-based therapeutics getting expelled into extracellular space [44, 47]. Once all these barriers are surmounted and nano-based therapeutics has reached the target cell, it will interact with the cell membrane, and the therapeutic agent either can directly penetrate the cell or can be subjected to endocytosis. If the nano-based therapeutics undergoes endocytosis, the entrapped nanosized objects that carry the therapeutic agents shall be released from the endosomes before their maturation. Further, the therapeutic agent is dissociated from the nano-based therapeutics inside the cell and is impacted by the molecules present endogenously. The dissociated therapeutic agent interacts with the cellular organelles of interest to initiate their mechanism of action. The complex process of nano-based therapeutic kinetics and dynamics can be explored by the quantum-level computational models and can also speed up and enhance the efficiency of the rational design of nano-based therapeutics. Nano-based therapeutics was administered through an intravenous route, and the fate is depicted in Figure 4.2.

In the case of pulmonary drug delivery, nano-based therapeutics are inhaled into the conducting airways and are deposited on the lining of the respiratory tract, which may be either the mucus layer or the surfactant lining fluid layer within the alveolar region. The deposited nanosized carrying objects may exhibit varying kinetics depending on the physicochemical properties of the administered nanosized carrying objects and place of deposition. Those are soluble in the mucus and the lining fluid is dissolved and rapidly absorbed by passive diffusion through the epithelial membrane, which is insoluble, and clearance is through mucociliary

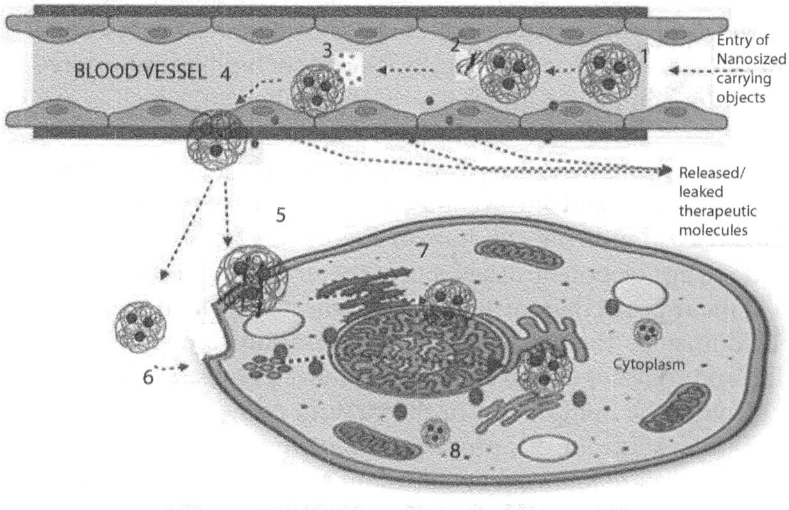

Figure 4.2 The pathway of intravenously administered nano-based therapeutic. (1) Entering the bloodstream, (2) interaction with biomolecules, (3) interaction with ions, (4) excavation into extracellular space, (5) direct penetration, (6) endocytosis, (7) entrapped nano-based therapeutic in cytoplasm, (8) release of drug molecules (not drawn to the scale).

transport and phagocytosis by alveolar macrophages after formation of protein corona with the lining fluid. The remaining molecules are translocated to epithelial and interstitial sites, blood circulation, and sometimes to lymphatic nodes. Upon reaching blood circulation, they are distributed throughout the body [48–50].

Oral administration is the most common route of administration of drugs but is the least preferred route for nano-based therapeutics. The number of biological barriers encountered by the orally administered nano-based therapeutic is far higher than other routes. Upon oral administration, they are initially exposed to saliva and then gastric fluid, which have two varying pH values and significantly affect the dissolution of nano-sized carrying objects. Interactions with stomach content such as food and beverages are unpredictable. Interactions with the gut mucosa may affect the stability and other cellular interactions. In the acidic medium of the stomach, the nanosized carrying objects may agglomerate or dissolve depending on the formation of protein corona and short residence time. From the stomach, they reach the intestine with an alkaline environment, where most of the nano-based therapeutics are intended to release the therapeutic molecules or drug payload, which is absorbed in the circulation through active or passive transport mechanisms. Absorption from the intestine through phagocytosis or pinocytosis is metabolized or dissolved in lysosomes and may be released in the cytosol of cells. The fate of orally administered medication is not complete without first pass metabolism, which is inevitable [51].

The transdermal nano-based therapeutics have gained attention as an alternative to oral and intravenous routes for achieving controlled and constant release utilizing the human body's largest organ, i.e., skin, for low-dose drugs or potent drugs. From the drug reservoir of transdermal patch of nano-based therapeutic, drug permeates through several layers of skin by diffusion. The mechanisms of permeation are intercellular, intracellular, and transappendageal. In intercellular permeation, drugs have to cross through the lipid matrix in the stratum corneum, which is suitable for lipophilic drugs. In this process, therapeutic molecules distributed in the lipid matrix are followed by diffusion into the inner side. Intracellular permeation is more suited for hydrophilic drugs, where they pass through a hydrophilic pathway created by the hydrated keratinized corneocytes into lipid matrices and then into the cytoplasm of cells. In the transappendageal route, permeation is through hair follicles, sweat glands, and oil glands [52]. Also, there are other routes of administration such as ocular and nasal routes for nano-based therapeutics.

The role of quantum computational modeling tools for the design, development, and optimization of nano-based therapeutics was inseparable starting from drug discovery to establishing clinical efficiency and evaluating post-marketing adverse reactions. There are various types of nano-based therapeutics, such as liposomes, dendrimers, carbon nanotubes, inorganic, and polymer-based nanoparticles, which have their own unique properties to suit the different routes of administration. The physicochemical characterization (size, morphology, deformability, surface charge, and composition changes) of nano-based therapeutics is an important step in the design, development, and optimization. However, these characterizations through experiments have intrinsic and practical limitations where the resolution of structural aspects is difficult, if not impossible. Other properties such as composition changes, size, and surface charges are tedious and expensive. As a result, it is difficult to correlate efficiency with the composition. Design and optimization of nano-based therapeutics only on the experimental research are unlikely. Also, failed *in vivo* results despite having successful *in vitro* results can be attributed to the insufficient mechanistic insight earned through trial-and-error-based experimental data [8]. The application of computational modeling methods enables one to screen the variables to predict the optimum conditions and minimize failure. However, computational simulations cannot completely replace experimental research and they complement and assist in the rational design, development, and optimization of novel nano-based therapeutics with enhanced efficiencies.

4.4 Design and Development of Nano-Based Therapeutics

When designing any kind of nano-based therapeutics, the objective will be to solve a specific set of problems and to achieve the advantages pertaining to it. It all starts with the characterization of physicochemical properties especially solubility and permeability. Once data on those parameters are obtained, depending on it, the selection of the suitable composition to achieve the best possible efficiency will be investigated. Simultaneously, the manufacturing process will be optimized to stabilize the formulation, followed by the evaluation of the pharmacokinetic and pharmacodynamic properties and finally the evaluation of the adverse reactions and toxicity [8, 34, 53, 54] (Figure 4.3).

4.4.1 Prediction of Solubility

The design of any therapeutic shall be based on the sound understanding of the solubility and permeability characteristics of the therapeutic agent. In order to make the therapeutic agent available to the site of action

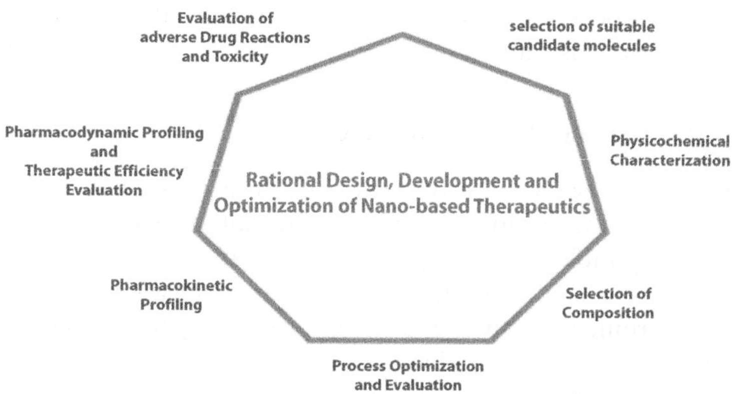

Figure 4.3 Steps in rational design, development, and optimization of nano-based therapeutic.

(attain bioavailability), the drug should be insoluble in aqueous biological medium and at the same time should be able to pass through the lipid bilayer membrane [18, 31, 55].

The low solubility of therapeutic agents is the cause of frequent clinical failure of contemporary drug molecules when formulated as oral dosage forms. Also, aqueous solubility is also crucial for absorption; it is a prerequisite to the design of the nano-based therapeutic to be soluble in a bio-environment. In this regard, computational simulation tools are efficiently able to explore the interaction of the drug molecule with the ingredients of the therapeutic system, which can predict the solubility and other methods for the enhancement of solubility [56–58]. For more details on the prediction of the aqueous solubility profile of drugs, see Refs. [31, 59].

Quantitative structure–property relationship tools can predict the fate of solvation characteristics and partition coefficients of organic chemicals in the biological environment, but with less accuracy in comparison with other quantum computational tools [60–62]. The optimal solubility of small molecules in the environment of polymeric nanosized carriers is a prior qualification for an efficient targeted delivery of therapeutics [31]. When it comes to solid dosage forms, the solubility of the therapeutic system in the biorelevant media plays a significant role in the effectiveness of the therapeutic. *In silico* models are available to predict the solubility of therapeutic agents in biorelevant fluids (simulated gastric fluids and simulated intestinal fluids) where additional factors impacting the dissolution, such as bile salts, are considered and developed based on quantitative structure–property relationships [63]. Using the molecular dynamics simulation model, the interaction of vinpocetine and excipients was simulated in the GROMACS package. HPMC was predicted for the nanoparticle formulation of vinpocetine to output smaller particles with enhanced solubility and improved dissolution rate [64].

4.4.2 Prediction of Permeability

The permeation (which involves entry of drug molecules into the cell membrane, its translocation is followed by exiting from the cell membrane) of drug molecules is imperative to the absorption of therapeutics and their bioavailability [65], including the permeation of oral drugs through gastrointestinal membranes, transdermal permeation of topical drugs, passage through the blood–brain barrier for drugs acting on the central nervous system, epithelial permeation of inhalation drugs, and permeation through corneal, conjunctival, scleral membranes for ocular drugs [66–71]. The permeation of small-molecule therapeutic agents, large molecular peptides, and bioactive molecules were studied using molecular dynamics simulation models [31, 72].

For the oral absorption of therapeutics, permeation through intestinal membrane is critical. The relation of lipophilicity, size of molecule, and capacities of hydrogen bonds was found to have an encouraging effect on the permeability. However, for the translocation of therapeutics, factors such as entrapment in the lipid membranes, non-specific binding to intracellular proteins, and specific binding to membrane-bound efflux transport proteins can play a role and affect the ease and time of translocation. The computational models employing molecular dynamics simulations and statistical approaches were able to predict the permeability of small molecular drugs through gastrointestinal membranes [73, 74]. The implementation of potential mean force technique permits estimation of free energy requirements to pass through the lipid membrane. This allows calculation of permeation coefficient and bestows the process of permeation of small molecular therapeutic agents. Using h-bonds with a portioned descriptor, simulation of about 49 weak basic drug like molecules was performed. The study also provides drug-membrane portioned complex information [75].

Understanding and predicting permeability of therapeutics and chemicals through the skin's barrier structure is beneficial for design of therapeutics and prediction of toxicities, respectively. Using atomistic molecular dynamics simulation, the transdermal permeability of water and seven other molecules consisting of solvents and small molecular drugs [29] and hormones [76] was predicted and data were validated against the permeability data in literature. Also, simulation for effect of permeation enhancers on the cell membrane permeability was also calculated by molecular dynamics simulations to identify and optimize the suitable enhancers for the therapeutic. The interactions between skin layers, therapeutics, and permeation enhancers and the transport mechanisms can be predicted through the basic quantitative structure–property relationship models, dissipative particle dynamic simulations, coarse-grained molecular dynamics, and COSMOL-based dermal transport modeling [77–82]. The permeation behavior of hydrophilic and hydrophobic drugs through the follicular pathway was simulated and was comparable to other experimental models [83]. Optimization of transdermal formulation based on the permeability of therapeutic agents developed a physiologically based transdermal platform called "BIOiSIM", which predicts the transdermal effectiveness of therapeutics [84].

In all the physiological barriers, the blood–brain barrier plays a crucial role in maintaining homeostasis in the central nervous system by the mechanisms of transport of essential nutrients, restricted transport of toxic entities, and removal of metabolites from brain. Thus, for the treatment of neurological disorders, brain tumors, and other neurodegenerative disorders, where the highly selective, semipermeable nature of BBB is affected, there is a dire need for prediction of interaction between the therapeutic and BBB. The mechanisms of crossing the blood–brain barrier via the paracellular pathway, passive diffusion, carrier-mediated transport, and receptor-mediated and adsorptive-mediated transcytosis were better understood by the computational models. They also help to find ways to enhance the permeation through various means of surface modification and coating with minimal experimentation [85]. The computational models of rule-based models, quantitative structure–activity relationship models, and microfluidic models were effectively used to understand the mechanism of drug transport through the endothelial cells of BBB [82, 85–88]. The application of external forces for the enhancement of permeability through BBB was also simulated for the magnetic nanoparticles and found to be suitable for the application [89]. There is a large molecular database available to aid in the prediction of BBB permeability of molecules [90]. The pulmonary availability of drugs was estimated by computational models in the epithelial lining fluids, intracellular retention, tissue retention for the alveolar and tracheobronchial regions, vascular space based on the permeability, and tissue retention predictions [91, 92]. For the drug delivery to the complex structural organ of the eye, the corneal membrane and conjunctival membranes are the major barriers to the permeation of ocular therapeutics. The corneal permeability of about 32 drugs were studied by Ramsay *et al.*, and they used these data in the quantitative structure–property relationship model with hydrogen bond donor, polar surface area, and halogen ratio of the molecules as molecular descriptors. The predicted conjunctival permeability results are comparable with those of corneal permeability [93]. Based on the physicochemical properties of drugs and corneal-PAMPA (parallel artificial membrane permeability assay) data, two QSPR models were developed for the investigation of corneal membrane permeability and membrane retention of therapeutics in less time with precision [94]. There are other computational models to study the flow behavior of ocular fluids inside different domains of the eye and its impact on permeability through corneal, conjunctival, and scleral tissues [71]. Considering the structural complexity of cornea (multi laminate structure), a unique model of quasi-3D was developed to successfully study the drug permeation across the cornea. The model has predicted the corneal permeabilities of both hydrophilic and hydrophobic drugs and the simulation was further extended to pharmacokinetic modeling of ocular drugs [94].

4.4.3 Selection of Components and Optimization of Formulation of Nano-Based Therapeutics

Even before studying the drug loading and release, the selection of composition of the nano-based therapeutic is required to be studied thoroughly in the design of nano-based therapeutics. The internal structural characteristics of lipid-based nano therapeutics of solid lipid nanoparticles, nanoemulsions, and nanostructured lipid carriers were investigated using coarse-grained molecular dynamics simulation with MARTINI. The simulation involves investigating the influence of solid to liquid mass fraction on the internal distribution of lipids in the matrix. The study involves evaluation of lipid-based nanocarriers composed of solid lipid (stearic acid), liquid lipid (oleic acid), and surfactant (sodium dodecyl sulfate) in water medium. The simulation study has concluded that, in the nanostructured lipid carriers, the mass ratio of solid to liquid lipids has limited influence on the distribution of lipids inside the lipid matrix [95]. The selection of suitable polymeric complexes for the stable nanoparticle formulation of a poorly soluble drug, i.e., fluticasone, to meet the requirement of higher dissolution rates and enhanced antibacterial potential was studied by molecular dynamics. The study concluded that combinations of hydroxy propyl methyl cellulose (HPMC)-polyvinyl pyrrolidone (PVP) and Eudragit-PVP were more suitable than the nanoparticles as the drug binding is better than the individual polymers. The combination was optimized and the experimental data were in good agreement with the simulation data [96]. A biomolecular simulation using molecular dynamics was reported in a study, where adsorption of gonadotrophin-releasing hormone (GnRH-I) and its cysteine-tagged modification (Cys-GnRH-I) to model silica surfaces followed by conjugating to the bovine serum albumin carrier protein (Cys-GnRH-I-SiNps) simulated its experimental results, and it was found that there was no immunological response. The simulation of testosterone with native peptide-silica nanoparticles attained a positive therapeutic effect. The study also found better antibody response with the silica nanoparticle conjugates than the conjugates of bovine serum albumin and was well correlated with the epitope presentation predicted by simulation. The research also mentioned the potential of biomolecular simulation for the rational design of vaccine systems [97]. A molecular dynamics simulation has provided insight into the peptide adsorption mechanisms, stabilization, aggregation, and sintering mechanisms of gold nanoparticles functionalized with cysteine-based peptides. The simulation results were not only confirmed by the experimental data but also provided information on aspects that cannot be scrutinized by experiments. This clearly shows the potential of biomedical simulation models to guide the development of complex biomedical nanosized objects [98].

4.4.3.1 Prediction of Therapeutic Loading

A deep insight into the pharmacokinetic behavior, i.e., the relation between a given drug and its concentration in plasma or tissues inside the body, has been provided by the use of computational models. To be specific, the therapeutic kinetics (absorption, distribution, and metabolism), drug loading, and drug release from the nanosized objects were simulated by researchers for the rational design of nano-based therapeutics. Investigating the drug binding and release properties of the carrying nanosized objects, it was found that those mechanisms operate at the molecular level [8, 24, 44].

Improved drug loading and stable and efficient encapsulation of therapeutic agent and release at the desired rate in desired location are the major expectations from nanosized carrying objects of nano-based therapeutics. In the design of nano-based therapeutics, considerations will be given to prevent the release of the therapeutic agent at the undesired locations coupled with prevention of drug accumulation at the targeted location. Thus, the encapsulation of drug or simply drug loading and drug release or drug leakage from the nanosized carrying objects shall be controlled and specifically targeted to obtain the benefits of efficiency and avoiding the harmful side effects [18].

Nano-based therapeutics usually includes the nanosized carrying objects that encapsulate the therapeutic agents within previous cores. The encapsulation and retention of therapeutic agents is governed by the colloidal interactions of therapeutic agents with nanosized carrying objects. This physical encapsulation resulting in loading efficiencies of about 20–30% implies that less than 30% of the mass of injected nano-based therapeutic provides therapeutic action and the remaining 70% mass does not have any therapeutic contribution. It is required to increase the loading efficiency and at the same time optimize the retention and release profiles along with the establishment of compatibility between the loaded therapeutic agent and nanosized carrying objects, which can be simulated by computational modeling methods. The computational modeling methods of molecular dynamics, Monte Carlo, and coarse-grained simulations were effectively able to analyze and optimize the loading efficiency and stability of the nano-based therapeutics such as nanoemulsions, dendrimers, polymeric micelles, carbon nanotubes, and liposomes. The outcome of the simulations and experimental data shows that the selection of suitable ingredients, solvents, and architecture of nanosized carrying objects augments the therapeutic encapsulation, retention, loading efficiency, and release profiles. There is the possibility of increasing the loading efficiency by the chemical encapsulation of therapeutics by directly linking them to nano-sized carrying objects via cleavable chemical bonds. However, this requires a prodrug and a formation of lipid or polymer, which eventually assembled to form the nano-based therapeutics. The molecular dynamics simulation of lipid-polyethylene glycol nanoemulsion helped to design and obtain a drug encapsulation efficiency of about 97% and higher drug loading capacities through optimized conjugation of the anti-cancer drug docetaxel with lauroyl moieties [18, 99]. Remote loading or active loading of therapeutic agents into the lipid-based nanosized carrying objects using the liposomes showed ion and/or proton gradients as the driving force to pull therapeutic agents into the liposomes. However, the drug candidate is required to be an amphipathic weak base. For other drug candidates, they can be chemically altered to the amphipathic weak base or acid prodrugs for increased loading efficiency. Steroid drugs were altered to their hemi succinate prodrugs, which are amphipathic weak acids to facilitate active loading and will be converted to steroids upon release. Also, the active loading is possible by complexation with cyclodextrins, which are amphipathic weak acids or bases themselves that can be actively loaded into liposomes as drug–cyclodextrin complexes. To predict the suitability of therapeutic agents for the active loading and estimate the loading capacity, which are capable of achieving desired loaded drug concentration, computational models based on QSPR were deployed [100]. Atomistic molecular dynamics simulation of PEGylated liposomes has shown that there is increased drug loading efficiency with PEGylation [101].

4.4.4 Prediction of Therapeutic Release or Leakage

Therapeutic molecules are transported from the encapsulated position in nanosized carrying objects to the outer surface and later on move into the biological medium. The mechanism of this transportation is by diffusion through water-filled pores, diffusion through a nanomaterial complex, surface erosion, bulk erosion, and osmotic pumping. The nanosized carrying objects release the loaded therapeutic molecules triggered by means of intracellular stimulus (such as pH, reactive oxygen, and elevated enzyme levels) or extracellular stimulus (such as heat, light, electrical, ultrasound induction, and magnetic induction). The stimuli may include physical, chemical, and biochemical means. The release profile of the hydrophobic drug molecule from dendrimers at various pH levels (neutral, acidic, and alkaline) were studied by fully atomistic molecular dynamics simulation and potential mean force calculations. There is a spontaneous release of drug at acidic pH, median release at neutral pH, and slower release at the alkaline pH. Also, the pH-induced conformational changes in dendrimers, ionization states, type of dendrimer, and pKa of the drug molecule were found to be the influential factors on stability of the formulation, drug loading, and release [2, 8, 19, 25, 102–106]. In protein-based therapeutics, the impact of pH and temperature on the drug release rate was simulated and compared with the experimental data performed with doxorubicin as the model drug. The simulated data showed consistency with the experimental data. In the carbon-based nano therapeutics, interaction of drug molecules with carbon nanocarriers as a function of various stimuli (pH, temperature, and external magnetic induction) and its impact on drug loading and release were explored with the help of molecular dynamics simulations. Using Monte Carlo simulation studies, drug release from the encapsulated emulsion systems and nanostructured lipid carriers (generally a combination of solid and liquid domains) was explored and the impact of size, location, and distribution of solid domains on the release rate of drug molecules from the nanostructured lipid carrier was noted. Drug release from the liposomal pores triggered by the electric field was simulated [8]. In a mechanistic modeling approach by the physiological-based biopharmaceutics model, the *in vitro* and *in vivo* correlation of drug release, lymphatic transportation, diffusion rate, and degradation of long-acting microsphere injection formulations of an anti-cancer drug was determined. In an experimental insight into the metal organic framework of an anti-cancer drug, using Monte Carlo simulations predicted the suitability of the pH-triggered controlled drug release profile for solid tumor targeting [19, 107].

4.4.5 Prediction of the Pharmacokinetic Profile of Nano-Based Therapeutics

A deep insight into the pharmacokinetic behavior, i.e., the relation between a given drug and its concentration in plasma or tissues inside the body, has been provided by the use of computational models. To be specific, the therapeutic kinetics (absorption, distribution, metabolism, and excretion) were simulated by researchers for the rational design of nano-based therapeutics. The kinetic property understanding is regarded as a necessary step both in the discovery and in the development of newer therapeutic agents and nano-based therapeutics. Investigating the drug binding and release properties of the carrying nanosized objects, it was found that those mechanisms operate at the molecular level [18, 31, 102].

4.4.5.1 *Prediction of Absorption*

When any nano-based therapeutic was administered orally, absorption into the gastrointestinal tract was via paracellular transport, transcytosis, and M cell uptake. It was the most complex mechanism of the pharmacokinetic profile that consists of absorption, distribution, metabolism, and excretion [35, 108, 109]. The absorption of orally administered therapeutics, using drug-specific physicochemical parameter prediction of absorbed therapeutic concentrations in the gastrointestinal compartments, was successfully performed by computational models such as the STELLA® software package way back in the 1990s [34, 110]. The other factors affecting oral absorption, such as insufficient dissolution, formation of insoluble complexes, unstable formulation in the gastrointestinal tract, and first-pass metabolism, are also estimated when predicting the absorption using mechanistic and computational models [111, 112].

4.4.5.2 *Prediction of Distribution (Protein Corona Formation)*

When it comes to parenteral administration of nano-based therapeutics, the voyage of nanosized carrying objects from the site of administration to the site of action is critical. The influence of the physiological environment and physicochemical characteristics of the nano-based therapeutic on the pharmacokinetic profile is very complex to understand. To simplify, the journey of nanosized carrying objects is split into vascular, transvascular, and interstitial phases. Upon injection of nano-based therapeutics into the bloodstream, the first barrier of blood protein interaction alters the biochemical properties of nanosized carrying objects resulting in the formation of protein-nanosized carrying object complex or corona. This corona flows through blood capillaries and reaches the target organs or tumorous tissues in the case of cancer. There is availability of continuum, particle-based and hybrid models to simulate this transit at the molecular scale. Upon reaching the capillary wall or extracellular scale, there will be cellular uptake of nanosized carrying objects. The whole process of distribution up to clearance is studied by pharmacokinetic models. The therapeutic efficacy and allied effects including adverse reactions are studied by pharmacodynamic models [35, 113]. In case of inhalation drugs, upon administration, nano-based carrying objects have to overcome the pre-absorption mucociliary clearance and cross the mucus and pulmonary endothelial barrier before absorbed in the blood and lymphatic circulations. The other non-intravenous routes of administration such as intraperitoneal, subcutaneous, and intramuscular required to minimize the distribution into irrelevant tissues and organs before reaching systemic circulation. In the dermal administration, the absorption through layers of skin is dependent on the size of nano-based carrying objects, along with other factors such as thickness of skin layers, physicochemical properties of nano-based carrying objects, and temperature [36].

When the nano-based therapeutic first interacts with proteins (usually human serum albumin, lipoprotein, and fibrinogens). The interaction of plasma proteins with the nanosized carrying objects was simulated by the coarse-grained simulation model. The amino acids within each protein are coarse-grained and represented as one bead for each amino acid to decrease the degrees of freedom. The force fields between these beads are calculated. In a study of simulating interaction of plasma proteins with nanosized carrying objects, the adsorption energies of the proteins and the orientation of adsorption were predicted, thereby effectively predicting which proteins form the corona with the nanosized carrying objects and their composition and thereby predicting its behavior [114]. The coarse-grained simulations allow to understand the impact of size and the surrounding environment on the corona formation. The interaction of protein is dependent on the size of the nano-based carrying object and there is also structural reorientation occurring during the corona formation in a slow manner. There is a competition between the components present in the surrounding environment, where non-proteinaceous substances competed with proteins to bind with ubiquitin. The aggregation tendency of ubiquitin was also observed [115]. In cancer therapeutics, the transportation of nanosized carrying objects in the microvasculature and extravasation into tumor tissues under the heterogenous blood flow was studied by computer simulations. In the blood flow, red blood cells move away, leaving a cell-free layer that traps the nanosized carrying objects in it. This process continues and increases the concentration at the endothelial surface layer, which later affects the shear-induced drug diffusion and caused slower drug transportation rate into the tumor tissue, which determines the excavation rate of nanosized carrying objects [116]. In a similar study, the impact of size of the nanoparticles on the transportation in the blood vessels was determined by a shear-induced nanoparticle diffusion model adopting finite volume-based equation [117].

In a study of the pharmacokinetic profiles of nano-based therapeutics, the utilization of physiologically based pharmacokinetic models may play a crucial role in developing and optimizing the nano-based therapeutics.

These models consider the physicochemical properties of modified nano-sized carrying objects in the biological medium such as morphology, topology, surface chemistry, and formation of protein corona, which impact the kinetic profile. The body's immune system identifies the nano-based therapeutics as the foreign bodies by the mononuclear phagocytes present in blood circulation and other tissues. Quantifying the mononuclear phagocytes within the target tissue could simulate a more realistic kinetic profile. The possibility of abrupt drug leakage from nanosized carrying objects is also taken into consideration [118, 119]. In case of inhalations, the nano-based therapeutics reaches the alveolar region and swallowed by the alveolar macrophages, which will either be cleared by mucociliary clearance at the tracheobronchial region or migrate to lymph nodes.

4.4.5.3 *Prediction of Metabolism*

The metabolism of nano-based therapeutics involves enzymatic degradation of components such as proteins, lipids, and other water-soluble materials, but often there might be non-degradable inorganic nano-based carrying objects and some organic coatings. The metabolism of nano-based therapeutics is mainly done through enzymatic degradation involving the liver. Silver nano-based therapeutics were studied for their metabolism by dissolution followed by precipitation of silver ions into silver sulfide using a physiologically based pharmacokinetic model [36].

4.4.5.4 *Prediction of Excretion*

The excretion of nano-based therapeutics from the body is predicted. The removal process is mainly through liver and kidneys via feces and urine, while the lungs and skin also contribute to excretion. In a PEGylated gold nano-based therapeutic study through a physiologically based pharmacokinetic model, hepatobiliary and urinary excretion was studied and their significance in the excretion process was determined. In the renal clearance, which involves glomerular filtration, tubular secretion, and urinary excretion, the rate of excretion is decreasing with larger nano-based therapeutics. There are studies predicting the urinary excretion of nano-based therapeutics occurring through the transcapillary pathway, which is dependent on the translocation rate of nano-based carrying objects from blood to urine, the permeability of capillary walls, the flux of blood through kidneys, and the total blood volume in the body. However, the portion of urinary clearance is smaller than the hepatobiliary clearance due to the smaller capillary pore size of kidneys. The biliary excretion of nano-based therapeutics, which travel from the extravascular compartment of the liver to the faecal compartment of the gut, is studied by several researchers. The rates of urinary excretion and biliary excretion have shown variation and are required to be studied in-depth using physiological based pharmacokinetic models [36]. The software packages available for the pharmacokinetic modeling are Monolix®, NanoCore 7.4, Stella®, GastroPlus™, PK-SIM®, and Simcyp®. For more information, see the work of Villa Nova *et al.* [120].

4.4.6 Understanding Protein Corona Formation

In the blood circulation, the nano-based therapeutics will be considered as foreign particles due to their size and will adhere to proteins in the blood. These protein agglomerates surrounding the nano-based carrying objects are referred to as 'protein corona' in the published literature. There is an impact of the coating or stealth sheath of nanosized carrying objects on the formation of protein corona and its properties. In the case of the PEGylation mechanism, its impact on the protein corona formation was found to vary; that is, PEGylation may either inhibit or accelerate the protein adhesion. The interaction between the protein corona, stealth sheath, and mechanism of stealth is not clearly understood. Mechanisms other than protein corona formation that contribute to the shielding of therapeutics were not clearly depicted. However, with the PEGylated nanoparticles, it was evident that protein corona is necessary for the stealth properties [53].

In the journey of nanosized carrying objects, they have to not only pass through the complement cascade but also steer through the hydrodynamic habitat of blood flow and deliver the drug payload through the cell membrane. Both of these processes are dependent on the surface properties of the nanosized carrying objects with a major influence on the size, shape, and rigidity or elasticity. The flow propensity of nanoparticles in the bloodstream can be compared to a laminar flow in a cylinder, where the particles are continuously pushed towards the direction of flow and also forced towards the walls or margination. The margination effect was more observed with spherical-shaped nanoparticles with more elasticity than the disk-shaped nanoparticles with more rigidity. This propensity was evident from the function and morphological attributes of leukocytes. The design of nanosized carrying objects considers a wide range of directions not limited to the properties required to deliver the drug payload through the biological barriers such as cell membrane to the target cells and sometimes to the specific organelle within the cell. Properties such as size, shape, elasticity, and charge of lipids are crucial in understanding the molecular mechanisms of interactions between nanosized carrying objects and the cell membrane. The permeation mechanisms involve either direct permeation through the diffusion or disruption of cell membrane. There are also triggers that can be applied either internally (pH change) or externally (optical, thermal, and magnetic) to release the drug payload at a controlled rate into the targeted tissues [45]. The delivery of the therapeutic agents to the specific targeted sites for a variety of nanosized carrying object types for many therapeutic areas such as anti-cancer agents, antibiotics, Alzheimer's drugs, anti-microbial peptides, diabetic drugs, immunomodulators, and immunosuppressants was investigated through computational simulation models.

Upon injection, the journey of nanosized carrying objects in the bloodstream comes across the hydrodynamic forces followed by the formation of protein corona by interacting with bloodstream proteins. A part of those bloodstream proteins may trigger the complement cascade where nanoparticles are specifically taken away by macrophages. The factors influencing the propensity of nanosized carrying objects in the bloodstream such as size and morphology of particles were better understood by the combination of mathematical calculation with computational fluid dynamics through investigation in abnormal tumor vasculature [121]. The mechanism of protein corona formation in the bloodstream was complex and critical for the intended application. The adsorption of bloodstream proteins to the nanoparticles was dependent on the surface properties. The protein corona shields the nanoparticles in the bloodstream and increases their residence time. The interaction of protective polymers PEG and poly-sarcosine with blood proteins was studied by molecular dynamics simulation and revealed that interactions are not specific to amino acids. The interactions are due to the general tendency depending on the charge and polarity of amino acid and the nature of water–polymer interaction. Researchers worked to utilize these mechanisms and contemplated on rationally designing and developing novel polymer materials that are superior in protective protein corona formation. Deploying the coarse-grained MARTINI model, the impact of PEGylation and PEG concentration that determines the surface charge on the PEGylated nano-based therapeutics (liposomes, dendrimers, and carbon nanotubes) that interact with bloodstream proteins was studied [32].

In an all atomic molecular dynamics simulation study, the interaction of bloodstream proteins with nanoparticles under the influence of electrostatic forces and hydrophobic surface properties was determined [46]. In liposomes, where the interaction between protective PEG corona and lipid bilayer was studied by all atomic molecular dynamics simulation, it was found that the properties of PEG are related to its amphiphilic nature and the ease of polymer electrolyte capabilities. Also, polyoxazolines as an alternative to PEG as a protective polymer was investigated [122]. In a molecular dynamics simulation, the impact of change in PEG length, branched structures, and functionalizing PEG with surface anchoring phospholipid, cholesterol, and cholane on liposomes was studied. For more insight into the PEGylation of nanosized carrying objects, see Sousa *et al.* [123].

4.4.7 Understanding the Interaction of Nanosized Carrying Objects with Bio-Membranes

The last barrier that nanosized carrying objects needs to surmount in the biological habitat to be efficacious is the cell membrane. For the hydrophobic nanosized carrying objects, traversing through the cell membrane involves direct translocation. The other way is through receptor-mediated endocytosis. To benefit from the endocytosis, nanoparticles are designed to release the drug payload at the acidic pH medium of endosomes. Molecular dynamics simulations provide a deep insight into both the mechanisms mentioned above and are applied to different types of nano-based therapeutics, such as carbon dots, graphene, dendrimers, nanocrystals, and gold nanoparticles. The other simulation methods used for simulating interaction of nanosized carrying objects with cell membrane are atomistic molecular dynamics, MARTINI, dissipative particle dynamics, and implicit solvent models [124]. In the direct transfection of nanosized carrying objects into the cell, the interaction with the cell membrane was investigated using molecular dynamics simulations. During the process of translocation, there are possibilities of disruption of the cell membrane followed by release of contents and pore formation. The simulation studies on the effect of size, shape, and surface morphology of nanosized carrying objects and the impact of PEGylation or any other polymer coatings and proteins on the cell membrane were studied by computational models of molecular dynamics [53, 125]. The receptor-induced endocytosis by the targeted binding of ligands was another complex mechanism studied using simulation models. This mechanism involves the impact of the distribution pattern of ligands on the surface of nanosized carrying objects, the extent of orientation, and their availability to the target receptors, which were studied by different computational techniques such as molecular dynamics simulation and Monte Carlo simulation. The degree to which the targeted ligand is exposed and the availability to the receptors were studied by atomistic simulations. A detailed review on the study of interactions between nanosized carrying objects and bio-membranes was provided by Ding *et al.* [108].

4.4.8 Prediction of the Pharmacodynamic Profile of Nano-Based Therapeutics

The physiological effects of the nano-based therapeutics and biochemical processes altered in the body are critical for efficacy and safety. The effects of nano-based therapeutics in the body were studied as pharmacodynamics/nano-pharmacodynamics and were further understood better by the application of computational models. As most of the nano-based therapeutics are administered intravenously, the drug effects in the blood circulation immediately upon administration and during the distribution of nano-based therapeutics were studied as hemodynamics. Cardiovascular drugs, where the effect of drugs on blood flow is crucial for the therapeutic effect, were studied by simulation of the drug transportation phenomenon using Buongiorno's model, where Brownian motion of nano-based carrying objects reduced the blood flow rate. The obtained results were further validated with the finite element method. The blood flow rate was directly proportional to the size of the nanosized carrying objects injected into the arteries and was not affected by the magnetic induction effect [126, 127]. The thrombolytic effect was studied with the activated platelet targeted tissue plasminogen activator-loaded nanovesicles in comparison with the free tissue plasminogen activator. The simulation shows that the thrombolytic efficiency of nanovesicles was greater even at lower doses (10% of the free tissue plasminogen activator) and that systemic exposure was limited, which mitigates the excessive bleeding complications [128].

The determination of the success of nano-based therapeutics is ultimately based on efficiency, and is more crucial in the treatment of cancer. The relationship between the concentration of therapeutic agents and their pharmacological effects can be determined by pharmacodynamic models. Binding of therapeutic molecules to the targeted biological receptors, enzymes, and DNA triggers the pharmacological responses depending on the administered dose. The pharmacological responses can be impacted by the limited quantity of receptors relative to the quantity of therapeutic molecules, and laws of mass action. The dose–response curve is linear initially and becomes nonlinear with increased concentrations of therapeutic molecules [129]. The mechanism of sensitivity to anti-cancer drugs is a valuable information, as the deep insight into the drug resistance mechanism enhances the potential to design new hypotheses or treatment strategies in the future to avoid or mitigate drug resistance. The drug resistance mechanisms of cancer drugs obtained from literature are genetic, epigenetic, posttranslational, cellular, pharmacokinetic, and microenvironmental. With molecular dynamics simulations, the genetic mechanisms of drug resistance are used to study the dynamic changes of the therapeutic agent and the conformational changes of the cancer cells [130]. Along with the desired therapeutic responses, there are undesired responses that cannot be avoided and are called side effects. These can be due to patient related factors such as age, gender, or disease stage, or due to related drug dosage. Even the nano-based therapeutics are designed to specifically target the intended sites; they also bind to other proteins, resulting in side effects. The prediction of side effects is currently performed through docking-based, network-based, and machine learning-based approaches [131]. The frequency of the side effects plays a significant role in the risk–benefit assessment of therapeutics. The current practice of determining the frequency of side effects by randomized clinical trials is expensive and time-consuming. In a simulation tool, the matrix decomposition algorithm estimated the frequencies of side effects both identified and yet to be identified. The approach helps to understand the therapeutic activity mechanism and shows that each component of the drug's latent signatures is related to different anatomical categories of drugs and their administration route [131, 132].

4.4.9 Prediction of Adverse Drug Reactions and Nanotoxicity

With the advent of the niche field of nanotechnology, advanced nano-sized objects and nano-based therapeutics were continuously being developed every day. The safety concerns related to their application has to be resolved for further sophisticated developments in the healthcare industry. The safety of any nano-based therapeutic is mainly dependent on the pharmacokinetic and pharmacodynamic profile. However, there is a slight gap between *in vitro* and *in vivo* testing results, but because of the advanced computational simulation models, they are able to achieve *in vitro* and *in vivo* correlation through various *in silico* simulation models. The computational simulation models enable us to better understand the correlation between pharmacokinetics and pharmacodynamics with the aid of Monte Carlo simulation, Bayesian models, and nano-quantitative structure–activity relationship models. The attribute of toxicity was investigated by various models with descriptors of more than one of the following: size, surface charge, relaxivity, lipophilicity, concentration, functional groups, and surface area, and studied in metallic, metal oxide nano-based therapeutics, dendrimers, and polymer-based therapeutics. The literature on prediction of cytotoxicity of nanosized carrying objects with more than 90% accuracy used quantity structure–activity relationship perturbation models [133]. The physicochemical properties of nanosized carrying objects and the pharmacokinetic parameters aid in the estimation of absorption, distribution, metabolism, distribution, and excretion properties of nano-based therapeutics. Another approach for predicting nanotoxicity was by extraction of comprehensive scientific literature and databases. By the application of multiple path particle dosimetry, a quantitative tool to estimate the accumulation of nanosized carrying objects in the respiratory tract and lungs for inhaled drugs and thereby ion toxicity can be predicted. These models can model the data to interpret occupational exposure limits without animal or human testing [130]. Immunotoxicity shows adverse drug reactions of foreign materials such as nano-based therapeutics on the immune system of the administered body. As the nano-based therapeutics are new to the healthcare industry, insight into the immunotoxic potential of the nanosized objects is limited. In the approval process, currently the data from the *in vivo* and *ex vivo* experiments were the gold standard in the immunotoxicity assessment. However, there is a need for computational simulation models to complement the experiments to minimize or avoid animal or human testing and meet ethical standards. There are models based on the databases and scientific literature to predict immunotoxicity on different types of immune cells [134]. For further details on toxicity prediction, see the work of Pérez Santín *et al.* [135].

4.4.10 Design of Target-Oriented Nano-Based Therapeutics

The design of nano-based therapeutics involves the fundamental considerations of targeted delivery of therapeutic agents at the optimum concentrations in such a way to obtain maximum efficiency with negligible or no toxicity. In case of chemotherapy, where the need for the nano-based therapeutics is far higher than other therapies, the design involves the delivery of chemotherapeutic agents that are usually cytotoxic in nature to the tumorous tissues with the highest specificity by avoiding drug distribution to the remaining non-tumorous tissues of the body. The nanosized carrying objects are designed to safeguard the therapeutic molecule, which is often referred to as the 'drug payload' of the nanosized carrying objects. The specific targeted delivery can be either active or passive. In active targeting, the surface of the nanosized carrying objects with specific functionalized ligands targeted the overexpressed receptors on the tumor cells. Examples of commonly used ligands are transferrin (protein), herceptin–rituxan (antibody), pegaptanib (aptamer), folate, and galactose (small molecules) [136]. In passive targeting, the physicochemical properties of the nanosized carrying objects facilitate the more selective targeting of the tumor cells in comparison with the healthy cells without any ligands. The phenomenon of passive targeting involves enhanced permeation and retention (EPR) effect and PEGylated liposomes, which will benefit from the vasculature of tumorous tissues. Nanosized carrying objects are designed to carry the therapeutic agents and shield them from the various biological barriers in the target site. The first barrier of cellular uptake from blood circulation by phagocytes and nanosized carrying objects was designed to have a surface characteristic that inhibits the cellular uptake and enables flow through the blood circulation to attain the desired concentration of therapeutic agents at the target site. This modified surface of nanosized carrying objects can be referred to as stealth sheath, and a well-established strategy is PEGylation of nanosized carrying object's surfaces. The alternative surface modifiers tested are polyoxazolines, PASylation, zwitterionic polymers, hydroxyethyl starch, and polypeptides. In the PEGylation process, an alternate polymer shield with a stealth nature will be created through functionalizing the polymer to an ingredient of the nanosized carrying object. The ingredient can be the nanosized carrying object itself or a therapeutic agent. PEGylation of boron nitride, folic acid, interferon, insulin, peptides, and protein drugs were a type of therapeutic agent and polymer. The PEGylation of proteins to modify their immunological properties was established more than four decades ago and was tried and tested with many kinds of nanosized carrying objects. The list includes PEGylated carbon nanotubes, inorganic nanoparticles, silver graphene nanoparticles, nano-graphene, lipid micelles, nanodiscs, liposomes, and dendrimers.

The alternate polymers for functionalization have been investigated for inorganic nanoparticles. For gold nanoparticles, functionalization was through the thiol group with hydrocarbons capped in a methyl group, hydroxyl group, carboxylic group, amine group, chlorine sulfate, and para mercaptobenzoic acid. Other investigations aim to navigate the nanoparticle to the targeted site where a hydrophobic-coated nanoparticle binds to the targeted hydrophobic core of the lipid bilayer. The coating composition was investigated for various moieties, and this has led to the development of amphiphilic inorganic nanoparticles to target the separation line between aqueous phase and lipid head groups. They consist of two kinds of coatings with the same coating moieties with polar coating placed on one-half of the particle and non-polar coating placed on the other half of the particle. The amphiphilic coating was investigated with gold nanoparticles, silver nanoparticles, graphene nanoflakes with ssDNA, and silica nanoparticles with hydrocarbons.

The functionalization of nano-sized carrying objects has been investigated by the computational modeling tools using the molecular dynamics model. The simulation modeling involves downloading the protein structures with estimated potentials followed by attaching the polymers to the protein and solvate in water to study its behavior. The simulation of human recombinant erythropoietin conjugated with polyethylene glycol and Pro-Ala-Ser was performed with molecular dynamics. The coarse-grained model was used in investigating site-specific PEGylation of protein lysozyme. The atomistic molecular dynamics simulation model was used to simulate the conjugation of PEG to bovine serum albumin, the impact of PEGylation on the stability and potency of interferon and insulin, and the steric shielding effect of PEGylated streptokinase [53]. For further details on protein–polymer conjugation, see the work of Lin *et al.* [137] and for bioinformatics tools, refer to the work of Malviya *et al.* [138].

4.5 Conclusion

While the advances in drug delivery technologies from the recent four decades are breathtaking, they are a drop in the ocean of technology that has not begun to be utilized in an authorized formulation or has yet to be discovered. As human life expectancy continues to increase, millions of people require long-term care for diverse diseases. Filling the present and future unmet demands requires nano-based therapeutic technologies to surmount age-old challenges, e.g., improving water solubility of poorly soluble therapeutic agents, surmounting biological barriers, and developing more efficient controlled release depot formulations. The research experiences from the past are vital assets for designing future drug delivery technologies carried out in therapy. As the development of COVID-19 vaccines demonstrated, meeting the unexpected crisis of the uncertain future calls for continuous cumulation of past disasters, knowledge, and technologies. Wide awakening efforts of supporting varied research subjects within the drug delivery area are urgently required more than ever. Nano-based therapeutics, which are a niche concept of targeted delivery of therapeutic agents with minimal side effects, were perfectly placed in the path of meeting those requirements. On the other hand, with less than four decades of practical development, skillsets with inculcated quantum computational modeling in the exploration of mechanistic insights into nano-based therapeutics form the basis of rational design, development, and optimization. The deep insights into nano-based therapeutics explored through quantum computational simulation models may contribute to reducing or even avoiding the need for expensive clinical trials. Thus, the quantum computational simulation is envisaged to be able to shorten the therapeutic development timelines and lower the risk of failure. In a dynamic world that is under unforeseen attack by novel biological threats, there is a constant need of working and updating the existing healthcare features in order to meet the demands of existing diseases and the ever-changing requirements of human healthcare, which we have witnessed with the recent COVID-19 crisis. It is concluded that the aforementioned capabilities of quantum computational modeling tools guide the healthcare research toward treating difficult-to-treat existing diseases and unforeseen hazards later on.

References

1. Prasad, M. *et al.*, Nanotherapeutics: An insight into healthcare and multi-dimensional applications in medical sector of the modern world. *Biomed. Pharmacother.*, 97, 1521–1537, Jan. 2018, doi: 10.1016/j. biopha.2017.11.026.
2. Patra, J.K. *et al.*, Nano based drug delivery systems: Recent developments and future prospects. *J. Nanobiotechnol.*, 16, 1, 1–33, Sep. 2018, doi: 10.1186/s12951-018-0392-8.
3. Yetisgin, A.A., Cetinel, S., Zuvin, M., Kosar, A., Kutlu, O., Therapeutic nanoparticles and their targeted delivery applications'. *Molecules*, 25, 9, 1–31, May 2020, doi: 10.3390/molecules25092193.
4. Chakravarty, M. and Vora, A., Nanotechnology-based antiviral therapeutics. *Drug Deliv. Transl. Res.*, 11, 3, 748–787, Jun. 2021, doi: 10.1007/s13346-020-00818-0.
5. Halwani, A.A., Development of Pharmaceutical Nanomedicines: From the Bench to the Market. *Pharmaceutics*, 14, 1, 1–21, Jan. 2022, doi: 10.3390/pharmaceutics14010106.
6. Ruman, U., Fakurazi, S., Masarudin, M.J., Hussein, M.Z., Nanocarrier-based therapeutics and theranostics drug delivery systems for next generation of liver cancer nanodrug modalities. *Int. J. Nanomed.*, 15, 1437–1456, 2020, doi: 10.2147/IJN.S236927.
7. Hafner, A., Lovrić, J., Lakŏ, G.P., Pepić, I., Nanotherapeutics in the EU: An overview on current state and future directions'. *Int. J. Nanomed.*, 9, 1, 1005–1023, Feb. 2014, doi: 10.2147/IJN.S55359.
8. Ramezanpour, M., Leung, S.S.W., Delgado-Magnero, K.H., Bashe, B.Y.M., Thewalt, J., Tieleman, D.P., Computational and experimental approaches for investigating nanoparticle-based drug delivery systems. *Biochim. Biophys. Acta, Biomembr.*, 1858, 7, 1688–1709, Jul. 2016, doi: 10.1016/j. bbamem.2016.02.028.

9. Anu Mary Ealia, S. and Saravanakumar, M.P., A review on the classification, characterisation, synthesis of nanoparticles and their application. *IOP Conf. Ser. Mater. Sci. Eng.*, 263, 032019, Nov. 2017, doi: 10.1088/1757-899X/263/3/032019.
10. Bhatia, S., Nanoparticles Types, Classification, Characterization, Fabrication Methods and Drug Delivery Applications, in: *Natural Polymer Drug Delivery Systems*, pp. 33–93, Cham: Springer International Publishing, Basil, Switzerland, 2016, doi: 10.1007/978-3-319-41129-3_2.
11. Rizwan, M., Shoukat, A., Ayub, A., Razzaq, B., Tahir, M.B., Types and classification of nanomaterials, in: *Nanomaterials: Synthesis, Characterization, Hazards and Safety*, pp. 31–54, Elsevier, Amsterdam, Netherlands, 2021, doi: 10.1016/B978-0-12-823823-3.00001-X.
12. Wang, Z., Hu, T., Liang, R., Wei, M., Application of Zero-Dimensional Nanomaterials in Biosensing. *Front. Chem.*, 8, 2, 1–19, 2020, doi: 10.3389/fchem.2020.00320.
13. Bashir, S. and Liu, J., Chapter 1 - Nanomaterials and Their Application, in: *Advanced Nanomaterials and their Applications in Renewable Energy*, J.L. Liu and S. Bashir (Eds.), pp. 1–50, Elsevier, Amsterdam, 2015, doi: https://doi.org/10.1016/B978-0-12-801528-5.00001-4.
14. Cai, S. and Yang, R., Two-Dimensional Nanomaterials With Enzyme-Like Properties for Biomedical Applications. *Front. Chem.*, 8, 1–17, 2020, doi: 10.3389/fchem.2020.565940.
15. Paramasivam, G., Palem, V.V., Sundaram, T., Sundaram, V., Kishore, S.C., Bellucci, S., Nanomaterials: Synthesis and Applications in Theranostics. *Nanomaterials*, 11, 12, 1–22, 2021, doi: 10.3390/nano11123228.
16. Mollazadeh, S., Sahebkar, A., Shahlaei, M., Moradi, S., Nano drug delivery systems: Molecular dynamic simulation. *J. Mol. Liq.*, 332, 115823, Jun. 2021, doi: 10.1016/j.molliq.2021.115823.
17. Khurana, A. *et al.*, Role of nanotechnology behind the success of mRNA vaccines for COVID-19. *Nano Today*, 38, 1–7, Jun. 2021, doi: 10.1016/j.nantod.2021.101142.
18. Shamsi, M., Mohammadi, A., Manshadi, M.K.D., Sanati-Nezhad, A., Mathematical and computational modeling of nano-engineered drug delivery systems. *J. Control. Release*, 307, 150–165, 2019, doi: https://doi.org/10.1016/j.jconrel.2019.06.014.
19. Kotzabasaki, M. and Froudakis, G.E., Review of computer simulations on anti-cancer drug delivery in MOFs. *Inorg. Chem. Front.*, 5, 6, 1255–1272, Jun. 2018, doi: 10.1039/c7qi00645d.
20. Andersson, M.P., Jones, M.N., Mikkelsen, K.V., You, F., Mansouri, S.S., Quantum computing for chemical and biomolecular product design. *Curr. Opin. Chem. Eng.*, 36, 100754, Jun. 2022, doi: 10.1016/j. coche.2021.100754.
21. Wang, W., Ye, Z., Gao, H., Ouyang, D., Computational pharmaceutics - A new paradigm of drug delivery. *J. Control. Release*, 338, 119–136, Oct. 2021, doi: 10.1016/j.jconrel.2021.08.030.

22. Nikezić, A.V.V., Bondžić, A.M., Vasić, V.M., Drug delivery systems based on nanoparticles and related nanostructures. *Eur. J. Pharm. Sci.*, 151, 105412, Aug. 2020, doi: 10.1016/j.ejps.2020.105412.
23. Yu, X., Trase, I., Ren, M., Duval, K., Guo, X., Chen, Z., Design of Nanoparticle-Based Carriers for Targeted Drug Delivery. *J. Nanomater.*, 2016, 1–15, 2016, doi: 10.1155/2016/1087250.
24. Sahai, N., Gogoi, M., Ahmad, N., Mathematical Modeling and Simulations for Developing Nanoparticle-Based Cancer Drug Delivery Systems: A Review. *Curr. Pathobiol. Rep.*, 9, 1, 1–8, Mar. 2021, doi: 10.1007/ s40139-020-00219-5.
25. Kashkooli, F.M., Soltani, M., Souri, M., Meaney, C., Kohandel, M., Nexus between *in silico* and *in vivo* models to enhance clinical translation of nanomedicine. *Nano Today*, 36, 101057, 2021, doi: https://doi.org/10.1016/j.nantod.2020.101057.
26. Thiel, W., Semiempirical quantum–chemical methods. *WIREs Comput. Mol. Sci.*, 4, 2, 145–157, Mar. 2014, doi: 10.1002/wcms.1161.
27. Zare, K. and Shadmani, N., Comparison of drug delivery systems: Nanotube and p-Sulphonatocalix[4]arene, by Density Functional Theory. *J. Nanostruct. Chem.*, 3, 1, 72, Dec. 2013, doi: 10.1186/2193-8865-3-72.
28. Al-Qattan, M.N., Deb, P.K., Tekade, R.K., Molecular dynamics simulation strategies for designing carbon-nanotube-based targeted drug delivery. *Drug Discov. Today*, 23, 2, 235–250, 2018, doi: https://doi. org/10.1016/j.drudis.2017.10.002.
29. Katiyar, R.S. and Jha, P.K., Molecular simulations in drug delivery: Opportunities and challenges. *Wiley Interdiscip. Rev. Comput. Mol. Sci.*, 8, 4, 1–18, Jul. 2018, doi: 10.1002/wcms.1358.
30. Liu, X., Shi, D., Zhou, S., Liu, H., Liu, H., Yao, X., Molecular dynamics simulations and novel drug discovery. *Expert Opin. Drug Discov.*, 13, 1, 23–37, 2018, doi: 10.1080/17460441.2018.1403419.
31. Salo-Ahen, O.M.H. *et al.*, Molecular dynamics simulations in drug discovery and pharmaceutical development. *Processes*, 9, 1, 1–63, 2021, doi: 10.3390/pr9010071.
32. Lee, H., Molecular Simulations of PEGylated Biomolecules, Liposomes, and Nanoparticles for Drug Delivery Applications. *Pharmaceutics*, 12, 6, 533, Jun. 2020, doi: 10.3390/pharmaceutics12060533.
33. Padding, J.T. and Briels, W.J., Systematic coarse-graining of the dynamics of entangled polymer melts: the road from chemistry to rheology. *J. Phys. Condens. Matter.*, 23, 23, 233101, Jun. 2011, doi: 10.1088/0953-8984/23/23/233101.
34. Haddish-Berhane, N., Rickus, J.L., Haghighi, K., The role of multiscale computational approaches for rational design of conventional and nanoparticle oral drug delivery systems. *Int. J. Nanomed.*, 2, 3, 315–31, 2007.
35. Li, M., Zou, P., Tyner, K., Lee, S., Physiologically Based Pharmacokinetic (PBPK) Modeling of Pharmaceutical Nanoparticles. *AAPS J.*, 19, 1, 26–42, Jan. 2017, doi: 10.1208/s12248-016-0010-3.

36. Utembe, W., Clewell, H., Sanabria, N., Doganis, P., Gulumian, M., Current Approaches and Techniques in Physiologically Based Pharmacokinetic (PBPK) Modelling of Nanomaterials. *Nanomaterials*, 10, 7, 1267, Jun. 2020, doi: 10.3390/nano10071267.
37. Manshadi, M.K.D. *et al.*, Delivery of magnetic micro/nanoparticles and magnetic-based drug/cargo into arterial flow for targeted therapy. *Drug Deliv.*, 25, 1, 1963–1973, Jan. 2018, doi: 10.1080/10717544.2018.1497106.
38. Shamsi, M., Sedaghatkish, A., Dejam, M., Saghafian, M., Mohammadi, M., Sanati-Nezhad, A., Magnetically assisted intraperitoneal drug delivery for cancer chemotherapy. *Drug Deliv.*, 25, 1, 846–861, Jan. 2018, doi: 10.1080/10717544.2018.1455764.
39. Hadjicharalambous, M., Wijeratne, P.A., Vavourakis, V., From tumour perfusion to drug delivery and clinical translation of *in silico* cancer models. *Methods*, 185, 82–93, Jan. 2021, doi: 10.1016/j.ymeth.2020.02.010.
40. Siepmann, J. and Siepmann, F., Mathematical modeling of drug release from lipid dosage forms. *Int. J. Pharm.*, 418, 1, 42–53, Oct. 2011, doi: 10.1016/j.ijpharm.2011.07.015.
41. Al-Anber, M.J., Theoretical Semiempirical Study of the Biomolecules Interaction with Carbon Nanotubes. *J. Macromol. Sci. Part B*, 50, 12, 2481–2487, Dec. 2011, doi: 10.1080/00222348.2011.557004.
42. Dama, J.F. *et al.*, The Theory of Ultra-Coarse-Graining. 1. General Principles. *J. Chem. Theory Comput.*, 9, 5, 2466–2480, May 2013, doi: 10.1021/ ct4000444.
43. Poon, W., Kingston, B.R., Ouyang, B., Ngo, W., Chan, W.C.W., A framework for designing delivery systems. *Nat. Nanotechnol.*, 15, 10, 819–829, Oct. 2020, doi: 10.1038/s41565-020-0759-5.
44. Salvati, A. *et al.*, Transferrin-functionalized nanoparticles lose their targeting capabilities when a biomolecule corona adsorbs on the surface. *Nat. Nanotechnol.*, 8, 2, 137–143, Feb. 2013, doi: 10.1038/ nnano.2012.237.
45. Li, Z. *et al.*, Emerging well-tailored nanoparticulate delivery system based on in situ regulation of the protein corona. *J. Control. Release*, 320, 1–18, Apr. 2020, doi: 10.1016/j.jconrel.2020.01.007.
46. Lee, H., Effects of Nanoparticle Electrostatics and Protein–Protein Interactions on Corona Formation: Conformation and Hydrodynamics. *Small*, 16, 10, 1906598, Mar. 2020, doi: 10.1002/smll.201906598.
47. Blanco, E., Shen, H., Ferrari, M., Principles of nanoparticle design for overcoming biological barriers to drug delivery. *Nat. Biotechnol.*, 33, 9, 941–951, Sep. 2015, doi: 10.1038/nbt.3330.
48. Rogueda, P.G. and Traini, D., The nanoscale in pulmonary delivery. Part 1: deposition, fate, toxicology and effects. *Expert Opin. Drug Deliv.*, 4, 6, 595–606, Nov. 2007, doi: 10.1517/17425247.4.6.595.
49. Liu, Q., Guan, J., Qin, L., Zhang, X., Mao, S., Physicochemical properties affecting the fate of nanoparticles in pulmonary drug delivery. *Drug Discov. Today*, 25, 1, 150–159, Jan. 2020, doi: 10.1016/j.drudis.2019.09.023.

50. Praphawatvet, T., Peters, J., II, Williams, R.O., Inhaled nanoparticles–An updated review. *Int. J. Pharm.*, 587, 119671, Sep. 2020, doi: 10.1016/j.ijpharm.2020.119671.
51. Pem, B. *et al.*, Fate and transformation of silver nanoparticles in different biological conditions. *Beilstein J. Nanotechnol.*, 12, 665–679, Jul. 2021, doi: 10.3762/bjnano.12.53.
52. Sharma, M., Transdermal and Intravenous Nano Drug Delivery Systems, in: *Applications of Targeted Nano Drugs and Delivery Systems*, pp. 499–550, Elsevier, Amsterdam, Netherlands, 2019, doi: 10.1016/B978-0-12-814029-1.00018-1.
53. Bunker, A. and Róg, T., Mechanistic Understanding From Molecular Dynamics Simulation in Pharmaceutical Research 1: Drug Delivery. *Front. Mol. Biosci.*, 7, 1–36, Nov. 2020, doi: 10.3389/fmolb.2020.604770.
54. Huynh, L., Neale, C., Pomès, R., Allen, C., Computational approaches to the rational design of nanoemulsions, polymeric micelles, and dendrimers for drug delivery. *Nanomed. Nanotechnol. Biol. Med.*, 8, 1, 20–36, Jan. 2012, doi: 10.1016/j.nano.2011.05.006.
55. Erlebach, A. *et al.*, Predicting Solubility of Small Molecules in Macromolecular Compounds for Nanomedicine Application from Atomistic Simulations. *Adv. Theory Simul.*, 3, 5, 2000001, May 2020, doi: 10.1002/ adts.202000001.
56. Larsen, A.S. *et al.*, Determining short-lived solid forms during phase transformations using molecular dynamics. *CrystEngComm*, 21, 27, 4020–4024, 2019, doi: 10.1039/C9CE00460B.
57. Chan, T. and Ouyang, D., Investigating the molecular dissolution process of binary solid dispersions by molecular dynamics simulations. *Asian J. Pharm. Sci.*, 13, 3, 248–254, May 2018, doi: 10.1016/j.ajps.2017.07.011.
58. Han, R. *et al.*, Insight into the Dissolution Molecular Mechanism of Ternary Solid Dispersions by Combined Experiments and Molecular Simulations. *AAPS PharmSciTech*, 20, 7, 274, Oct. 2019, doi: 10.1208/ s12249-019-1486-9.
59. Johnson, S.R. and Zheng, W., Recent progress in the computational prediction of aqueous solubility and absorption. *AAPS J.*, 8, 1, E27–E40, Mar. 2006, doi: 10.1208/aapsj080104.
60. Mathieu, D., QSPR versus fragment-based methods to predict octanol-air partition coefficients: Revisiting a recent comparison of both approaches. *Chemosphere*, 245, 125584, Apr. 2020, doi: 10.1016/j.chemosphere.2019.125584.
61. Duarte Ramos Matos, G., Kyu, D.Y., Loeffler, H.H., Chodera, J.D., Shirt, M.R., Mobley, D.L., Approaches for Calculating Solvation Free Energies and Enthalpies Demonstrated with an Update of the FreeSolv Database. *J. Chem. Eng. Data*, 62, 5, 1559–1569, May 2017, doi: 10.1021/ acs.jced.7b00104.
62. Nedyalkova, M.A., Madurga, S., Tobiszewski, M., Simeonov, V., Calculating the Partition Coefficients of Organic Solvents in Octanol/Water and Octanol/Air. *J. Chem. Inf. Model.*, 59, 5, 2257–2263, May 2019, doi: 10.1021/acs.jcim.9b00212.

63. Bergström, C.A.S. and Larsson, P., Computational prediction of drug solubility in water-based systems: Qualitative and quantitative approaches used in the current drug discovery and development setting. *Int. J. Pharm.*, 540, 1–2, 185–193, Apr. 2018, doi: 10.1016/j.ijpharm.2018.01.044.
64. Li, C.-X., Wang, H.-B., Oppong, D., Wang, J.-X., Chen, J.-F., Le, Y., Excipient-Assisted Vinpocetine Nanoparticles: Experiments and Molecular Dynamic Simulations. *Mol. Pharm.*, 11, 11, 4023–4035, Nov. 2014, doi: 10.1021/mp500045t.
65. Venable, R.M., Krämer, A., Pastor, R.W., Molecular Dynamics Simulations of Membrane Permeability. *Chem. Rev.*, 119, 9, 5954–5997, May 2019, doi: 10.1021/acs.chemrev.8b00486.
66. Bennion, B.J. *et al.*, Subscriber access provided by UB + Fachbibliothek Chemie | (FU-Bibliothekssystem) Predicting a Drug's Membrane Permeability: A Computational Model Validated With *In Vitro* Permeability Assay Data. *J. Phys. Chem. B*, 121.20, 5228–5237, 2017. Available: http://pubs.acs.org.
67. Dahan, A., Beig, A., Lindley, D., Miller, J.M., The solubility–permeability interplay and oral drug formulation design: Two heads are better than one. *Adv. Drug Deliv. Rev.*, 101, 99–107, Jun. 2016, doi: 10.1016/j. addr.2016.04.018.
68. Ferreira, M., Silva, E., Barreiros, L., Segundo, M.A., Costa Lima, S.A., Reis, S., Methotrexate loaded lipid nanoparticles for topical management of skin-related diseases: Design, characterization and skin permeation potential. *Int. J. Pharm.*, 512, 1, 14–21, Oct. 2016, doi: 10.1016/j. ijpharm.2016.08.008.
69. Nair, M., Jayant, R.D., Kaushik, A., Sagar, V., Getting into the brain: Potential of nanotechnology in the management of NeuroAIDS. *Adv. Drug Deliv. Rev.*, 103, 202–217, Aug. 2016, doi: 10.1016/j.addr.2016.02.008.
70. Sheng, R. *et al.*, Novel 1-Phenyl-3-hydroxy-4-pyridinone Derivatives as Multifunctional Agents for the Therapy of Alzheimer's Disease. *ACS Chem. Neurosci.*, 7, 1, 69–81, Jan. 2016, doi: 10.1021/ acschemneuro.5b00224.
71. Bhandari, A., Ocular Fluid Mechanics and Drug Delivery: A Review of Mathematical and Computational Models. *Pharm. Res.*, 38, 12, 2003–2033, Dec. 2021, doi: 10.1007/s11095-021-03141-6.
72. Lomize, A.L. *et al.*, PerMM: A Web Tool and Database for Analysis of Passive Membrane Permeability and Translocation Pathways of Bioactive Molecules. *J. Chem. Inf. Model.*, 59, 7, 3094–3099, Jul. 2019, doi: 10.1021/acs.jcim.9b00225.
73. Soliman, M.E. *et al.*, Simulation Models for Prediction of Bioavailability of Medicinal Drugs—the Interface Between Experiment and Computation. *AAPS PharmSciTech*, 23, 3, 86, Apr. 2022, doi: 10.1208/ s12249-022-02229-5.
74. Bemporad, D., Essex, J.W., Luttmann, C., Permeation of Small Molecules through a Lipid Bilayer: A Computer Simulation Study. *J. Phys. Chem. B*, 108, 15, 4875–4884, Apr. 2004, doi: 10.1021/jp035260s.
75. Dickson, C.J., Hornak, V., Bednarczyk, D., Duca, J.S., Using Membrane Partitioning Simulations To Predict Permeability of Forty-Nine Drug-Like

Molecules. *J. Chem. Inf. Model.*, 59, 1, 236–244, Jan. 2019, doi: 10.1021/acs.jcim.8b00744.
76. Orsi, M. and Essex, J.W., Permeability of drugs and hormones through a lipid bilayer: insights from dual-resolution molecular dynamics. *Soft Matter*, 6, 16, 3797, 2010, doi: 10.1039/c0sm00136h.
77. Lundborg, M., Wennberg, C.L., Narangifard, A., Lindahl, E., Norlén, L., Predicting drug permeability through skin using molecular dynamics simulation. *J. Control. Release*, 283, 269–279, Aug. 2018, doi: 10.1016/j.jconrel.2018.05.026.
78. Pecoraro, B., Tutone, M., Hoffman, E., Hutter, V., Almerico, A.M., Traynor, M., Predicting Skin Permeability by Means of Computational Approaches: Reliability and Caveats in Pharmaceutical Studies. *J. Chem. Inf. Model.*, 59, 5, 1759–1771, May 2019, doi: 10.1021/acs. jcim.8b00934.
79. Gupta, R., Dwadasi, B.S., Rai, B., Mitragotri, S., Effect of Chemical Permeation Enhancers on Skin Permeability: *In silico* screening using Molecular Dynamics simulations. *Sci. Rep.*, 9, 1, 1456, Dec. 2019, doi: 10.1038/s41598-018-37900-0.
80. Otto, D., Combrinck, J., Otto, A., Tiedt, L., de Villiers, M., Dissipative Particle Dynamics Investigation of the Transport of Salicylic Acid through a Simulated *In Vitro* Skin Permeation Model. *Pharmaceuticals*, 11, 4, 134, Dec. 2018, doi: 10.3390/ph11040134.
81. McLean, K. and Zhan, W., Mathematical modelling of nanoparticle-mediated topical drug delivery to skin tissue. *Int. J. Pharm.*, 611, 121322, Jan.2022, doi: 10.1016/j.ijpharm.2021.121322.
82. Bagchi, S., Chhibber, T., Lahooti, B., Verma, A., Borse, V., Jayant, R.D., *In-vitro* blood-brain barrier models for drug screening and permeation studies: an overview. *Drug Des. Devel. Ther.*, 13, 3591–3605, Oct. 2019, doi: 10.2147/DDDT.S218708.
83. Yu, F., Tonnis, K., Kasting, G.B., Jaworska, J., Computer Simulation of Skin Permeability of Hydrophobic and Hydrophilic Chemicals – Influence of Follicular Pathway. *J. Pharm. Sci.*, 110, 5, 2149–2156, May 2021, doi: 10.1016/j.xphs.2020.12.020.
84. Maharao, N., Antontsev, V., Hou, H., Walsh, J., Varshney, J., Scalable *in silico* Simulation of Transdermal Drug Permeability: Application of BIOiSIM Platform. *Drug Des. Devel. Ther.*, 14, 2307–2317, Jun.2020, doi: 10.2147/DDDT.S253064.
85. Lombardo, S.M., Schneider, M., Türeli, A.E., Günday Türeli, N., Key for crossing the BBB with nanoparticles: the rational design. *Beilstein J. Nanotechnol.*, 11, 866–883, Jun. 2020, doi: 10.3762/bjnano.11.72.
86. Miranda, A., Cova, T., Sousa, J., Vitorino, C., Pais, A., Computational modeling in glioblastoma: from the prediction of blood–brain barrier permeability to the simulation of tumor behavior. *Future Med. Chem.*, 10, 1, 121–131, Jan. 2018, doi: 10.4155/fmc-2017-0128.

87. Shityakov, S. and Förster, C.Y., Computational simulation and modeling of the blood–brain barrier pathology. *Histochem. Cell Biol.*, 149, 5, 451–459, May 2018, doi: 10.1007/s00418-018-1665-x.
88. Jiang, L., Li, S., Zheng, J., Li, Y., Huang, H., Recent Progress in Microfluidic Models of the Blood-Brain Barrier. *Micromachines*, 10, 6, 375, Jun.2019, doi: 10.3390/mi10060375.
89. Gkountas, A.A., Polychronopoulos, N.D., Sofiadis, G.N., Karvelas, E.G., Spyrou, L.A., Sarris, I.E., Simulation of magnetic nanoparticles crossing through a simplified blood-brain barrier model for Glioblastoma multiforme treatment. *Comput. Methods Programs Biomed.*, 212, 106477, Nov.2021, doi: 10.1016/j.cmpb.2021.106477.
90. Meng, F., Xi, Y., Huang, J., Ayers, P.W., A curated diverse molecular database of blood-brain barrier permeability with chemical descriptors. *Sci. Data*, 8, 1, 289, Dec. 2021, doi: 10.1038/s41597-021-01069-5.
91. Eriksson, J., Sjögren, E., Thörn, H., Rubin, K., Bäckman, P., Lennernäs, H., Pulmonary absorption – estimation of effective pulmonary permeability and tissue retention of ten drugs using an *ex vivo* rat model and computational analysis. *Eur. J. Pharm. Biopharm.*, 124, 1–12, Mar. 2018, doi: 10.1016/j.ejpb.2017.11.013.
92. Bäckman, P. *et al.*, iBCS: 2. Mechanistic Modeling of Pulmonary Availability of Inhaled Drugs versus Critical Product Attributes. *Mol. Pharm.*, 19, 7, 2040–2047, May 2022, doi: 10.1021/acs.molpharmaceut.2c00112.
93. Ramsay, E. *et al.*, Corneal and conjunctival drug permeability: Systematic comparison and pharmacokinetic impact in the eye. *Eur. J. Pharm. Sci.*, 119, 83–89, Jul. 2018, doi: 10.1016/j.ejps.2018.03.034.
94. Pak, J., Chen, Z.J., Sun, K., Przekwas, A., Walenga, R., Fan, J., Computational modeling of drug transport across the *in vitro* cornea. *Comput. Biol. Med.*, 92, 139–146, Jan. 2018, doi: 10.1016/j.compbiomed.2017.11.009.
95. Khalkhali, M., Mohammadinejad, S., Khoeini, F., Rostamizadeh, K., Vesicle-like structure of lipid-based nanoparticles as drug delivery system revealed by molecular dynamics simulations. *Int. J. Pharm.*, 559, 173–181, 2019, doi: https://doi.org/10.1016/j.ijpharm.2019.01.036.
96. Ahmed, S. *et al.*, Experimental and molecular modeling approach to optimize suitable polymers for fabrication of stable fluticasone nanoparticles with enhanced dissolution and antimicrobial activity. *Drug Des. Devel. Ther.*, 12, 255–269, Feb. 2018, doi: 10.2147/DDDT.S148912.
97. Connell, D.J. *et al.*, Rationalising drug delivery using nanoparticles: a combined simulation and immunology study of GnRH adsorbed to silica nanoparticles. *Sci. Rep.*, 8, 1, 17115, Nov. 2018, doi: 10.1038/ s41598-018-35143-7.
98. Monti, S., Barcaro, G., Sementa, L., Carravetta, V., Ågren, H., Dynamics and self-assembly of bio-functionalized gold nanoparticles in solution: Reactive molecular dynamics simulations. *Nano Res.*, 11, 4, 1757–1767, Apr. 2018, doi: 10.1007/s12274-017-1704-2.

99. Decuzzi, P., Facilitating the Clinical Integration of Nanomedicines: The Roles of Theoretical and Computational Scientists. *ACS Nano*, 10, 9, 8133–8138, Sep. 2016, doi: 10.1021/acsnano.6b05536.
100. Cern, A., Marcus, D., Tropsha, A., Barenholz, Y., Goldblum, A., New drug candidates for liposomal delivery identified by computer modeling of liposomes' remote loading and leakage. *J. Control. Release*, 252, 18–27, Apr. 2017, doi: 10.1016/j.jconrel.2017.02.015.
101. Dzieciuch, M. *et al.*, PEGylated Liposomes as Carriers of Hydrophobic Porphyrins. *J. Phys. Chem. B*, 119, 22, 6646–6657, Jun. 2015, doi: 10.1021/acs.jpcb.5b01351.
102. Mahdavi, M., Rahmani, F., Nouranian, S., Molecular simulation of pH-dependent diffusion, loading, and release of doxorubicin in graphene and graphene oxide drug delivery systems. *J. Mater. Chem. B*, 4, 46, 7441–7451, 2016, doi: 10.1039/C6TB00746E.
103. Kamaly, N., Yameen, B., Wu, J., Farokhzad, O.C., Degradable Controlled-Release Polymers and Polymeric Nanoparticles: Mechanisms of Controlling Drug Release. *Chem. Rev.*, 116, 4, 2602–2663, Feb. 2016, doi: 10.1021/acs.chemrev.5b00346.
104. Taylor, K., Tabish, T.A., Narayan, R.J., Drug Release Kinetics of DOX- Loaded Graphene-Based Nanocarriers for Ovarian and Breast Cancer Therapeutics. *Appl. Sci.*, 11, 23, 11151, Nov. 2021, doi: 10.3390/ app112311151.
105. Jain, V., Maingi, V., Maiti, P.K., Bharatam, P.V., Molecular dynamics simulations of PPI dendrimer–drug complexes. *Soft Matter*, 9, 28, 6482, 2013, doi: 10.1039/c3sm50434d.
106. Arifin, D.Y., Lee, L.Y., Wang, C.-H., Mathematical modeling and simulation of drug release from microspheres: Implications to drug delivery systems. *Adv. Drug Deliv. Rev.*, 58, 12–13, 1274–1325, Nov. 2006, doi: 10.1016/j.addr.2006.09.007.
107. Gao, G.F. *et al.*, Predicting drug release and degradation kinetics of long-acting microsphere formulations of tacrolimus for subcutaneous injection. *J. Control. Release*, 329, 372–384, Jan. 2021, doi: 10.1016/j. jconrel.2020.11.055.
108. Dogra, P. *et al.*, Mathematical modeling in cancer nanomedicine: a review. *Biomed. Microdevices*, 21, 2, 40, Jun. 2019, doi: 10.1007/ s10544-019-0380-2.
109. Feitosa, R.C., Geraldes, D.C., Beraldo-de-Araújo, V.L., Costa, J.S.R., Oliveira-Nascimento, L., Pharmacokinetic Aspects of Nanoparticle-in-Matrix Drug Delivery Systems for Oral/Buccal Delivery. *Front. Pharmacol.*, 10, 1–15, Sep. 2019, doi: 10.3389/fphar.2019.01057.
110. Grass, G.M., Simulation models to predict oral drug absorption from *in vitro* data. *Adv. Drug Deliv. Rev.*, 23, 1–3, 199–219, Jan. 1997, doi: 10.1016/S0169-409X(96)00436-X.
111. Dressman, J.B., Thelen, K., Jantratid, E., Towards Quantitative Prediction of Oral Drug Absorption. *Clin. Pharmacokinet.*, 47, 10, 655–667, 2008, doi: 10.2165/00003088-200847100-00003.

112. Huang, W., Lee, S.L., Yu, L.X., Mechanistic Approaches to Predicting Oral Drug Absorption. *AAPS J.*, 11, 2, 217–224, Jun. 2009, doi: 10.1208/s12248-009-9098-z.
113. Dogra, P. *et al.*, A mathematical model to predict nanomedicine pharmacokinetics and tumor delivery. *Comput. Struct. Biotechnol. J.*, 18, 518–531, 2020, doi: 10.1016/j.csbj.2020.02.014.
114. Lopez, H. and Lobaskin, V., Coarse-grained model of adsorption of blood plasma proteins onto nanoparticles. *J. Chem. Phys.*, 143, 24, 243138, Dec. 2015, doi: 10.1063/1.4936908.
115. Tavanti, F., Pedone, A., Menziani, M.C., A closer look into the ubiquitin corona on gold nanoparticles by computational studies. *New J. Chem.*, 39, 4, 2474–2482, 2015, doi: 10.1039/C4NJ01752H.
116. Xu, Z. and Kleinstreuer, C., Heterogeneous blood flow in microvessels with applications to nanodrug transport and mass transfer into tumor tissue. *Biomech. Model. Mechanobiol.*, 18, 1, 99–110, Feb. 2019, doi: 10.1007/s10237-018-1071-2.
117. Xu, Z. and Kleinstreuer, C., Direct nanodrug delivery for tumor targeting subject to shear-augmented diffusion in blood flow. *Med. Biol. Eng. Comput.*, 56, 11, 1949–1958, Nov. 2018, doi: 10.1007/s11517-018-1818-z.
118. Moss, D.M. and Siccardi, M., Optimizing nanomedicine pharmacokinetics using physiologically based pharmacokinetics modelling. *Br. J. Pharmacol.*, 171, 17, 3963–3979, Sep. 2014, doi: 10.1111/bph.12604.
119. Yuan, D., He, H., Wu, Y., Fan, J., Cao, Y., Physiologically Based Pharmacokinetic Modeling of Nanoparticles. *J. Pharm. Sci.*, 108, 1, 58–72, Jan. 2019, doi: 10.1016/j.xphs.2018.10.037.
120. Villa Nova, M. *et al.*, Nanomedicine Ex Machina: Between Model-Informed Development and Artificial Intelligence. *Front. Digit. Health*, 4, 1–12, Feb. 2022, doi: 10.3389/fdgth.2022.799341.
121. Gao, Y. *et al.*, Simulation study of the effects of interstitial fluid pressure and blood flow velocity on transvascular transport of nanoparticles in tumor microenvironment. *Comput. Methods Programs Biomed.*, 193, 105493, Sep. 2020, doi: 10.1016/j.cmpb.2020.105493.
122. Magarkar, A., Róg, T., Bunker, A., A computational study suggests that replacing PEG with PMOZ may increase exposure of hydrophobic targeting moiety. *Eur. J. Pharm. Sci.*, 103, 128–135, May 2017, doi: 10.1016/j.ejps.2017.03.008.
123. Sousa, S.F., Peres, J., Coelho, M., Vieira, T.F., Analyzing PEGylation through Molecular Dynamics Simulations. *ChemistrySelect*, 3, 29, 8415–8427, Aug. 2018, doi: 10.1002/slct.201800855.
124. Tian, F., Yue, T., Li, Y., Zhang, X., Computer simulation studies on the interactions between nanoparticles and cell membrane. *Sci. China Chem.*, 57, 12, 1662–1671, Dec. 2014, doi: 10.1007/s11426-014-5231-7.

125. Róg, T., Girych, M., Bunker, A., Mechanistic understanding from molecular dynamics in pharmaceutical research 2: Lipid membrane in drug design. *Pharmaceuticals*, 14, 10, 1–99, Oct. 2021, doi: 10.3390/ph14101062.
126. Tripathi, J. *et al.*, A review on recent advancements in the hemodynamics of nano-drug delivery systems. *Nanosci. Technol.: Int. J.*, 11.1, 2020. DOI: 10.1615/NanoSciTechnolIntJ.2020033448
127. Vasu, B., Dubey, A., Bég, O.A., Gorla, R.S.R., Micropolar pulsatile blood flow conveying nanoparticles in a stenotic tapered artery: NON-Newtonian pharmacodynamic simulation. *Comput. Biol. Med.*, 126, 104025, Nov. 2020, doi: 10.1016/j.compbiomed.2020.104025.
128. Gu, B. *et al.*, Multiphysics Modelling and Simulation of Thrombolysis via Activated Platelet-Targeted Nanomedicine. *Pharm. Res.*, 39, 1, 41–56, Jan. 2022, doi: 10.1007/s11095-021-03161-2.
129. Kim, T.H., Shin, S., Shin, B.S., Model-based drug development: application of modeling and simulation in drug development. *J. Pharm. Investig.*, 48, 4, 431–441, Jul. 2018, doi: 10.1007/s40005-017-0371-3.
130. Sun, X. and Hu, B., Mathematical modeling and computational prediction of cancer drug resistance. *Brief. Bioinform.*, 19, 6, 1382–1399, Nov. 2018, doi: 10.1093/bib/bbx065.
131. Sachdev, K. and Gupta, M.K., A comprehensive review of computational techniques for the prediction of drug side effects. *Drug Dev. Res.*, 81, 6, 650–670, Sep. 2020, doi: 10.1002/ddr.21669.
132. Galeano, D., Li, S., Gerstein, M., Paccanaro, A., Predicting the frequencies of drug side effects. *Nat. Commun.*, 11, 1, 4575, Dec. 2020, doi: 10.1038/s41467-020-18305-y.
133. Singh, A.V. *et al.*, Artificial Intelligence and Machine Learning in Computational Nanotoxicology: Unlocking and Empowering Nanomedicine. *Adv. Healthc. Mater.*, 9, 17, 1901862, Sep. 2020, doi: 10.1002/adhm.201901862.
134. Schrey, A.K. *et al.*, Computational prediction of immune cell cytotoxicity. *Food Chem. Toxicol.*, 107, 150–166, Sep. 2017, doi: 10.1016/j. fct.2017.05.041.
135. Pérez Santín, E. *et al.*, Toxicity prediction based on artificial intelligence: A multidisciplinary overview. *WIREs Comput. Mol. Sci.*, 11, 5, 1–32, Sep. 2021, doi: 10.1002/wcms.1516.
136. Attia, M.F., Anton, N., Wallyn, J., Omran, Z., Vandamme, T.F., An overview of active and passive targeting strategies to improve the nanocarriers efficiency to tumour sites. *J. Pharm. Pharmacol.*, 71, 8, 1185–1198, Jul. 2019, doi: 10.1111/jphp.13098.
137. Lin, P. and Colina, C.M., Molecular simulation of protein–polymer conjugates. *Curr. Opin. Chem. Eng.*, 23, 44–50, Mar. 2019, doi: 10.1016/j. coche.2019.02.006.

138. Malviya, R., Sharma, P.K., Sundram, S., Dhanaraj, R.K., Balusamy, B., *Bioinformatics tools and big data analytics for patient care*, vol. 1, p. 240, Routledge Taylor & Francis Group, Florida, United States, Aug. 2022.

5

Application of Quantum Computational Simulation in Drug Delivery Strategies with Carbon Nanotubes

Rupali Sharma[1*], Yashomita Mehta[1], Shekhar Sharma[2], Jagriti Narang[3] and Shabnam Thakur[1]

[1]*Amity Institute of Pharmacy, Amity University Haryana Amity Education Valley Manesar, Gurugram, Haryana, India*
[2]*Llyod Institute of Management and Technology, Knowledge Park II, Greater Noida, U.P., India*
[3]*Department of Biotechnology, Jamia Hamdard University, New Delhi, India*

Abstract
One of the most promising applications of nanotechnology is in the field of target-specific delivery of medications and nano-biosensors utilizing carbon nanotubes (CNTs). Because of their distinctive features, carbon nanotubes (CNTs) have attracted much interest from scientists lately. The crucial characteristics of CNTs used in the medical industry include their diminutive size, lightweight, high surface area-to-volume ratio, and affinity to restrain substances. Despite some CNT deficiencies, they are not dispersible in various aqueous solvents and biological media, so it is very difficult to develop an efficient drug delivery system with them. Functionalization could thereby increase their biocompatibility and solubility. The many forms of functionalization procedures, including non-covalent, covalent, defect modification, and encapsulation, are important. The covalent and noncovalent drug adsorption on the external surface of functionalized carbon nanotubes (fCNTs) could be investigated using quantum chemical calculations. For the evolution of any new drug delivery system, we need to understand the system's basic nature, properties, and methods of preparation. An effective method for analyzing drug delivery systems is quantum computing. Researchers are urged to apply computational approaches such as quantum mechanics (QM), molecular dynamics (MD) simulation studies, and computer-aided molecular design (CAD)

**Corresponding author*: rsharma9@ggn.amity.edu; rupali6kalra@gmail.com

Rishabha Malviya, Sonali Sundram and Dhanalekshmi Unnikrishnan Meenakshi (eds.)
Drug Delivery Systems Using Quantum Computing, (141–170) © 2024 Scrivener Publishing LLC

tools to provide a suitable low-cost method for this optimization. Various computational simulation approaches determined the behavior of nanostructures of CNTs, how drugs are delivered, the significance of maintaining a balance between therapeutic effectiveness and cytotoxicity, and how nanomaterials interact with cell membranes. The creation of CNT-based nano-biosensors will benefit from the knowledge gained from the quantum computational calculations that were carried out. The interaction between CNT and the biomolecule was investigated using the density functional theory approach, to assess the synthesized nano-biosensor's stability and reactivity. In other words, the computation performed in this manner can assist researchers in creating and using targeted medications and minimize the laboratory's trial-and-error procedure.

***Keywords*:** Carbon allotropes, biosafety, quantum mechanic, molecular dynamics, monte carlo, drug delivery, nano-biosensors

5.1 Introduction

The materials that have at least one exterior dimension and a size range of around 1–100 nm are referred to as nanomaterials (NMs). Nanomaterials are now widely employed in many branches of science and technology due to the rapid growth of nanotechnology. On the line separating molecules from nanomaterials is the carbon molecule C 60, which exclusively has hexagonal and pentagonal interatomic bonding networks [1].

Carbon nanomaterials (CNs), which range from zero-dimensional (0-D) fullerenes and carbon nano-onions (CNOs) to one-dimensional (1-D) CNTs and carbon nanoscrolls (CNSs), two-dimensional (2-D) graphene, and three-dimensional (3-D) diamond, have promising uses in a variety of fields, including smart drug delivery, toxicological studies, energy storage, and field emission display, biomedical applications, and biological and chemical sensors [2]. The first smallest stable carbon nanostructure is called fullerene, initially reported in 1985 [3]. After six years, in 1991, Iijima made the first experimental TEM observation of carbon nanotubes, leading to the discovery of the material (transmission electron microscopy). CNTs are made of sheets of graphite that can be rolled up into a nanoscale tube with a diameter of up to 100 nm and a length in the micrometer scale. The key component of artificial carbon allotropes based on honeycomb networks is carbon nanotubes (CNTs), which are new, useful nanomaterials [4] as shown in Figure 5.1.

The carbon nanotube, which is made of a hexagonal mesh and simulates a sheet of graphite, contains a carbon atom at the apex of each mesh. This graphene sheet is continually attached along its two edges as it is

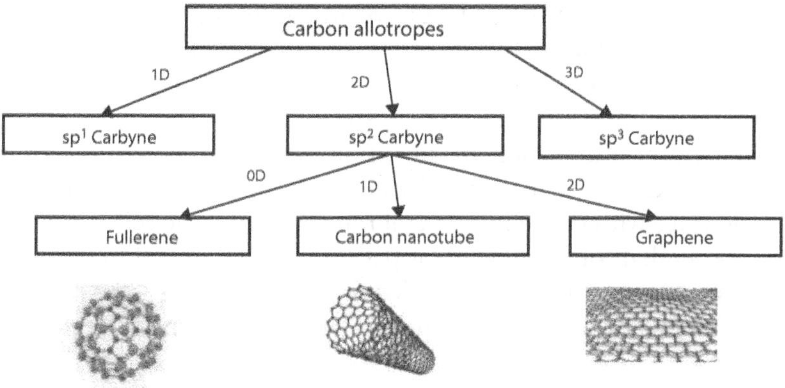

Figure 5.1 Types of carbon allotropes.

rolled. Three basic forms are single-, double-, and multi-walled CNTs, respectively, based on the number of the rolled graphite sheet. The kind garnering the greatest interest is the single-walled CNTs, whose diameter of 0.4–2 nm justifies the name of the nanotube [5].

The hexagonally organized, sp2-hybridized carbon atoms that make up CNTs are encased in cylindrical sheets of graphene [6]. Single-walled carbon nanotubes (SWCNTs) are single cylinders, whereas multi-walled carbon nanotubes (MWCNTs) have many cylinders with coincident centers. MWCNTs' internal diameter of the cylindrical tube is approximately 1–2 nm and the external diameter of the cylindrical tube is approximately 2–100 nm with lengths starting from 1 mm; SWCNTs have a diameter span of 0.4 to 3.0 nm and 20 to 1000 nm.

MWCNTs can be divided into two groups based on how the graphite layer is structured: "Russian doll" models having numerous sheets of graphene placed within a concentric structure, and "parchment-like" structures with a sheet of graphene rolled up around itself [7].

The extraordinary characteristic of CNTs is aspect ratios and exceptionally strong carbon atom bonds, making them special and distinctive. The innovative materials that demonstrate property advantages at lower concentrations than conventional nanotechnology might be imagined because of the extraordinary properties of CNTs. The development of CNT-based target drug delivery or composites that are interesting for novel applications is constrained by the fact that CNTs are precipitated in solvents and that they are not easily manipulated in solvents due to their lack of solubility [8].

CNTs' bare or pristine surfaces are hydrophobic by nature, making them insoluble in biological mediums and unsuitable for drug delivery. Functionalization could be used to increase CNTs' solubility and their compatibility with biological media. Encapsulation, non-covalent, covalent, and defect modification are all options for functionalization. To facilitate the uptake of pharmaceutically active medicines, the functionalized CNTs interface with the cellular membrane and enter the cells via an endocytosis-dependent or -independent pathway. In summary, functionalized CNTs offer new prospects in the fields of nanotechnology and nanomedicine by lowering drug resistance, increasing solubility, and limiting cellular penetration [9].

5.2 Properties of CNTs

A family of materials having unanticipated properties like electrical, mechanical, and thermal properties is now possible thanks to carbon nanotubes (CNTs). One way to visualize the structure of CNTs is rolled into a cylindrical shape by one or more sheets of graphene with sp^2-hybridized carbon atoms and arranged in a skep pattern. Regardless of whether they are single- or multi-walled carbon nanotubes, they exhibit a number of remarkable qualities, such as ultralight weight, incredible strength, good conductivity, and significant electronic properties range from metallic to semiconducting (MWCNTs) [10].

5.2.1 Electrical Properties of CNTs

Depending on their diameter, length, and chirality, carbon nanotubes can either be metallic or semiconducting. It depends on their chiral vector, and the carbon atoms in CNTs can be arranged around the outside of the nanotube in an armchair, zigzag, or chiral pattern. CNTs are presented by the two integers (n, m). The geometrical structure of a 2-dimensional sp2 graphene sheet has been characterized in terms of a chiral vector (C),

$$\text{where } C = na_1 + ma_2 \tag{5.1}$$

(n,m) with integer n and m. The letters a1 and a2 represent the unit vectors in the hexagonal lattice sheets of graphene [6, 11] as shown in Figure 5.2.

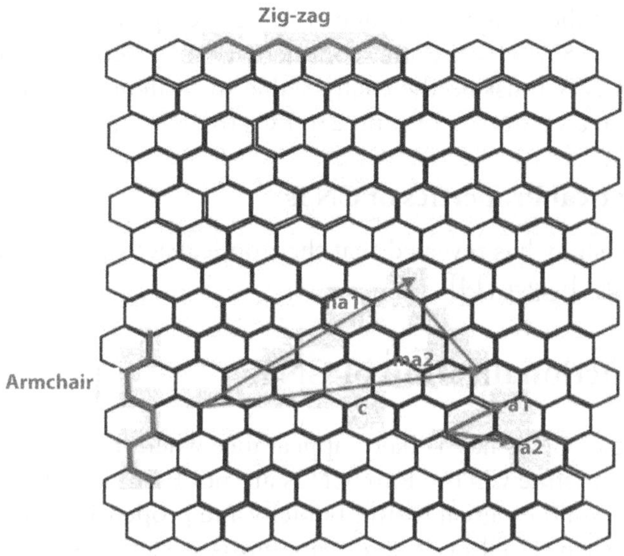

Figure 5.2 Generating the CNT chirality vector from the graphene sheet.

As a rule, these nanotubes are referred to as armchair ones when n = m, zigzag ones when m = 0, and chiral ones for other ranges. If n m is greater than or equal to 3 times any value, the carbon nanotube is likely to be metallic or exceedingly conductive; otherwise, it may be a semi-metallic type or a semiconductor. The armchair type can typically be described as metallic, whereas the other types can be classified as semiconductors [12, 13].

5.2.2 Elastro-Mechanical Properties of CNTs

A two-dimensional arrangement of carbon atoms in graphene sheets that permits significant out-of-plane distortions while maintaining remarkable strength against any in-plane deformation or fracture shows that carbon nanotubes have very good elasto-mechanical properties. Nanotubes' structural and material properties suggest that they could be used to create the next generation of composite materials that are incredibly light but both robust and highly elastic [14].

5.2.3 Thermal Properties of CNTs

CNTs have very good thermal conductors that exhibit a property called ballistic conductors. Both SWNTs and MWNTs are good insulators with a lateral axis of the carbon nanotube [15]. At a temperature of around 3600 Wm^{-1} K^{-1}, the SWNT's thermal conductivity may be measured along its tube axis. CNTs must be thermally stable in ordinary dry air at around 750°C and in a vacuum at around 2800°C [16].

5.2.4 Optical Properties of CNTs

A literature study has revealed that their optical activity declines as chiral nanotubes get bigger [14].

5.3 Functionalization of CNTs

The surface of CNTs has a "bald" appearance when they are in their most pure form (pristine CNTs). Pure CNTs are not used directly in biological fluid because of their great hydrophobicity and propensity to have harmful side effects such as cell-cycle arrest, apoptosis, and necrosis [17, 18].

Increasing the absorption and excretion from biological systems can also be made possible by functionalization, in addition to hydrophilicity, increased drug-carrying capacity, improved cellular uptake from biomembrane, and internalization. Enhancing the water solubility of CNTs is mostly dependent on physical characteristics like wettability and a reduction in the interbond attraction. Few conventional methods for improving the solubility of CNTs are the addition of polar functional groups, the use of surfactants, and PEGylation. Due to a delay in the opsonization process, PEGylation also ensures enhanced circulation of functionalized carbon nanotubes in the biological fluid in addition to improving water solubility. Additionally, it has been demonstrated that functionalization enhances the cellular absorption and internalization of CNTs [19].

Before exposing the CNT surface to the biological fluid, functionalization of the CNTs on that surface is of utmost importance. The CNT structure is functionalized into a usable component of a therapeutic drug suitable for clinical application. By adding functionalities to CNT, researchers were able to increase the material's biocompatibility and solubility while reducing toxicity [20]. Non-covalent, covalent, and defect modification and the encapsulation of bioactive compounds or medications inside CNTs are important functionalization techniques [9, 21].

5.3.1 Covalent Functionalization

The stable functionalized CNT that results from this covalent functionalization of CNTs through the addition of carbenes or nitrene as well

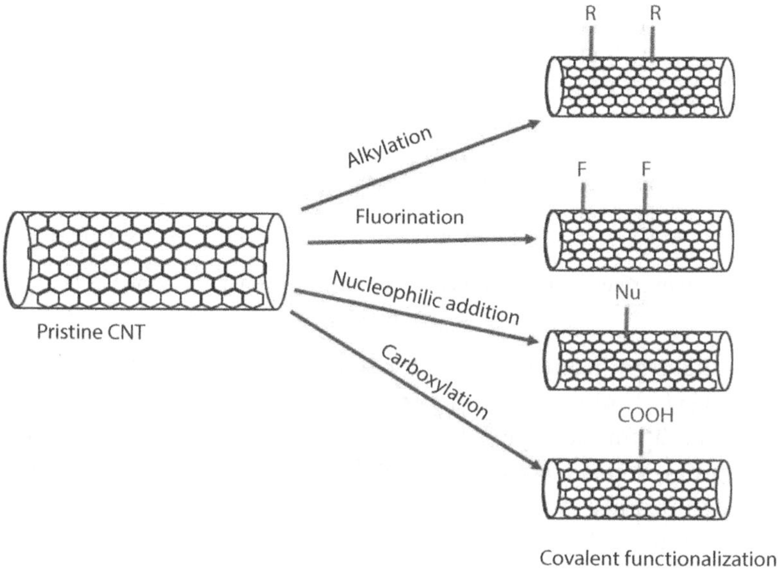

Figure 5.3 An overview of both covalent and non-covalent approaches for functionalizing CNTs.

as amidation (-NH2), halogenation (F/Cl/Br), thiolation (SH), and other processes is very suitable for use in drug delivery vehicle [22]. Functionalization with these groups causes the hybridization of carbon atoms to shift hybridization from sp2 to sp3, preventing electron transfer, which has been proven to be helpful for the administration of therapeutic compounds since it causes the progressive release of pharmaceuticals [19].

The benefit of this covalent functionalization is an improvement in solubility and biological media compatibility. The creation of defects on CNTs is one of the main disadvantages of this sort of functionalization [16, 23, 24] as shown in Figure 5.3.

5.3.2 Noncovalent Functionalization of CNTs

Through van der Waals forces, CNTs may be functionalized non-covalently. The attachment of polymers, biomolecules, or amphiphilic surfactants at the supramolecular level is often how this functionalization is accomplished. The bulk of carbon surfaces, including graphene sheets, are used for CNTs and are wrapped by using the hydrophobic or "π–π" stacking interaction. With the exception of slight length reduction, the physical characteristics of the CNTs can be essentially conserved in the network,

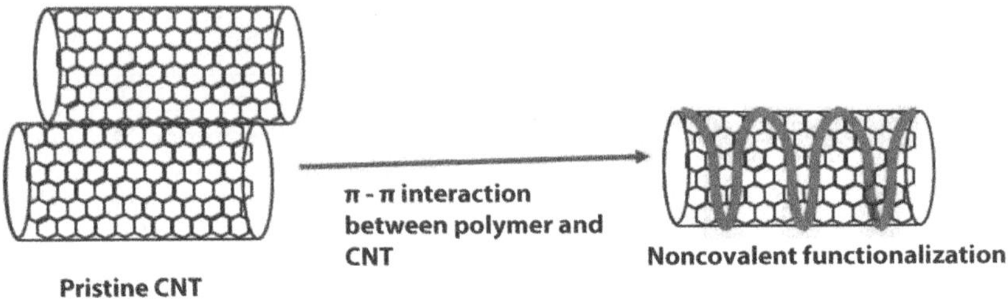

Figure 5.4 An overview of non-covalent approaches for functionalizing CNTs.

which holds significant potential for imaging and photothermal ablation [16]. However, unlike covalent functionalization, noncovalent functionalization has a limited capacity to deliver the drugs due to the existence of weak van der Waals forces. The solubility and dispersibility of CNTs can also be significantly increased by noncovalent functionalization [24, 25] as shown in Figure 5.4.

5.3.3 Encapsulation Inside CNTs

There are various examples that demonstrate that the bounding with the nanocarrier, with biological fluid, or the enclosing environment can calculate the inactivation mechanism or excretion technique of these molecules, even though many large or small biomolecules are adsorbed or interact onto the outer area of nanodevices that can be noticed to display an enhanced therapeutic activity such as enhanced hydrophilicity, reduced cytotoxicity, and improved bioavailability. Because of this, these nanosystems are especially well suited for further applications like material storage, compound production, and medicine administration. Recently, CNTs have been used to encapsulate molecules [26].

5.3.4 "Defect" Functionalization

Defect functionalization is the term typically used to describe oxidizing functionalization. By interacting CNTs with oxidizing agents, for example, $KMnO_4$, H_2SO_4, O_2, HNO_3, $K_2Cr_2O_7$, and H_2SO_4, they can functionalize the graphene sheets of carbon concurrently, becoming soluble in organic and polar solvents and acids without the use of techniques such as sonication, by adding surfactants, or any other method.

CNT fabrication processes also produce contaminants. Amorphous carbon and catalyst nanoparticles are the major by-products. The methods used for the raw material's purification, like acid oxidation, cause the tube caps to open and the sidewalls to develop holes [27]. The final products are pieces of nanotubes having oxygenated functionalities, primarily carboxylic and carbonyl groups, on their ends and sidewalls. Using IR spectroscopy, thermogravimetry, and other methods, numerous groups have investigated the chemical makeup of these moieties [20, 26].

5.4 Significance of CNT in Drug Delivery

1. Carbon nanotubes (CNTs) are proposed as an extremely lightweight, large surface area-to-volume ratio packing material of gas molecules adsorbed on the surface or they can be encapsulated in the interior side of nanoscale containers for molecular drug delivery [4, 14]. They are hollow, caged molecule cylinders formed with rolling sheets of graphene into consistent cylinders.
2. CNTs have a strong level of compatibility with biological systems due to their physical similarity to physiological molecules and subsequent functionalization with active molecules [5].
3. CNTs could diffuse throughout the circulation and enter cells into a biological fluid by an endocytosis-independent mechanism, promoting cellular uptake of medicinal compounds [28].
4. As a result of their low charge carrier density, they make excellent nanomaterial for use in biosensing transducers that can identify and target the precise cell that is specifically used in cancer treatment [29, 30].
5. Because of their chemical inertness and low toxicity, functionalized CNTs help to minimize medication resistance.
6. CNTs are inactive until they reach the places where they can assist medications to work more effectively.
7. When carbon nanotubes are inserted into a lipid membrane, the hydrophobic tails around the sheets of carbon nanotube become more arranged than the remainder of the lipid tails, increasing the order parameter of the lipid tails and thickening the biological biomembrane, which has the least impact on the membrane [1].
8. CNTs can be employed for target medication delivery because of their various distinctive parameters such as size,

shape, and capacity to functionalize with various amino acids, proteins, drug molecules, and other biopolymers [31].

5.5 Overview of CNT-Based Drug Delivery

For the transport of diverse biological materials, including proteins, peptides, etc., and chemotherapeutics, CNTs have been used as a drug carrier nanomaterial. Many terrible diseases, primarily malignancies, have been researched using CNTs as therapeutic carriers [19]. There are many ways to administer CNTs, including intravenous, intramuscular, transdermal, subcutaneous, inhalation, and intraperitoneal injection. The internalization cycle of CNTs is influenced by several factors, including the interaction of the biological medium with the biomolecules and different surface characteristics, such as functionalization, hydrophobicity or hydrophilicity, heterogeneity, and chemical composition, shape, porosity, and surface crystallinity [7].

There are several ways that cell membranes can internalize CNTs. The first approach is dependent on the endocytosis approach, which can be mediated by either a receptor or a non-receptor. In this mechanism, the biological cytoplasmic biomembrane engulfs the CNTs by enveloping them in cellular vesicles inside the body cell through energy-dependent active transport. Lysosomal degradation of the various drug molecule with CNT delivery mechanism can result from the internalization of CNTs by the development of endosomes within the cell. The second is based on a distinct endocytosis pathway also known as the direct penetration and nano penetration pathway, which comprises the direct transport of the extracellular material into the cell membrane by the fusion or diffusion technique [10]. This direct method for penetration has the benefit of preventing endosomal trapping and the possibility of lysosomal destruction.

5.6 Pharmacokinetics of CNTs

Understanding the outcome of CNTs on the interior side of the healthy human body is essential for the use of CNTs in the design of novel pharmaceutical therapeutic products. Excellent pharmacokinetic parameters are necessary for a drug carrier, including adequate and proper absorption, distribution to the intended area, entry to the site for target drug delivery, circulation time inside the blood, half-life clearance, etc. [16]. The key determinant of CNT accumulation in the tissue and toxicity is the route of delivery of the CNTs. There are several ways to provide CNT, including orally and intravenously as well as subcutaneously, abdominally, and via injection. When CNTs are given in different ways, there are several modes of absorption and transportation [23].

5.6.1 Absorption

The first crucial stage for drug carriers to finish their drug delivery job after administration is absorption. Various data are established by the literature on CNT absorption from their administration sites; whole and diminutive fiber length carbon nanotubes (CNTs) were used to create CNTs for oral administration. These findings suggest that the short fiber length of CNTs and the aggregation of CNTs are not essential for the oral delivery of medications and that CNTs are capable of being absorbed on their own. By using a transmission electron microscope, another research has shown that physically reduced CNTs can be ingested and absorbed by intestinal mucous membrane columnar cells. The capability of the intestinal tract to absorb CNTs is extremely important because it allows the development of oral formulation with CNTs as a nanocarrier [32].

Although injected subcutaneously and intravenously, some CNTs remain in the tissues permanently whereas others are absorbed by the lymphatic canal. It is a lymphatic target effect when drugs absorbed through the lymphatic canal CNTs travel down and collect in the lymph node. CNTs can directly enter the bloodstream and distribute throughout numerous internal organs, including the heart, kidney, liver, and spleen, when given intravenously. A literature review showed that surface modification is largely responsible for the measurement of the volume of plasma from which a CNT is completely removed per unit time of intravenously injected CNTs. The use of polyethylene glycol (PEG), which coats the carriers' surfaces to reduce immunogenicity and inhibit nonspecific phagocytosis by the reticuloendothelial system (RES), is thought to be one of the most crucial methods for extending the half-life of CNTs in blood circulation [25].

5.6.2 Distribution

Distribution refers to the numerous areas in which ingested CNTs might circulate and be eliminated from the body in the clinical investigation of CNTs as drug nanocarriers. It was interesting to learn that these CNTs' biodistribution and blood vasculature are influenced by the length of the PEG chains. Investigations have focused on both *ex vivo* and *in vivo* biodistributions. Long PEG-modified CNTs have a circulation time that is much longer than that of short PEG-modified CNTs because the former absorbs less than the latter.

In a different study, the effects of PEG chain lengths on biological cellular uptake of gluconated CNTs were evaluated. It was found that adsorption of shorter-chain PEG to SWCNTs was ineffective at preventing the degradation of CNTs from macro phagocytosis both *in vitro* and *in vivo*, while adsorption of long-chain PEGs significantly reduced their *in vivo* nonspecific uptake of CNTs.

Furthermore, it was shown that PEG-covalently altered CNTs had a longer blood circulation half-life than PEG-noncovalently altered CNTs with equal chain length [33].

5.6.3 Metabolism and Excretion

The potential of their effective application in clinical settings is made possible by the biodegradability and expulsion from the body. The animal body appears to be able to metabolize functionalized CNTs. It has been demonstrated that SWCNTs with carboxylated surfaces may endure 3 months of breakdown in a phagolysosomal simulant length decrease and the accumulation of ultrafine carbonaceous debris. Aryl sulfonate and unmodified SWCNTs do not deteriorate in the same conditions [34]. The observed metabolism phenomenon may be caused by the peculiar chemistry of carboxylation, in which the addition of reactive, modifiable functional groups such as COOH, and OH onto CNTs outer surfaces also causes bond degradation of the rolled graphene backbone in the form of nearby active sites that act as an attack for the further oxidative degradative reaction [35].

5.7 Biosafety of Carbon Nanotube

Numerous studies have been conducted in the last decade to ascertain and investigate the toxicity of CNTs. The biosafety of CNTs, which are used to carry the drugs in pharmaceutical medication delivery, is the main concern. Following administration, carbon nanotubes begin to circulate in the biofluids and interact with structural proteins, ribosomes, mitochondria, channels, and nuclear pores [19]. Numerous *in vitro* and *in vivo* experimental studies have been conducted to understand the various factors that contribute to CNT toxicity, including the route of administration (such as oral, transdermal, subcutaneous, and intraperitoneal), surface area, shape, composition, and residual catalyst metal impurities (i.e., cobalt, nickel, molybdenum, etc.). When it enters the gastrointestinal tract, it has been seen to cause Crohn's disease and colon cancer. When it enters the systemic circulation, it has been seen to cause blood clotting and numerous heart diseases. When pure CNTs penetrate the respiratory system, it causes asthma, bronchitis, and lung cancer. When CNTs are deposited in the eye, they can cause glaucoma, change the aqueous humor, and create inflammatory reactions on the epidermal layer. When CNTs are used to encapsulate leftover metal, oxidative stress and mitochondrial breakdown can lead to cell death. The accumulation of CNTs in tissues and cells is detrimental to living cells. SWCNTs aggregate more readily due to their large surface area and stronger van der Waals forces, whereas MWCNTs aggregate less due to their reduced surface area and numerous active defective sites on the side wall [7, 21].

5.7.1 Mechanism of CNT Toxicity

Free radical creation, the production of radical oxygen species (ROS), a rise in inflammatory reactions, and the formation of granuloma are the key mechanisms underlying the toxicity of CNTs. The creation of CNT aggregates in the cells, prolonged exposure to long CNTs, and the trapping of metal impurities within the CNT wall are the causes of ROS production. ROS assault the mitochondrial membrane system through peroxidation [21]. The damaged mitochondrial membrane then allows the free radicals or ROS to travel to the lysosomes, where they cause the lysosome membrane to rupture. Injured lysosomes produce digesting enzymes, which cause damage or death to whole cells. However, when ETCs (electron transmission chains) are blocked, mitochondria are unable to produce ATP, which depletes the energy available for the living processes of the cell and damages or kills the entire cell [25, 32].

5.7.2 Scenario to Bypass Carbon Nanotube Toxicity

Functionalization of CNTs, as opposed to virgin CNTs, is the first important strategy to decrease toxicity and increase biocompatibility. PEG-CNTs were reported to improve pharmacokinetic features, which include the lengthening of blood circulation half-life while reducing cytotoxicity and reticuloendothelial system (RES) capture in multiple investigations employing PEG functionalization of CNTs. The second method uses ultrasonication to purify CNTs while reducing metal catalyst residue impurities [36]. The third strategy involves using chemicals to control the distribution of CNTs within organelles, thus reducing or avoiding the disadvantage of CNTs. Fourthly, because there are dosage differences between the therapeutical and toxicological effects of CNTs, toxicological consequences may be avoided by managing CNT doses [7, 25].

5.8 Quantum Computational

Quantum computational technologies use fundamentally different computing approaches. These computational (QC) technologies are based on the laws of quantum mechanics. These laws permit few calculations to be done swiftly besides proficiently than conventional computing does. These computing methods are now game changers in many pharmaceutical sectors. With these quantum computing methods, the simulation of various molecules, solids, and other atomic particles can be done more precisely. These techniques help overcome many limitations such as scaling one of the conventional computational methods. These methods provide us with much easier and correct equations for large complex molecules. In terms of quantum computation, we usually call these equations Schrödinger equations. With these methods of quantum mechanics, the calculations of the interaction between the molecules become more precise and accurate. Quantum computational methods are used to explain the biological and chemical properties of molecules in *in silico* drug design applications. These procedures are also helpful in determining intermolecular or intramolecular forces and conformational states of molecules. These methodologies have very wide applications in the drug discovery process because they provide very sophisticated and precise processes termed algorithms. These are used for scheming wave functions of macromolecular systems. In this chapter, we have explained the promising part played by quantum computational techniques in all characteristics of *in silico* DD and all the significant recent advances.

Quantum computational methods are categorized into two categories:

5.8.1 Structure-Based Drug Design Methods (SBDD)

The structure-based drug design methods (SBDD) explicate the broad insights of protein and the protein–ligand structure will be explained. With these structure-based drug design methods, it became very easy to explain more new features with novel descriptors. Linear scaling quantum computing methods (LSQM) have made accessibility of targeted receptors easy. This makes it therapeutically very effective in the treatment of many diseases, for example, novel QM-derived descriptors, for instance, electrostatic potential (ESP) maps of molecules. This method is also used to determine the local hardness and softness. Various other techniques are used to determine or probe ligand complexes such as Fukui indices and frontier orbital analysis.

Linear scaling quantum mechanics (LSQM) is also helpful in calculating the binding affinity of small molecules and between proteins. This mechanic is also helpful in determining chemical interactions. Recently, researchers developed a new approach, i.e., the divide and conquer (D&C) approach. This divide and conquer (D&C) approach is useful in determining the binding affinities of ion-mediated and metal-containing systems. The SBDD method requires a structure of the ligand in the complex molecule. The structure is confirmed by X-ray crystallography or NMR. The most important impact of SBDD methods is refinement.

The two renowned researchers, Rajamani and Reynolds, explained very well how linear scaling quantum mechanics (LSQM) works. They applied this set of rules of QM in the computerized database called DivCon, to form protonation states of enzymes such as b-secretase in catalytic aspartates. From these studies, it has been concluded that enzyme aspartates favor such states in the presence of the inhibitor. If the inhibitor is not there, it prefers to remain in a di-deprotonated state. Raha and Merz worked on the same computer program DivCon, and they found the proton affinity of the various enzymes that are catalytic such as aspartates of HIV-1 protease. These catalytic enzymes show quite a good affinity for proteases whether the inhibitors are present or not. From this, they calculate the binding affinities.

5.8.2 Ligand-Based Drug Design Methods (LBDD)

In LBDD methods, the designing of the analogs takes place with a vast acquaintance of various properties such as structure–activity relationship and potency of the compounds. There are a few methods that can be used for the ligand-based drug design such as the Topliss method or a simple analog design. This Topliss method relies on structural likeness properties. In computational QM methods, pharmacophore models or the shape of the compounds plays a vital role in design purposes. For this, if a repository of substantial size that has a quite fair strength is available, then only a quantitative structure–activity relationship (QSAR) model can be used. We can use these models for prediction purposes. Calculation of free energy of binding via various computational methods generally practices the ramification of energy, which is known as an index of scoring or scoring function. This scoring function calculates the parameters directly or indirectly, which is correlated to the binding free energy.

It is known that carbon nanotubes dominate with exceptional features in terms of their mechanical and physical strength. Due to this, carbon nanotubes have a wide-ranging use in the pharmaceutical world. Some of the leading and all possible applications are explained very well here as examples that are very useful on the nanoscale, as well as nanodevices for delivery of drugs such as fluid transport and fluid storage. It is clear from all conventional experimental approaches that it is very difficult to perform the experiments at the micromillimeter, millimicron, and nanometer scales. However, with quantum simulation techniques such as LBDD, it became an easier task for the researchers to copy these experiments at smaller scales. It provides promising results in calculating various indices such as mechanical property, wave propagation, and resonant frequency. By using various parameters, it has been concluded by various researchers that the vibration parameter of a molecule is calculated best. This chemistry of quantum mechanics accomplished such a big feat in drug delivery by scheming the electronic properties of biological systems [36].

5.9 Various Simulation Approaches in Drug–CNTs Interaction

Quantum computing enables the simultaneous delineation of numerous compounds and the immersion of their conformations, the creative solutions of suitable active ingredients among thousands of compounds, the effectiveness of the various approaches of comparing the effects of various active ingredients and their interactions, predicting the chemical reaction mechanisms, drug repurposing, and the development of techniques for analyzing large-scale molecules that may lead to the development of the pharmaceutical formulations. It is discovered that each of these efficient strategies works incredibly well in obtaining distinct and focused results.

Additionally, this permits the simultaneous delineation of numerous compounds and the immersion of their conformations, the inventive identification of a promising active drug candidate among thousands of compounds, the efficacy of that drug candidate compared to the effects of these medications and their interactions, the prediction of various physiochemical reaction mechanisms, drug repurposing, and the development of techniques for analyzing large-scale molecules that may aid in treatment.

It becomes clearer that quantum computing also allows for the simultaneous mapping of several biomolecules and the modeling of a biomolecule such as protein, amino acid, DNA, and RNA interactions and helps

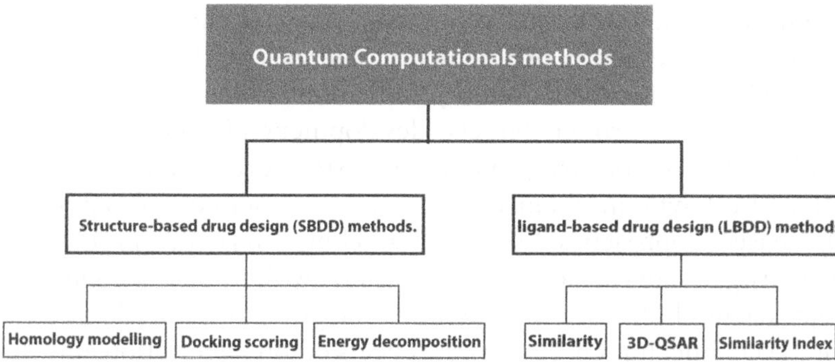

Figure 5.5 Quantum computational methods in *in silico* DD.

researchers develop methods for analyzing a wide range of molecules that may be able to find the outcome in the development of more potent therapeutics [37].

Few of the computative approaches maneuver the classical or QM-based approach. These quantum mechanic methods are systematized into semiempirical and *ab initio* procedures. All the computative approaches append various approaches such as molecular mechanics (MM), Monte Carlo (MC), and molecular dynamics (MD). It has been proven that these computative approaches are employed in studying the interactivity of molecules and atoms. With these pragmatic approaches, results found are very promising. However, the inter-atomic potentials between atoms and molecules are described by certain characteristics, which affect how accurate this type of computational method is [35, 36].

All of these molecular-based approaches have been proposed for drug design, but QM-based approaches can solve a broad range of biological issues, providing a deep knowledge of electronic structures of lots of molecules or atoms, describing receptor–target binding or structures of receptors and their interacting sites to calculate the strong or weak binding affinities, and figuring out the addition of a proton to an atom, molecule, or ion in the binding mechanisms of pharmaceuticals [37], as shown in Figure 5.5.

5.9.1 QM Approaches

There are many QM methods used in drug discovery, and the results we get from these techniques are very precise, specific, and detailed. The most exact calculations are done using QM methods, which are also the most expensive and time-consuming to compute [38]. We can easily differentiate one from another. Evaluation of system behaviors, analysis of cleavage energies for amide bonds or interaction energies, optimization of molecule composition, identification of protonation or electron transfer states, prognostication of protein targets, development of compound models, evaluation of receptor–drug interactions, structure–activity relationship (SAR), and catalytic mechanisms of enzymes are all possible applications of QM-based approaches [39–41]. Modeling enzymatic processes and rational drug design can advance significantly with the use of sophisticated QM approaches [42, 43]. When defining the physicochemical characteristics of the channel lumen, such as the effects of ion penetration and charge transfer, one can use QM-based methods. Using QM approaches to predict ligand affinities that assess drug therapeutic and pharmacological profiles, including their adverse effect, can provide drug research topics with a reasonable foundation and hasten the creation of more efficient therapies. Using QM methods, various therapeutics can be quickly screened followed by an evaluation of the specific binding affinities [44].

Zinc-mediated ligand binding has been assessed using the QM-based calculating practical data and the free energy of binding has been computed. The method has made it possible to calculate the binding energies for receptor–enzyme complexes in addition to estimating binding affinities, and the outcome was in good agreement with the practical data.

Additionally, using QM approaches, it is possible to anticipate the properties of nanoparticle-based medication delivery systems. The creation

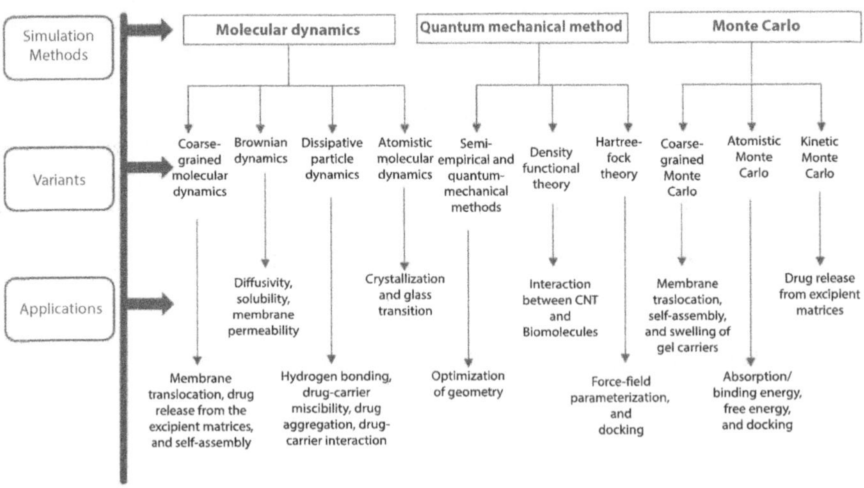

Figure 5.6 Schematic representation of various simulation methods and their application.

of bio-degradable polymers based on the delivery of drugs has led to the description of bond breakage at the polymer–solvent interface using a QM model [44].

A few methods that are mentioned here with their explicit features are shown in Figure 5.6.

5.9.1.1 Ab Initio Approach

The phrase "*ab initio*" refers to a computational technique that is directly generated from theoretical concepts without the need for experimental data. The complicated approach, which can simulate hundreds of atoms without input–output data, may be used to calibrate higher-level models and assess the creation or breaking of bonds. It has an emphasis on the electron density and is based on their electronic-wave function, i.e., *ab initio* (DFT) [45].

5.9.1.2 Semiempirical Approach

On the other hand, semiempirical means nimble and, in turn, the least accurate. This chapter also focuses on the approaches used in the pharmaceutical world. In these segments, the authors also enlighten us on various principles of these methods. Only for the system's valence electrons are QM calculations used in semiempirical approaches like AM1 and PM3 [38].

5.9.1.3 Hartree–Fock (HF) Approach

The Hartree–Fock semblance is used to calculate the prospect of determining an electron and is thought to be unaffected by the possibility of calculating additional electrons. This has to be done because when electrons are in the proximity of another electron, they experience the presence of one another due to some force. Thus, we can say that this HF theory gives a very clear picture of interelectronic or interelectronic interactions by using while excluding the electron correlation, which includes the quantum mechanical exchange term [46].

5.9.2 Molecular Dynamics (MD) Approaches

Molecular dynamics (MD) has offered fresh perspectives on the processes underlying medication diffusion or absorption [9]. The use of CNTs as a nanocarrier and molecular dynamics (MD) has produced new discernment into the mechanics bolstering drug diffusion or immersion [9]. It is possible to copy the processes regulating the biological activity of carbon nanomaterials and assay relations between specifics and carriers or media using traditional MD and coarse-granulated MD (CGMD) approaches [47]. Calculating experimental data and CGMD combination have been enforced for predicting the conformation kinetics of the biomolecule nimbus that is perhaps useful for designing case-specific carbon nanoparticles for delivery of a variety of rectifiers.

MD approaches for simulating and assessing the prolixity and target of medicines include designing and optimizing the drug delivery or finding solutions to challenging exploration issues in experimental contexts. Multi-scale models can give examples of the micromeritic properties of carbon nanotube on medicine, drug release design, cellular uptake, and distribution through the tiny blood vessel, including capillaries, venules, and arterioles as well as enable prognosticating mechanisms of action of carbon nanocarriers. It is possible to predict the behaviors of action of carbon nanocarriers by using multi-scale models to better understand the impact of the external properties, size, and morphologies of carbon nanotubes on drug loading and their distribution through the tiny blood vessel [48].

5.9.3 Monte Carlo (MC) Simulation Approaches

These MC simulations allow for the evaluation of how surface modification affects NP binding to endothelial cells as well as CNT interactions with normal and malignant cells.

5.9.4 Hybrid Approaches

5.9.4.1 MD and QM Approaches

The occurrence of more successful therapeutical pharmaceutical formulations, the selection of suitable drug carriers, and the quick identification of various excipients that provide insights into the molecular action of additive and drug targets and their binding in the biological media will be able to determine drug loading of the nanocarriers, and the pivotal assets of dosage formulations and their stability have all been credited to MD and QM simulations as a dynamic and energetic tool [49]. To examine the production and drug release profiles from nanocarriers, their accumulation in untargeted organs should be prevented and bioavailability should be modified; a combination of MD and QM algorithms can be used [50]. By using this combination, drug nanocarriers can be outlined for transporting the second carriers for the second-generation tyrosine kinase inhibitors, which resulted in increased activity and decreased side effects, which have been developed using MD and QM approaches [51].

5.9.4.2 MM and QM Approaches

When developing new medications, QM/MM is an invaluable way of deciphering the ligand–protein crystal structure and assessing the impact of different substituents on binding modes [52]. The mechanisms of enzyme-catalyzed reactions can be simulated using QM/MM-based methods, which also provide a deeper understanding of binding within the target and effective sites of enzymes, and through these actions, ligands are stabilized in complex formation, drug repositioning (the technique by which ligands bind to proteins), binding between the biomolecule and other compounds, and ligand–target optimization. These methods could also make a distinction between ligands and determine the binding between active therapeutic drugs and their ligands. Additionally, it might be used to specify the geometry of an enzyme's active site and create the various structures of a biomolecule such as proteins and tiny molecules. To research the active sites of proteins and mimic their targets, QM/MM techniques have been created [53].

5.10 Applications of Quantum Computational Methods

5.10.1 Applications of Quantum Computational Methods in DDS

In this chapter, we have covered every aspect of quantum computational technology in drug discovery and drug delivery. The approach based on QM can impact various dimensions of drug design. These methods overcome the various problems of researchers using SBDD and LBDD methods. Quantum computational methods rely on the variational principle, with which it calculates the interaction or binding affinities between ligand and receptor. They do perform it to generate various descriptors. For handling QM methodologies, there must be a vast knowledge of validation studies. These QM methodologies have been proven to provide good results compared to classical methods. From a few studies, it has been reported that in classical methods, two properties, i.e., entropic and dynamic, play an important role in ligand and structure binding but QM methods overcome these hurdles.

1. There are a few examples that have shown very promising results for many compounds reaching clinical trials, and a few of them have been approved by the FDA. There are a few examples of anti-AIDS drugs. The target for anti-AIDS is HIV-1 (Human Immunodeficiency Virus 1) protease. A few examples of drugs that were developed using QM methods are Saquinavir in Figure 5.7 and Amprenavir in Figure 5.8.

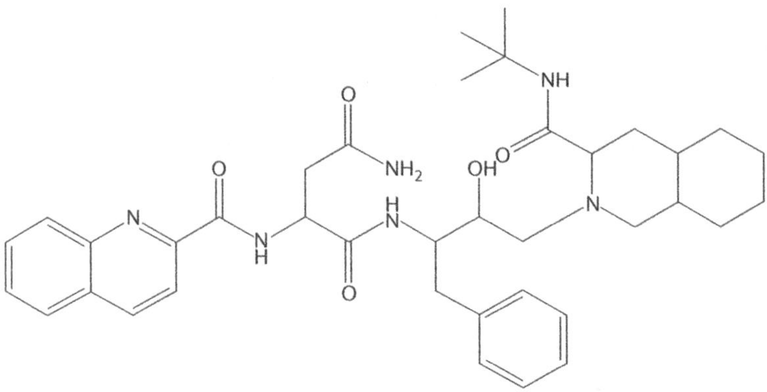

Figure 5.7 Saquinavir.

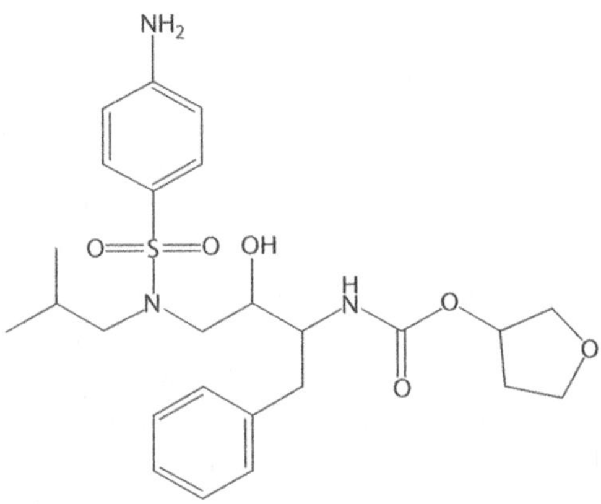

Figure 5.8 Amprenavir.

The two anti-HIV drugs, Saquinavir and Amprenavir, were developed by using the principles of SBDD, in which the novel descriptors are used for determining the binding affinities. One more interesting drug example belongs to the category of carbonic anhydrase II inhibitor: Dorzolamide. This drug was developed by the SBDD tools.

2. The QM approach may be used to estimate the nanocarrier's capacity for drug loading. This can be a reliable indicator of the stability of the effective and target drug delivery system, and binding energies can be a sign of drug loading. The binding sites that are more crucial are identified in the most stable configurations (higher '-' binding energies) [50]. The projected high drug loading is caused by the substantial negative values of the estimated binding energies in the nanoparticle and drug system.

3. One of the most crucial methods for improving a medicine's solubility is choosing the proper drug delivery technology [51]. It will be crucial to determine the solvation energies for the drug delivery systems because of this. The solubility of a given medicine in an aqueous solution ought to be increased by a suitable carrier. When a nanocarrier is present, the drug's E_{solv} value decreases, increasing the drug's solubility. According to the quantum theory of atom in the molecule (QTAIM), hydrogen bonds that develop between the drug and the carrier are the main cause of the strong binding and enhancement in solubility.

$$\Delta E_{solv} = E_{aq} - E_{gas} \tag{5.2}$$

4. Strong hydrogen bonds between the active drug candidate and nanocarrier can allow for the selection of an appropriate drug delivery method. The medication loading and physical stability increase with the strength of the hydrogen bonding. The next generation of drug carriers is anticipated to be based on hydrogen-bond-based drug conjugations [52, 53]. By forming hydrogen bonds, CNTs with different functional groups, like -COOH, -OH, and active -NH, can be an intelligent choice for drug administration. Resolving the type, strength, and characteristics of hydrogen bonding in drug delivery systems will therefore motivate scientists to create efficient systems [54].

An example is the energy gap between the lower unoccupied molecular orbital (LUMO) and the higher occupied molecular orbital (HOMO) for both phases, as well as other quantum molecular descriptors like electrophilicity power (ω) and global hardness (η).

When the drug's E_g and η are equal, the drug with the nanocarrier are the same. To put it another way, there is little to no transfer of charge between the medication and the nanocarrier. That can be absolute for a target delivery of drugs since the drug can simply be released from the nanocarrier's external surface.

The fact that E_g and η values in aqueous solution are greater indicates that certain arrangements are more stable. A rise in system reactivity is suggested by the drop in E_g.

As toxicity is assessed using electrophilicity power (ω), it may be concluded that the presence of the nanocarrier either slightly reduces or does not affect the drug's toxicity. It is analyzed to determine the electrophilicity index (ω).

$$\omega = (I+A)^2/8\eta \tag{5.3}$$

where I = ionization potential; here, $I = -E_{HOMO}$

A = electron affinity; here, $A = -E_{LUMO}$

Global hardness (η) is the ability of one particle to resist changes to its electrical state [55].

$$\eta = (I\text{-}A)/2 \tag{5.4}$$

5. It will be possible to determine whether a product created with a nanocarrier through covalent or noncovalent bonding is stable using quantum simulation calculations. Relative energies for reactants, products, and transition states are Gibbs free energy (ΔGz), activation enthalpy (ΔHz), and activation energy (ΔEa). The product's energy values show that the reaction was exothermic ($H < 0$) and spontaneous ($G < 0$). For instance, using a quantum mechanical technique, covalent or noncovalent interactions associated with the orientations of the functional groups, such as NO, and NH2 of various medications, specifically anticancer agents, towards the nanocarrier were studied (DFT).

The covalent bond between the drug and the nanocarrier may form as the temperature rises. Examined are four mechanisms using the functional groups NH2 (NH2 mechanism), NO (NO mechanisms), and integrating N-atoms (N mechanism). When we compare the two mechanisms, the NO mechanism's products form more quickly than the NH2 mechanism's because the former has lower activation energies (kinetic product). Because it is a thermodynamic product, the result of the NH2 process is more stable [56, 57].

5.10.2 Applications of Quantum Computational Methods in Nanobiosensors

CNTs and other biomolecules that significantly immobilize amino acids offer new possibilities for the recognition, diagnosis, and medication of diseases including tumors, kidney damage from illness, etc. Quantum mechanical computations were carried out to comprehend how these amino acids and other biomolecules interacted with CNT. It offered helpful information to develop the CNT-based nano-biosensors based on calculation findings. The interaction between CNT and the amino acid was studied using the notion of density functional theory, in which the focal point is the density of the electron rather than the wave function [4]. The stability and reactivity of the developed nano-biosensor were assessed using the concept of DFT-based criteria that were directly concerned with the relative energies of the highest occupied molecular orbitals (HOMOs) and the lowest unoccupied molecular orbitals (LUMOs).

$$E_{ADS} = E_{AA/CNTs} - E_{CNTs} - E_{AA} \tag{5.5}$$

where $E_{AA/CNTs}$ is the total energy of the CNTs with a biomolecule, E_{CNTs} is the total energy of CNTs, and E_{AA} is the total energy of biomolecules in relaxing geometry. '+' adsorption energy alluded to local minima, whereas '-' adsorption energy denoted the stable produced complexes [2].

5.11 Conclusion

Nowadays, nanotechnology, particularly CNTs, is the industry with the greatest number of applications across the spectrum. The most adaptable and promising biomedical materials are CNTs. The solubility and toxicity of carbon nanotubes are improved and decreased by functionalization with biocompatible chemicals. The development of carbon nanotubes and their industrial uses depend heavily on functionalization. Excellent mechanical, electrical, and optical characteristics of CNTs assist in the expansion of their application scenario. Drug delivery, biosensors, bioimaging, tissue engineering, and cancer therapy are just a few uses for CNT.

Aside from these developments, one of the most challenging issues facing CNT economics is conserving purity and producing it in large quantities at cheaper costs, which is necessary due to its numerous uses. Recently, methods like CVD have been developed to solve this problem, producing purer and defect-free CNTs at significantly cheaper costs. It is anticipated that improved technologies will be created in a few years that will fulfill our demands. Many researchers are concentrating on devising new procedures to boost purity while sustaining moderate manufacturing costs. Numerous difficult and time-consuming jobs have been simplified, expedited, and cost-effectively completed thanks to computational studies that make use of different simulation methodologies. It has assisted in the identification of several novel applications in this sector over the past 10 years and is useful in many facets of CNT biological applications. It will set the stage for novel findings and applications in the fields of biomedicine and materials research. It can satisfy our needs for years.

References

1. Dinga, H.M. and Ma, Y., Computational approaches to cell–nanomaterial interactions: keeping a balance between therapeutic efficiency and cytotoxicity. *Nanoscale Horiz.*, 3, 6–27, 2018.
2. Yoosefian, M. and Etminan, N., Leucine/Pd-loaded (5,5) single-walled carbon nanotube matrix as a novel nanosensors for *in silico* detection of protein. *Amino Acids*, 50, 653–661, 2018.
3. Georgakilas, V., Perman, J.A., Tucek, J., Zboril, R., Broad Family of Carbon Nanoallotropes: Classification, Chemistry, and Applications of Fullerenes, Carbon Dots, Nanotubes, Graphene, Nanodiamonds, and Combined Superstructures. *Chem. Rev.*, 115, 11, 4744–4822, 2015.
4. Yoosefian, M. and Etminan, N., Density functional theory (DFT) study of a new novel bionanosensor hybrid; tryptophan/Pd doped single-walled carbon nanotube. *Physica E.*, 81, 116–121, 2016.
5. Anzar, N., Hasan, R., Tyagi, M., Yadav, N., Narang, N., Carbon nanotube - A review on Synthesis, Properties and plethora of applications in the field of biomedical science. *Sens. Int.*, 1, 100003, 2020.
6. Monajjemi, M., Falahati, M., Mollaamin, F., Computational investigation on alcohol nanosensors in combination with carbon nanotube: A Monte Carlo and *ab initio* simulation. *Ionics*, 19, 155–164, 2013.
7. N.Al-Qattan, M., Deb, P.K., Tekade, R.K., Molecular dynamics simulation strategies for designing carbon-nanotube-based targeted drug delivery. *Drug Discov. Today*, 2, 235–250, 2018.
8. Rahman, M.M. and Asiri, A.M., Introductory Chapter: Carbon Nanotubes and Their Applications, in: *Carbon Nanotubes*, 2018.
9. Yoosefian, M. and Moslem, J., A molecular study on drug delivery system based on carbon nanotube for the novel norepinephrine prodrug, Droxidopa. *J. Mol. Liq.*, 284, 258–264, 2019.
10. Singh, S., Patel, H., Goyani, M., Patel, P., Carbon nanotubes- A Novel Drug Delivery System. *Multidiscip. Int. Res. J. Guj. Technol. Univ.*, 2, 2581–8880, 2012.

11. Cheung, W., Pontoriero, F., Taratula, O., Chen, A.M., He, H., DNA and carbon nanotubes as medicine. *Adv. Drug Deliv. Rev.*, 62, 633–649, 2010.
12. Jagadeesan, A.K., Thangavelu, K., Dhananjeyan, V., Carbon Nanotubes: Synthesis, Properties and Applications, in: *Century Surface Science*, 2020.
13. Srivastava, D., Menon, M., Cho, K., Computational Nano Technology with Carbon nanotubes and fullerenes. *Comput. Sci. Eng.*, 3, 42–55, 2001.
14. Meyyappan, M. and Srivastava, D., Carbon Nanotubes. *Nanosci. Eng. Technol.*, 12, 205–216, 2003.
15. Prajapati, V., Sharma, P.K., Banik, A., Carbon Nanotubes and its Applications. *I.J.P.S.R.*, 2, 5, 1099–1107, 2011.
16. Jha, R., Singh, A., Sharma, P., Fulori, N., Smart carbon nanotubes for drug delivery system: A comprehensive study. *J. Drug Deliv. Sci. Technol.*, 58, 101811, 2020.
17. Ketabi, S. and Rahmani, L., Carbon nanotube as a carrier in drug delivery system for carnosine dipeptide: A computer simulation study. *Mater. Sci. Eng. C.*, 73, 173–181, 2017.
18. Rungnima, C., Rungrotmongkol, T., Hannongbua, S., Hisashi, O., Replica exchange molecular dynamics simulation of chitosan for drug delivery system based on carbon nanotube. *J. Mol. Graph. Model.*, 39, 183–192, 2013.
19. Meher, J.G., Kesharwani, P., Chaurasia, M., Singh, A., Chourasia, M.K., Carbon Nanotubes (CNTs): A Novel Drug Delivery Tool in Brain Tumor Treatment, in: *Nanotechnology-Based Targeted Drug Delivery Systems for Brain Tumors*, pp. 375–396, 2018.
20. Tasis, D., Tagmatarchis, N., Bianco, A., Prato, M., Chemistry of Carbon Nanotubes. *Chem. Rev.*, 106, 1105–1136, 2006.
21. Vashist, S.K., Zheng, D., Pastorin, G., Al-Rubeaan, K., Luong, J., Sheu, F.S., Delivery of drugs and biomolecules using carbon nanotubes. *Carbon*, 49, 4077–4097, 2011.
22. Madani, S.Y., Naderi, N., Dissanayake, O., Tan, A., Seifalian, A.M., A new era of cancer treatment: carbon nanotubes as drug delivery tools. *Int. J. Nanomed.*, 6, 2963–2979, 2011.
23. Kushwaha, S.K.S., Ghoshal, S., Rai, A.K., Singh, S., Carbon nanotubes as a novel drug delivery system for anticancer therapy: A review. *Braz. J. Pharm. Sci.*, 49, 629–643, 2013.
24. Liu, Z., Tabakman, S., Welsher, K., Dai, H., Carbon Nanotubes in Biology and Medicine: *In vitro* and *in vivo* Detection, Imaging and Drug Delivery. *Nano Res.*, 2, 85–120, 2009.
25. Zhang, W., Zhang, Z., Zhang, Y., The application of carbon nanotubes in target drug delivery systems for cancer therapies. *Nanoscale Res. Lett.*, 6, 555, 2011.
26. Pastorin, G., Crucial Functionalizations of Carbon Nanotubes for Improved Drug Delivery: A Valuable Option. *Pharm. Res.*, 26, 746–769, 2009.
27. Zaidi, B., Smida, N., Althobaiti, M.G., Aldajani, A.G., Almdhaibri, S.D., Polymer/Carbon Nanotube Based Nanocomposites for Photovoltaic

Application: Functionalization, Structural, and Optical Properties. *Polymers*, 14, 6, 1093, 2022.
28. He, H., Pham-Huy, L.A., Drama, P., Xiao, D., Zuo, P., Pham-Huy, C., Carbon Nanotubes: Applications in Pharmacy and Medicine. *BioMed. Res. Int.*, 578290, 12, 2013.
29. Hossen, S., Hossain, M.K., Basher, M.K., Mia, M.N.H., Rahman, M.T., Uddin, M.J., Smart nanocarrier-based drug delivery systems for cancer therapy and toxicity studies: A review. *J. Adv. Res.*, 15, 1–18, 2019.
30. Pinillos-Madrid, J.F. and Gallardo, C., Carbon Nanotubes: A viable drug delivery platform for the treatment of cancer. *J. Appl. Pharm. Sci.*, 5, 09, 143-152, 2015.
31. Zhang, Y., Pettibone, D., Xu, Y., Mahmood, M., Karmakar, A., Casciano, D. *et al.*, Toxicity and efficacy of carbon nanotubes and graphene: The utility of carbon-based nanoparticles in nanomedicine. *Drug Metab. Rev.*, 46, 232-246, 2014.
32. Amenta, V. and Aschberger, K., Carbon nanotubes: potential medical applications and safety Concerns. *WIREs Nanomed. Nanobiotechnol.*, 7, 371–386, 2020. ISSN:1939-0041.
33. Liu, Z., Davis, C., Cai, W., He, L., Chen, X., Dai, H., Circulation and long-term fate of functionalized, biocompatible single-walled carbon nanotubes in mice probed by Raman spectroscopy. *PNAS*, 105, 5, 1410–1415, 2008.
34. Hassanzadeh, P., Towards the quantum-enabled technologies for the development of drugs or delivery systems. *J. Control. Release.*, 324, 260–279, 2020.
35. Bravyi, S., Smith, G., Smolin, J.A., Trading Classical and Quantum Computational Resources. *Phys. Rev. X*, 6, 21043, 2016.
36. Cavalli, A., Carloni, P., Recanatini, M., Target-Related Applications of First Principles Quantum Chemical Methods in Drug Design. *Chem. Rev.*, 106, 9, 3499, 2006.
37. Wallrapp, F., Masone, D., Guallar, V., Electron Transfer in the P450cam/PDX Complex. The QM/MM e-Pathway. *J. Phys. Chem.*, 112, 12989–12994, 2008.
38. Honarparvar, B., Govender, T., Kruger, H.G., Maguire, G.E., Soliman, M.E., Kruger, H.G., Integrated approach to structure-based enzymatic drug design: molecular modeling, spectroscopy, and experimental bioactivity. *Chem. Rev.*, 114, 1, 493–537, 2014.
39. Arodola1, O.A. and Soliman, M.E.S., *Quantum mechanics implementation in drug-design workflows: does it help?*, vol. 11, pp. 2551–2564, Dove press, London, 2017.
40. Berhane, N.H., Rickus, J.L., Haghighi, K., The role of multiscale computational approaches for the rational design of conventional and nanoparticle oral drug delivery systems. *Int. J. Nanomed.*, 2, 3, 315–331, 2007.
41. Njafa, J.P.T., Quantum Mechanics and Applications. *J. Lasers Opt. Photonics*, 5, 33, 2018. ISSN: 2469-410X.
42. Galindo-Murillo, R. and Aguilar-Suarez, L.E., Barroso-Flores J. A Mixed DFT-MD Methodology for the *In Silico* Development of Drug Releasing

Macrocycles. Calix and Thia-Calix[N]Arenes as Carriers for Bosutinib and Sorafenib. *J. Comput. Chem.*, 37, 940–946, 2016.

43. Kolosnjaj-Tabi, J., Hartman, K.B., Boudjemaa, S., Ananta, J.S., Morgant, G., Szwarc, H. *et al.*, *In Vivo* Behavior of Large Doses of Ultrashort and Full-Length Single-Walled Carbon Nanotubes after Oral and Intraperitoneal Administration to Swiss Mice. *ACS Nano.*, 4, 1481–1492, 2010.
44. Riviere, J.E., Pharmacokinetics of nanomaterials: an overview of carbon nanotubes, fullerenes, and quantum dots. *WIREs Nanomed Nanobiotechnol.*, 1, 26–34, 2009.
45. Monajjemi, M., Baei, M.T., Mollaamin, F., Quantum mechanic study of hydrogen chemisorptions on nanocluster vanadium surface. *Russ. J. Inorg. Chem.*, 53, 1430–1437, 2008.
46. Kaddi, C.D., Phan, J.H., Wang, M.D., Computational nanomedicine: modeling of nanoparticle-mediated hyperthermal cancer therapy. *Nanomedicine*, 8, 8, 1323–1333, 2013.
47. Staples, M., Daniel, K., Cima, M.J., Langer, R., Application of Micro- and Nano-Electromechanical Devices to Drug Delivery. *Pharm. Res.*, 23, 847–863, 2006.
48. Blumberger, J., Lamoureux, G., Klein, M.L., Peptide Hydrolysis in Thermolysin: *Ab Initio* QM/MM Investigation of the Glu143-Assisted Water Addition Mechanism. *J. Chem. Theory Comput.*, 3, 1837–1850, 2007.
49. Bottini, M., Rosato, N., Bottini, N., PEG-Modified Carbon Nanotubes in Biomedicine: Current Status and Challenges Ahead. *Biomacromolecules*, 12, 3381–3393, 2011.
50. Lin, S.H., Cui, W., Wang, G.L., Meng, S., Liu, Y.C., Jin, H.W. *et al.*, Combined computational and experimental studies of molecular interactions of albuterol sulfate with bovine serum albumin for pulmonary drug nanoparticles. *Drug Des. Devel. Ther.*, 10, 2973–2987, 2016.
51. Fahr, A. and Liu, X., Drug delivery strategies for poorly water-soluble drugs. *Expert Opin. Drug Deliv.*, 4, 4, 403–416, 2007.
52. Kim, B.S., Park, S.W., Hammond, P.T., Hydrogen-bonding layer-by-layer-assembled biodegradable polymeric micelles as drug delivery vehicles from surfaces. *ACS Nano.*, 2, 386–392, 2008.
53. Senthilkumar, T., Lv, F., Zhao, H., Liu, L., Wang, S., Conjugated Polymer Nanogels Binding Anticancer Drug Through Hydrogen Bonds for Sustainable Drug Delivery. *ACS Appl. Bio Mater.*, 2, 6012–6020, 2019.
54. Hasanzade, Z. and Raissi, H., Density functional theory calculations and molecular dynamics simulations of the adsorption of ellipticine anticancer drug on graphene oxide surface in aqueous medium as well as under controlled pH conditions. *J. Mol. Liq.*, 255, 269–278, 2018.
55. Rahbar, M., Morsali, A., Bozorgmehr, M.R., Beyrmabadi, S.A., Quantum chemical studies of chitosan nanoparticles as effective drug delivery systems for a 5-fluorouracil anticancer drug. *J. Mol. Liq.*, 302, 112495, 2020.

56. Avarand, S., Morsali, A., Heravi, M.M., Beyramabadi, S.A., A quantum chemical study on the magnetic nanocarrier-tirapazamine drug delivery system. *Nanosyst. Phys. Chem. Math.*, 12, 2, 167–174, 2021.
57. Malviya, R., Sharma, P.K., Sundram, S., Dhanaraj, R.K., Balusamy, B., Bioinformatics tools and big data analytics for patient care, in: *Routledge Taylor & Francis Group*, vol. 1, p. 240, 2022.

6

Quantum Computation Approach for Nanotechnology-Based Targeted Drug Delivery Systems

Smriti Ojha[1*], Sudhanshu Mishra[1], Anubhav Anand[2], Amrita Singh[1] and Palak Gupta[3]

[1]Department of Pharmaceutical Science and Technology, Madan Mohan Malviya University of Technology, Gorakhpur, Uttar Pradesh, India
[2]Department of Pharmaceutics, Hygia Institute of Pharmaceutical Education and Research, Lucknow, Uttar Pradesh, India
[3]Department of Chemistry, University of Florida, Gainesville, FL, USA

Abstract
Nanotechnology is a promising platform for targeted drug delivery in an effective manner. Drug development and research with quantum computing (QC) emerges as a game changer for pharmaceutical industries. The main objective of the present chapter is to review various QC techniques that are used to effectively simulate the problems at molecular and atomic level to find out exceptional results. QC devices use mathematical models and quantum gates for computational implementation. The mode of action of nanomaterials and their interactions with the active ingredient, biological membranes, and other biological tissues or molecules have been improved with complementary experimental and computational studies. Several strategies, like 3-dimensional protein structures and active ingredient interactions with various systemic proteins, can have an effect on the development process. An absolute QC model is needed for the representation of actual physiological conditions for nanocarriers loaded with active ingredients and their biodistribution under a given vascular geometry and hemodynamic conditions. These virtual tools help to design potent nanocarriers with minimal side effects.

***Keywords*:** Quantum computing, nanotechnology, drug development, virtual tools, engineered nanomaterials, subatomic level

**Corresponding author*: smritiojha23@gmail.com

Rishabha Malviya, Sonali Sundram and Dhanalekshmi Unnikrishnan Meenakshi (eds.)
Drug Delivery Systems Using Quantum Computing, (171–194) © 2024 Scrivener Publishing LLC

6.1 Introduction

The evolution of computer technology and concepts of quantum theory are interlinked and are the primary goal of the field of quantum computing (QC) [1]. The two biggest challenges facing the pharmaceutical industry today are technological advancements and productivity growth. High research and development costs as well as a stagnant number of new chemical entities are the main obstacles [1]. The nature and behavior of matter and its energy at the quantum level are explained by quantum theory [2]. Unlike conventional computers, which only work with binary data, quantum computers can process a variety of different types of data [3]. The ability of quantum computers to analyze data between 0 and 1 bit might open up new possibilities for simulation and calculation [4]. The basic information unit in quantum operations is the spin direction, which can be either up or down and is analogous to the 0 or 1 bit in a classical computer system. Due to quantum entanglement, electron spin can simultaneously adopt the values of 0 and 1, which considerably improves the efficiency of complex computations [5]. QC fundamentals are based on the basic principles of quantum physics, so it is regarded as a different computing paradigm in related areas, which permits certain computations to be conducted significantly more swiftly and efficiently than classical computing does and hence holds huge potential for the pharmaceutical sector [6]. Big pharma corporations are creating quantum task forces and devoting significant sums of money to investigate the application of quantum computers and software to chemistry and biology because they recognize how game-changing this approach is [7]. Nowadays, QC is unitized in pharmaceutical industries in a broad manner, one of which is the application of nanotechnology for targeted drug delivery. To carry out drug synthesis, it is very important to identify new chemical entities, those having the potential to treat diseases with safety and efficacy [8]. The area is related to the investigation and exploitation of coherent quantum effects in engineered nanostructures and nanocarriers using quantum mechanics theory. Because quantum technologies rely on the manipulation of matter at extremely small scales, they are reliant on and supported by advances in nanotechnology. The concept of quantum mechanics and its dependency on the size of matter is related to the physics of electron properties in materials with significantly smaller particle sizes. When observing and calculating from macro to micro dimensions, this phenomenon is not present. However, it takes over once it reaches the nanoscale size range. Engineered nanomaterials go beyond mere shrinkage, whether created top-down (where a bulk material is shrunk to a nanoscale pattern or structure) or bottom-up (where larger structures are developed or grown atom by atom or molecule by molecule). It becomes dominant as it reaches the nanoscale size range.

Nanotechnology deals with the nano-scale devices—components of those devices, their production processes, and their engineering to build a usable system. We frequently develop and build nanoscale devices while taking into account the quantum effect of the system. We are dealing with a single particle or energy level here [9, 10]. Now, in quantum computing, we execute computation based on the phenomena of quantum mechanics with the devices that accept spins, oscillation, or a single charge to build the most fundamental basis of computation, a qubit or a two-level energy system. Thus, when we want to create a two-level system, we will look into nanostructure or nano-material or nano-scale electronic devices, such as semiconductor quantum dots, Josephson junction, cQED elements, etc. [11]. Figure 6.1 explains various stages for targeted drug delivery and drug discovery for the same.

At the lower end of the nanoscale (smaller size), quantum effects can start to influence the physical as well as chemical properties of matter, further affecting electrical, optical, and magnetic behavior when open to a suitable environment. Nanoscale materials can be fabricated in one dimension, two dimensions, or all three dimensions (such as nanoparticles and quantum dots) by using various techniques for their production. Any material's bulk qualities are just the culmination of all quantum forces operating on its atoms. As things get smaller and smaller, eventually the average stops working and researchers have to deal with the unique behavior of individual atoms or molecules, which might be very different from

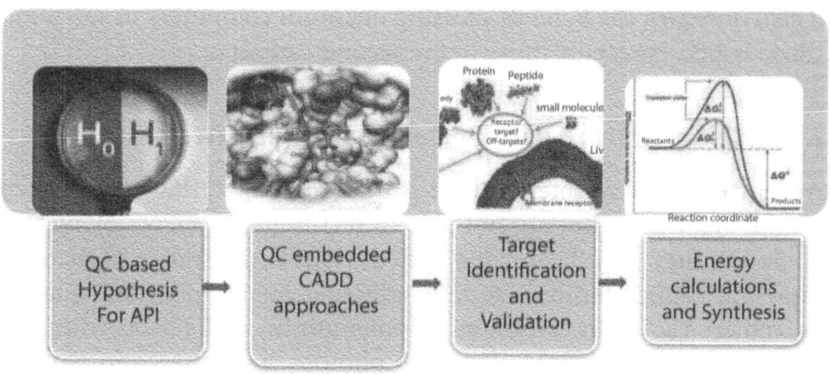

Figure 6.1 QC-aided targeted drug delivery and drug development.

how individual atoms behave when they are assembled into a bulk material [12–14].

Materials lowered to the nano-size range can carry different properties from those observed in the macro-size range [15]. Concisely, nanotechnology can be thought of as one of the applications of quantum theory. Scientists hypothesize that single-molecule sensors, sophisticated memory storage, and neural-like networks that can be accomplished with a very small number of molecules might all be developed depending on the nature of quantum variance. These devices will replace larger microelectronic equipment and be small, light, self-contained, and energy-efficient [16]. Quantum theory and mechanics explain how energy and matter interact at the atomic and subatomic scales. On the atomic and sub-atomic levels, quantum theory and quantum mechanics describe the interaction of energy and matter [17].

6.2 The Types of Quantum Computers

6.2.1 Scalable Quantum Computers

A QC program can apply quantum algorithms almost error-free owing to a common set of gates that operate on quantum bits (qubits) [18]. Very promising algorithms are being used for many works, like chemical interactions and process optimization, so that QC can handle them more precisely than any other old methods [19]. Although a QC can potentially target an accelerated speed, this benefit is not always relevant. The vast majority of optimization methods fall into the second group, while the first covers a few niche topics like factorization, discrete logarithms, and the simulation of quantum many-body systems. The higher overhead demands will also diminish the benefits of using quantum computers to solve some of those issues, especially those in the second category.

6.2.2 Noisy Intermediate-Scale Quantum Devices

Devices called Noisy Intermediate-scale Quantum (NISQ) can execute quantum algorithms but are not able to rectify minor errors or mistakes [20]. So far, they have been constructed using a wide range of materials, comprising cations, anions, superconducting qubits, photons, quantum dots, centers vacant in carbon atom polymorphs, and silicon with phosphorus inserted into it [21]. Although photons and cold atoms will soon follow, the first two are the most complex [22]. Variational algorithms, which apply to optimization issues in general, such as locating a string of bits, are a very active area of research with NISQ devices [21, 23].

6.2.3 Analog Quantum Devices

Another category of devices falls under what is referred to as analog quantum simulators [24]. These can be thought of as analog quantum computers that, despite not being able to construct a universal set of gates or fix faults, are more effective than classical devices at solving some particular problems [25]. They are particularly well adapted to issues with quantum many-body difficulties that are prevalent in chemistry, high-energy physics, and condensed materials [26].

6.3 Role of QC in Computer-Aided Drug Design

The cost of computing increases exponentially as the accuracy of molecular simulations is increased in the computer-aided drug design (CADD) process [27, 28]. How we conceive of solids, molecules, atoms, nuclei, and subatomic particles may alter as a result of quantum computing. By bypassing the scaling restrictions of conventional computational techniques, it could assist in numerically solving the Schrödinger equation for larger and more complicated molecular systems [29]. We anticipate quantum computing to have an impact on all facets of the drug-development pipeline as quantum hardware advances in terms of quantum bit quality, count, and volume. NISQ1 devices and variational algorithms enable the early adoption and commercialization of quantum chemistry and QML [30], positioning pharma to be one of the technology's primary benefactors even though fault-tolerant quantum devices are still at least 10 years away.

Currently, pharmaceutical companies process molecules using computer-assisted drug discovery, which uses non-QC methods like MD and DFT [31]. However, because of the severe limitations of conventional computers, even straightforward calculations to anticipate the behavior of medium-sized drug molecules could take a very long time to complete. Methods may exhibit how the development process would be accelerated with QC-enhanced CADD. By assisting in the precise prediction of molecular properties, QC could increase the efficiency of current CADD technologies. This can have a variety of effects on the research process, including mimicking protein folding and the interactions between drug candidates and biologically significant proteins [32, 33]. In this case, QC might make it possible for researchers to evaluate computational libraries simultaneously against a wide range of potential target architectures [1]. The structural flexibility of the target molecule is typically constrained by present techniques due to a lack of computing resources and time. Finding the optimal treatment alternatives may be more challenging due to these restrictions.

The main technological challenge facing drug research and development today is that traditional computers are ineffective at solving so-called optimization issues, which require attempting every possible solution in order to discover the best one [24]. This process consumes a lot of time and energy. Theoretically, quantum computers excel at resolving optimization issues. They can describe all potential outcomes of a problem simultaneously, including quantum interactions at the particle level, by using parallel states of quantum superposition. Theoretically, quantum computers should be able to process enormous amounts of data quickly once they attain their promised processing speeds [34].

ONIOM is one of the most well-known, effective, and straightforward hybrid quantum mechanics and/or molecular mechanics (QM/MM) methods for dealing with complex molecular systems [35]. In hybrid QM/MM approaches, the lower cost of MM methods integrated with an accuracy of QM methods is employed [36]. The ONIOM method includes a large amount of straightforward linear extrapolation. It permits its extension to two-layer ONIOM (QM1:QM2) [37], three-layer ONIOM (QM1:QM2:MM), and, in principle, any n-layer n-level-of-theory approaches. The fixed-charge calculation of electrostatic interactions is one of the most important topics in molecular mechanics (MM) force fields [38]. Since QM has shown to be useful in illustrating the mechanistic and structural states of drug interactions with carriers and receptors, it is appropriate to incorporate QM inside a hybrid QM/MM [35, 39] technique as the initial stage in computational elaboration. The likelihood of a reaction can be estimated using the energy difference between the products and the reactants [38].

This technology has the potential to significantly improve society whether used independently or in conjunction with other technologies. If the right measures are adopted to assure ethical research, development, and application, the ripple effects of the quantum revolution could enhance the lives of many people [40]. Therefore, it is crucial that we investigate the effects that quantum computing might have on the pharmaceutical sector if we want to guarantee that its potential to assist in the discovery of treatments for difficult-to-cure diseases like cancer and Alzheimer's becomes a benefit that is distributed fairly across the world [41, 42].

On a longer time scale, QC is utilized for hypothesis planning in collaboration with machine learning, which further helps to develop the structure–activity relationship of the potent compounds and also helps in the interpretation of the 3D structure of ligand-protein responsible for binding [43]. A number of development trials could be significantly reduced by integrating QC-based optimization of a quality lead compound. Current approaches that are already used in CADD could be improved for giving better results in a definite time duration when QC is utilized in pre-existing tools; it also increases the accuracy and precision of results related to ligand identification [44].

6.4 Development of Molecular Formulations

By assisting in the precise prediction of molecular properties, QC could increase the efficiency of current CADD technologies [40]. This can have a variety of effects on the research process, including mimicking protein 3-dimensional structures and drug candidate interactions with biologically important proteins. In this situation, QC may allow researchers to simultaneously assess computational libraries against a large number of potential target structures [45]. The goal of the molecular simulation is to identify a compound's ground state, which is its most stable state. Researchers use M&S systems to simulate the interactions between each electron in each atom of that chemical to predict how they will react to one another [46]. This is a rather easy technique if the molecules under evaluation are simple enough. Consequently, even the most potent supercomputers of today can only simulate molecules with up to a few hundred atoms, limiting their calculations to a relatively tiny subset of all known compounds [47]. The structural flexibility of the target molecule is typically constrained by existing techniques due to a lack of computing resources and time. Finding the optimal treatment alternatives may be more challenging due to these restrictions [48].

There are approximately 300,000 materials and 15 million chemical structures. Numerous useful substances still need to be discovered. Nature creates materials every day with amazing properties that industrial production methods cannot duplicate [49]. For instance, spider silk is produced at optimum rather than being forged in a furnace, making it stronger by weight than steel. The only by-product of its production is water. Given that spider silk is a protein produced by DNA, the greater subatomic modeling capabilities of quantum computers may one day allow us to produce a similar or improved substance in an equally eco-friendly manner [50].

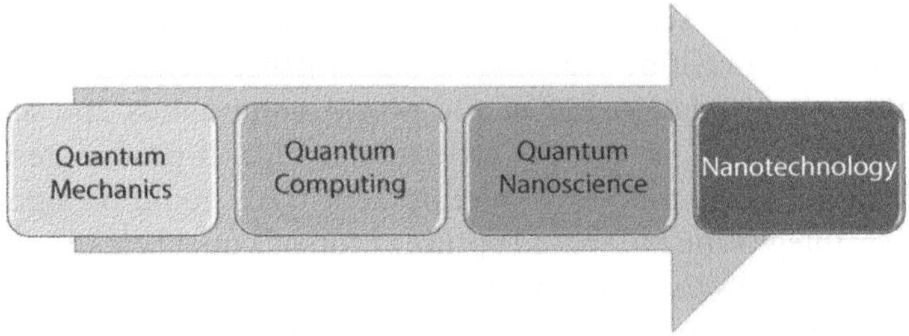

Figure 6.2 Schematic representation showing the role of QC in nanotechnology.

The industrial sector's anticipated applications for quantum computing could benefit the auto, aerospace, and electronics industries by [51, 52]:

- Better strength-to-weight ratios in the materials
- Batteries with a vastly increased energy density
- Catalytic and synthetic methods are more effective and helpful with energy production.

Computer simulation is being used to design and test a lot of items. Hardware parts and subparts for the automotive and aerospace industries are 3D-modeled with particular engineering safety margins. These margins can add up, resulting in over-engineered, overweight, or expensive items that can hinder their potential to be commercially viable [53, 54].

It is projected that, in the future, quantum computers will be able to simulate the interactions between the component systems more extensively and precisely by computing the system pathways, signal-to-noise ratio, and load vibration. This thorough study can minimize the overall effects of numerous separate safety margins, reduce costs without sacrificing system performance, and optimize the production of individual components within the context of the entire system [55]. Figure 6.2 is a representation of the development of various nanotechnological devices with the help of QC.

6.4.1 QC-Based Development of Nanocarriers

Innovative technologies, including those enabled by quantum mechanics, have demonstrated tremendous potential over the past 10 years for developing innovative designs with functional legends, quickly detecting ailments, and investigating unique strategies for treating rare hereditary diseases and disorders, better treatment approaches or drug design processes, and more precise comparisons of large-scale molecules [56, 57]. The appearance of one of the most notable technological advances in recent years has been quantum computing science. Along with securing data and decoding of data, QCs can be utilized for executing complicated computations, expedited calculations, integration of big datasets identifying patterns, or abnormalities in massive databases and coming up with answer security, the use of quantum computers (QCs) for extremely complex calculations, expedited calculations, analysis, processing, integration, and integration of large datasets discovering answers for difficult problems by seeing anomalies or patterns in massive databases challenges, the development of innovative classifiers capable of producing extremely complex data maps, estimating molecular energy, accurately simulating chemical systems or complicated phenomena levels, investigation of the catalytic processes for creating innovative, sophisticated simulations (such as precise representation of quantum behavior, molecular behaviors, electronic cloud for the development of novel therapeutic agents, including anti-cancer ones), solving selection and optimization issues, computing a vast number of protein 3D structures and its folding, for discovery of potent active ingredient, learning more about DNA transcription, prediction of matter characteristics and properties their molecular and cellular level interactions, and quick identifications of new medications [58, 59].

Quantum molecular design is a complex, multi-step process, but here we will look at it in three main steps [59].

Step 1: Extract binding information from a target. In the beginning, the quantum computer has some idea of what the final molecule needs to be like. For example, if the goal is to create a molecule in the Janus family of kinases, then the algorithm searches for an example of a kinase in that family.

Step 2: Search in a larger database. Here, the algorithm searches that target against a database of scaffolds published in the literature, theoretical scaffolds created combinatorically, and even combines some of them. The result is a list of matches that have the same binding pattern at their most basic level.

Step 3: Create a novel chemical space by varying functional groups.

The fundamental informational unit in traditional computing is the bit, which has just two potential states: 0 or 1 [60]. The fundamental component of quantum information, however, is the qubit. A qubit can represent more information than a conventional bit because it can be in a superposition of 0 and 1. A quantum computer's exponentially expanded state space is then produced through the entanglement of qubits with each other. Artificial intelligence (AI) and machine learning (ML) can benefit from quantum computing. AI's most effective feature, machine learning (ML), interacts with both supervised and unstructured data. Learning is connected to data analysis and mining. Nanocomputing devices execute ML paradigms to address challenging analytical issues [14]. ML approaches offer the potential to forecast drug interactions and uncover new targets or treatments against several illnesses, in addition to cross-validation and offering a deeper grasp of the structural properties of the complicated molecules. As a supervised learning algorithm, the regularized least squares classifier successfully predicted interactions between the drugs and their targets. ML approaches such as the random forest, support vector machine, k closest neighbor, Naive Bayes classifier, or Gaussian process could be beneficial for predicting the pharmacological profiles of the effectiveness of drug delivery, or potential treatment solutions. Fuzzy logic, decision trees, and genetic programming approaches offer the chance to overcome the obstacles connected to the development of contemporary therapeutic artificial intelligence (AI) simulation for sophisticated problem solving, decision making, and search offering the best answers and knowledge or insight [9]. ML deals with increased accuracy, learning from the experience for maximizing machine performance, and providing self-learning algorithms and knowledge. By enhancing ML methods, quantum computing and speeding AI could lead to more efficient completion of difficult activities like processing the sensory data or investigating nonlinear linkages and influences. Nanoparticles (NPs) can vary in size or stability. Modern factory control methods push the boundaries of advanced analytics, particularly when using machine learning and analyzing numerous factors.

QC may assist in uncovering novel correlations in data, improving pattern recognition, and advancing categorization. The integration of QC with machine learning is projected to have a considerable impact in various areas. Semiconductor chip production also incorporates machine learning. However, due to a computational bottleneck, more sophisticated analyses cannot be performed using classical computing's limited number of components. Quantum computing is anticipated to examine extra interaction elements and procedures to boost output [10, 61].

It takes a lot of computing power to simulate and optimize production flows and robotics scheduling for complicated items like autos. Faster optimization runs and more dynamic optimizations in production may be made possible by quantum computing [62, 63].

Future QC should be able to assess software systems that are much more complicated than what is now possible for classical computers to evaluate [64]. Supply chains are developing from a linear model to an organic model that is more responsive and dependent on changing real-time market demands and the current availability of critical components [65, 66]. Quantum computing may help Industry 4.0's digital supply chain by enhancing risk management, accelerating decision-making, and lowering operating expenses in addition to reducing lost sales due to out-of-stock or discontinued product situations [67].

In the past years, many synthetic polymers have been investigated to apply nanomedicine exclusively as drug delivery systems. Polylactic-co-glycolic acid (PLGA) is one of the most valuable polymers because of its biodegradability, the ability to self-assemble into nanometric micelles capable of trapping the same molecules like drugs, and the ability to reach them into the body in a time-controlled release. These PLGA-derived systems can be loaded with various drugs, peptides, and proteins, which are released slowly into the body.

There were two steps to these modelings. First, molecule structures were optimized, and then ONIOM2 was used to compute molecular energies (DFT: UFF) [35]. The thiazolo [5,4-d] pyrimidinone derivative, which Bondock has synthesized, has been evaluated as an anti-microbial and anti-cancer agent. The quantitative structure–activity relationship of these novel thiazoline derivatives with anti-cancer activity has been investigated with the density functional theory by Madadi *et al.* A study was carried out to consider the complex's geometry and binding energy between PLGA copolymers and the thiazoline derivatives. The ONIOM methodology with the Gaussian 03 program was used to accomplish the whole geometry optimization and frequency computations utilizing a combined QM and MM method to optimize the thiazoline derivatives and PLGA polymer. The ONIOM binding energies revealed that these complexes have relatively low stability, indicating that the PLGA polymer can be employed for drug delivery. The more potent binding interaction between the drug and the polymer is due to substitution at 1-methyl-1H-tetrazole-5-yl and 1-phenyl-1H-tetrazol-5-yl [2].

6.4.2 QC in Biosensor

In contrast to the medical communities that require drug sensors to identify and remove an overdose of a substance that may pose a grave hazard, law enforcement agencies must be able to identify illicit drugs [68]. While these medications or biomolecules can be found using a variety of laboratory methods, such as chromatography, mass spectroscopy, infrared (IR), and Fourier-transformed IR (FT-IR) spectroscopy, etc., electrochemical sensors are crucial for quick responses and mass applications [69].

In most cases, an electrochemical sensor transforms the information from a physical response between the biomolecule and the sensor surface into a recognizable electronic signal. The majority of these electrochemical sensors consist of a receptor, where biomolecule contact occurs, a transducer, which turns these data into an electrochemical signal, and an amplifier, which amplifies the signal into a detectable one. Due to its numerous benefits in the field of drug sensing, LDNs, particularly 2D nanomaterials, have drawn a lot of attention from the scientific community in recent years [70]. In 2D nanomaterials, which consist of a single layer of atoms designated as nanosheets, electrons are permitted to travel more freely than in 3D structures without interlayer electron interaction, increasing their sensitivity to medicines. Due to their exceptional mechanical strength and hardness, 2D nanomaterials (graphene) can be used to create some of the most durable sensors. It was discovered that the field-effect transistor coated with a monolayer of deformed graphene demonstrated an ultra-sensitivity toward nucleic acids and other biomolecules with a structure in the millimeter range by considerably enhancing the local conductivity. Ozkan *et al.* also employed core-shell nanoparticles with boron nitride nanosheets (BNNS) to create a sensor for the anticancer medication etoposide, which demonstrated extremely high sensitivity with a detection limit (LOD) of 0.00025 ng/ml [71].

6.4.3 QC-Based Targeted Drug Delivery

The following QC use cases apply to various aspects of drug development and will appear at various times over a long period. All of them, on the other hand, may make it easier to develop targeted compounds with greater precision and efficiency [72, 73]. Despite their best efforts, many biologically important proteins have yet to be crystallized, whether due to their size, solubility (for example, membrane proteins), or inability to express and purify in sufficient quantities. Pharmaceutical corporations occasionally produce medications without even understanding the structure of a protein, taking the risk of a trial-and-error method in later phases of drug development since the commercial case for a certain drug has the potential to be so great [74].

Folic acid (FA)-conjugated graphene quantum dots (GQD) were used for targeting the cancer cell. The NH2 group of the FA and the COOH groups of the GQDs reacted to form an amide bond, which then joined the GQDs and FA. The GQDs were planar nanocrystals of graphene with <0.8 nm thickness and 1–4 nm diameter [13].

Salmas *et al.* employed several *in silico* techniques to focus on the interactions of the medication molecules risperidone, clozapine, aripiprazole, olanzapine, ziprasidone, and quetiapine in the active site of the D2R [75]. *In silico* docking experiments and molecular dynamics (MD) simulated studies were employed to determine the conformations of drug molecules (ligands) at the D2R's binding site [76]. In a second method, in addition to the chemical interaction energies, the deformation and desolvation energies of ligands with neighboring amino acid residues, within a radius of 5 of the protein-bound ligand, were also considered for determining the ligand binding affinities. Despite the fact that Asp114's main interaction with the drugs is an H-bond that predominates over amino acid in the binding cavity, the exclusion of weak interactions leads to an overestimation of binding affinity. When all drug interactions with amino acids within a 5 radius of the ligand at the binding site are taken into consideration, the estimated binding energies are in the same energy range. The latter technique results in improved experimental and calculated binding energy agreement, which together show that all drugs bind to the catalytic domain of the D2R favorably and roughly equally well [74, 77].

By adjusting the synthesis conditions, it is possible to alter the quantity and kind of nitrogen (N) functionalities in graphene quantum dots, which leads to a variety of applications. The properties of nitrogen-doped graphene quantum dots (N-GQDs) can be easily altered, and they are very biocompatible. Through the use of molecular dynamics (MD) simulations and density functional theory (DFT) calculations, it was determined how different N-functionalities affected the effectiveness of N-GQDs as drug delivery systems. All of the ΔG_{ad} values were negative, indicating that the GC drug's spontaneous adsorption on the pristine GQDs and N-GQDs had taken place. However, compared to pristine GQDs and edge N-GQDs, the center N-GQDs' DGad values for adsorption of GC drugs were higher. The MD data showed that the GC drug was appropriately loaded onto the nanocarrier surfaces and that the only conditions in which the center N-GQDs could release the medication favorably were the acidic environments found in cancer tissues [12].

6.4.4 Target Identification

The paradigm through which one views quantum chemical simulation may change as a result of quantum computing. Quantum computers can conduct crucial quantum chemistry and machine learning tasks in ways that are beyond the capabilities of classical computers by effectively preparing highly entangled states that are otherwise impossible to describe on classical computers [78]. High-throughput screening (HTS) on sizable molecular libraries utilizing *in vitro* activity assays has historically been used to find hits. These searches typically yielded few results and necessitated the synthesis of several chemicals, which in turn required a significant time and resource commitment. In the early stages of drug discovery, evidence of therapeutic effects on the activation or inhibition of a particular biological pathway linked to a disease is gathered [79]. The discovery of a new compound is one of the most complicated aspects of the drug scheming process. Once a lead compound for a novel therapeutic has been revealed, then toxicological studies needed to be performed before the instigation of its clinical trials. The alteration in the chemical structure of a lead compound is a vital process in order to develop its analogs and prodrugs. The effectiveness of random screening has amplified significantly through automated High-Throughput Screening systems [80]. These systems utilize cell culture organization with associated receptor molecule resultant through gene cloning and enzyme assay. The surfacing of new technologies has made it promising to partition a large number of libraries of peptides and nucleic acid resultant from combinatorial chemistry measures [81, 82].

The biological component known as the target, which is typically a protein, is what causes such a reaction. The phrase could alternatively, in a more general sense, refer to the RNA or genes connected to the protein. The phrase "druggable" refers to a target's ability to be accessed by a drug candidate and to produce a biological response that can be measured both *in vitro* and *in vivo*. The possibility of success during the discovery phase is increased by the identification of suitable targets and their accompanying confirmation by studies of the mechanism of action, which also enables one to anticipate adverse effects related to the modulation of the target. Chemical proteomic methods, including affinity chromatography, biochemical fractionation, and radioactive ligand binding assays, are used in conventional target identification strategies [83]. These techniques separate the target protein from a mixture of other proteins by using a tiny chemical with known action [84].

A method of target identification that is more comprehensive and uses computational models to find target relationships and their mechanisms within intricate biological systems has been inspired by the growth of systems biology. In order to identify common patterns of biological response and therapeutic activity, many of these methods match the available biological data on regulatory networks, molecular pathways, and cell phenotypes with the same kind of profiles for bioactive compounds and targets [1]. To carry out structured-based drug discovery, the tertiary structure of the target protein must be described after target identification and confirmation. On a protein crystal, which is often difficult to obtain due to the experimental challenges of purifying and crystallizing proteins, this can be accomplished utilizing NMR and X-ray crystallography techniques [7]. The identification of binding sites, or those areas of the protein where the drug candidate is more likely to bind and have the desired therapeutic effect, is an important component of target characterization [85]. In co-crystallized structures of the target or related proteins, small-molecule binding sites are frequently known; however, if the binding sites are unknown or if the goal is to find new binding sites for novel applications, a few computational programs, including Ligsite, Qsite finder, and CASTp, can be used [86].

6.4.5 Target Validation

After hit compounds are found, they move onto an optimization phase to create a smaller group of superior possibilities, known as leads. The set of leads is further optimized through an iterative process involving CADD development, *in vitro* testing, and animal testing. Structure-based methods concentrate on calculating the candidate molecules' binding affinities, whose precise and effective calculation is thought to be the ultimate goal of CADD [27, 61].

6.4.6 Identification of Hit and Its Validation

When generating and validating results, QC's capacity to analyze complicated phenomena concurrently would be extremely useful. Pharma companies may currently only use CADD on small to medium-sized drug candidates, and usually sequentially [87]. The bottleneck lies in computing power. Pharma companies might broaden these use cases to include specific biologics, such as fusion proteins or semi-synthesized biologics, and conduct *in silico* search and validation studies in a more high-throughput show with adequate QC [82]. In the future, this use case would effectively encompass all of known biology in addition to protein identification. This step would already deliver potential lead molecules that are much easier and faster to optimize if a robust enough hit-generation and validation approach was used. Unfortunately, the majority of methods that have been used to calculate binding affinities are based on MM approximations that do not adequately account for phenomena like polarization, charge transfer, and other quantum mechanical factors that are necessary to reach such accuracy [4]. Due to these restrictions, quantum mechanical (QM) methods must be used, which, by their very nature, include such effects.

However, due to the approximations that must be used to reduce the intrinsic computing complexity of exact QM simulations on classical computers, the use of QM calculations in drug discovery is not widely used in practice. The already complex simulation of the receptor–ligand complex is further complicated by the requirement to combine explicit and implicit solvation techniques in order to correctly account for solvation effects. In this case, when quantum computers advance in size and accuracy that makes it possible to simulate big molecules, the development of quantum algorithms that can deal with quantum simulation effectively and precisely may help to increase the applicability of FES techniques to drug discovery [59, 87].

6.4.7 Optimization of Lead

QC may provide for improved ADME, more precise activity and toxicity estimates for organ systems, dosage and solubility optimization, and other safety issues during lead optimization, which is one of the top three parameters to increase R&D productivity [1].

6.5 Data Generation, Interpretation, and Co-Relation

The metalevel of research and development heavily emphasizes the integration of essential data, such as by creating logical relationships between data points through effective (semantic) administration. The complexity of the biological information that can be handled increases with the complexity of the graphs that guide the drug development research process. The goal of "topological data analysis," which is currently the subject of research, is to identify "holes" and "connections" in massive datasets. This may eventually allow R&D professionals to pinpoint actual scenarios and "business verticals" where such algorithms are applicable, such as when identifying connections between brain cells in response to medication [34].

Additionally, QC may be used to create fake data by using ML algorithms to "deepfake" missing data points during the research phase. This may be particularly useful in cases where there is a dearth of data, as in the case of rare diseases, where the lack of data can be lessened by using generated datasets. Here, QC will set a new bar for the amount of starting data required, the speed at which ML models can be trained, and the degree of accuracy [63, 84].

6.6 Role of QC in Clinical Trials

Clinical trials might benefit from population pharmacogenetic modeling, patient stratification, and the identification of patients. QC could choose the best trial sites during the planning and execution of the trial [88]. QC could be used to supplement causality analysis for adverse events to enhance active safety surveillance. Even though QC in pharma R&D has a ton of potential, it will almost probably be utilized further down the value chain. QC can help calculate reaction rates, optimize catalytic processes, and finally offer significant efficiencies in the creation of new product formulations in the synthesis of active substances. QC in the pharmaceutical industry could include supply chain enhancements in the business-related value pools and logistical optimization (such as optimizing on-site flows of materials, heat, and trash in production facilities). Finally, as we get closer to market access and commercialization, QC might enable automatic drug recommendations [89].

6.7 Future Prospects

A new chemical entity typically takes 10 years and $2 billion in today's notoriously expensive and time-consuming drug research and development process. In the next 5 to 10 years, we anticipate that the first QC tools that pharmaceutical companies adopt will be hybrid ones that combine traditional algorithms with QC subroutines where they can bring value [82]. The majority of resources devoted to quantum computing research concentrate either on the principles behind the construction of quantum computers or on applications in STEM fields. Protein folding, quantum annealing, and molecular comparison and design are a few issues that need to be resolved [45].

6.8 Conclusion

Recent years have seen significant advancements in both hardware and algorithms, which has accelerated the development of quantum computing. The eventual commercial utility of quantum computers is being brought closer by these developments. These innovative machines will find many applications in the exciting field of drug discovery. In contrast to current quantum chemistry techniques, the quantum simulation will make it possible to characterize molecular systems more quickly and accurately [2]. Additionally, algorithmic advances in quantum machine learning provide intriguing substitutes for traditional machine learning methods, which may potentially be helpful for the biochemical efforts required in the initial stages of drug discovery. In the meantime, quantum hardware is scaling up quickly to a point where a precise simulation is challenging even if the biggest supercomputers in the world are used [60]. We examine the potential paradigm shift that these most recent developments can bring about in the way people approach drug discovery, highlighting both the benefits and drawbacks of each breakthrough. We specifically point out how, while fault-tolerant, error-corrected quantum computers are still in the development stage, hybrid quantum-classical techniques to quantum simulation and quantum machine learning could result in significant advancements employing noisy intermediate-scale quantum devices [20].

References

1. Koksharov, Y.A., Gubin, S.P., Taranov, I.V., Khomutov, G.B. and Gulyaev, Y.V., Magnetic nanoparticles in medicine: Progress, problems, and advances. *J. Commun. Technol. Electron.*, 67, 2, 101–116, 2022.
2. Dopfer, O. and Fujii, M., Probing solvation dynamics around aromatic and biological molecules at the single-molecular level. *Chem. Rev., 116*, 9, 5432–5463, 2016.
3. Christensen, A.S., Kubar, T., Cui, Q., Elstner, M., Semiempirical quantum mechanical methods for noncovalent interactions for chemical and biochemical applications. *Chem. Rev., 116*, 9, 5301–5337, 2016.
4. McKinsey, McKinsey's Hamburg office, 2022 Jul 2, Available from: https://www.mckinsey.com/industries/life-sciences/our-insights/pharmas-digital-rx-quantum-computing-in-drug-research- and-development.
5. Sliwoski, G., Kothiwale, S., Meiler, J., Lowe, E.W., Computational methods in drug discovery. *Pharmacol. Rev., 66*, 1, 334–395, 2014.
6. Cao, Y., Romero, J., Aspuru-Guzik, A., Potential of quantum computing for drug discovery. *IBM J. Res. Dev., 62*, 6, 6–1, 2018.
7. Arcon, J.P., Turjanski, A.G., Martí, M.A., Forli, S., Biased Docking for Protein–Ligand Pose Prediction, in: *Protein-Ligand Interactions and Drug Design*, pp. 39–72, Humana, New York, 2021.
8. Kumar, N., Srivastava, R., Prakash, A., Lynn, A.M., Structure-based virtual screening, molecular dynamics simulation and MM-PBSA toward identifying the inhibitors for two-component regulatory system protein NarL of Mycobacterium Tuberculosis. *J. Biomol. Struct. Dyn., 38*, 11, 3396–3410, 2020.

9. Pircalabioru, G.G., Bleotu, C., Curutiu, C., Mihaescu, G., Chifiriuc, M.C., Nanodrug delivery systems in cancer, in: *Biomedical Applications of Nanoparticles*, pp. 31–62, William Andrew Publishing, Faculty of Applied Chemistry and Materials Science, University Politehnica of Bucharest, Romania, 2019.
10. Nanotechnology in Quantum Computing, 2022 Jul 2. Available from: https://www.azonano.com/article.aspx?ArticleID=3251.
11. Ulbrich, K., Hola, K., Subr, V., Bakandritsos, A., Tucek, J., Zboril, R., Targeted drug delivery with polymers and magnetic nanoparticles: covalent and non-covalent approaches, release control, and clinical studies. *Chem. Rev.*, *116*, 9, 5338–5431, 2016.
12. Lim, P.F., Leong, K.H., Sim, L.C., Abd Aziz, A., Saravanan, P., Amalgamation of N-graphene quantum dots with nanocubic like TiO2: an insight study of sunlight sensitive photocatalysis. *Environ. Sci. Pollut. Res.*, *26*, 4, 3455–3464, 2019.
13. Wu, G., Cheng, Y., Yang, Z., Jia, Z., Wu, H., Yang, L., Li, H., Guo, P., Lv, H., Design of carbon sphere/magnetic quantum dots with tunable phase compositions and boost dielectric loss behavior. *Open Chem. Eng. J.*, *333*, 519–528, 2018.
14. Szefler, B., Nanotechnology, from quantum mechanical calculations up to drug delivery. *Int. J. Nanomed.*, *13*, 6143, 2018.
15. Chaturvedi, S., Dave, P.N., Shah, N.K., Applications of nanocatalyst in new era. *J. Saudi Chem. Soc.*, *16*, 3, 307–325, 2012.
16. Nanotechnology Examples and Applications, 2022 Jul 2. Available from: https://www.nanowerk.com/nanotechnology-examples-and-applications.php.
17. Aspuru-Guzik, A. and Walther, P., Photonic quantum simulators. *Nat. Phys.*, *8*, 4, 285–291, 2012.
18. Gyongyosi, L. and Imre, S., Scalable distributed gate-model quantum computers. *Sci. Rep.*, *11*, 1, 1–28, 2021.
19. Stassi, R., Cirio, M., Nori, F., Scalable quantum computer with superconducting circuits in the ultrastrong coupling regime. *NPJ Quantum Inf.*, *6*, 1, 1–6, 2020.
20. Klco, N., Dumitrescu, E.F., McCaskey, A.J., Morris, T.D., Pooser, R.C., Sanz, M., Solano, E., Lougovski, P., Savage, M.J., Quantum-classical computation of Schwinger model dynamics using quantum computers. *Phys. Rev. A.*, *98*, 3, 032331, 2018.
21. Babukhin, D.V. and Pogosov, W.V., The effect of quantum noise on algorithmic perfect quantum state transfer on NISQ processors. *Quantum Inf. Process.*, *21*, 1, 1–18, 2022.
22. Bharti, K., Cervera-Lierta, A., Kyaw, T.H., Haug, T., Alperin-Lea, S., Anand, A., Degroote, M., Heimonen, H., Kottmann, J.S., Menke, T., Mok, W.K., Noisy intermediate-scale quantum (NISQ) algorithms, in: *arXiv preprint*, 2021, arXiv:2101.08448.

23. Preskill, J., Quantum computing in the NISQ era and beyond. *Quantum, 2,* 79, 2018.
24. Beebe, N.H., A Complete Bibliography of Publications. *J. Math. Phys.*, 2005–2009, 2019.
25. Jiang, M., Zhang, Z.K., Dong, D.Y., Liu, B., Tarn, T.J., The improved quantum switching mechanism based on contention. *Chaos, Solitons Fractals, 39,* 4, 1936–1942, 2009.
26. Zak, M. and Fijany, A., Quantum resonance for simulating combinatorial problems. *Phys. Lett. A, 345,* 1-3, 167–172, 2005.
27. Sun, H., Chen, H.F., Chen, R., Genome comparisons as a tool for antimicrobial target discovery, in: *In Silico Models for Drug Discovery*, pp. 31–38, Humana Press, Totowa, NJ, 2013.
28. Nikolaeva Koleva, M. and Fernandez-Ballester, G., In silico approaches for TRP channel modulation, in: *TRP Channels*, pp. 187–206, Humana Press, New York, 2019.
29. Wille, R., Fowler, A., Naveh, Y., Computer-aided design for quantum computation. *IEEE/ACM International Conference on Computer-Aided Design, z, ICCAD*, 2018 Nov 5.
30. Drug & Alcohol Screening - QML Pathology, 2022 Jul 2, Available from: https://www.qml.com.au/commercial/drug- and-alcohol-screening/.
31. Yu, W. and MacKerell, A.D., Computer-aided drug design methods, in: *Antibiotics*, pp. 85–106, Humana Press, New York, 2017.
32. Helstrom, C.W., Quantum detection and estimation theory. *J. Stat. Phys., 1,* 2, 231–252, 1969.
33. O'Malley, P.J., Babbush, R., Kivlichan, I.D., Romero, J., McClean, J.R., Barends, R., Kelly, J., Roushan, P., Tranter, A., Ding, N., Campbell, B., Scalable quantum simulation of molecular energies. *Phys. Rev. X, 6,* 3, 031007, 2016.
34. Esposito, C., Landrum, G.A., Schneider, N., Stiefl, N., Riniker, S., GHOST: adjusting the decision threshold to handle imbalanced data in machine learning. *J. Chem. Inf. Model., 61,* 6, 2623–2640, 2021.
35. Lundberg, M., The Oniom method and its applications to enzymatic reactions, in: *Multi-scale quantum models for biocatalysis*, pp. 21–55, Springer, Dordrecht, 2009.
36. Sailor, M.J., Computationally Enabled Sensors. *ACS Sens., 6,* 6, 1988–1989, 2021.
37. Chung, L.W., Sameera, W.M.C., Ramozzi, R., Page, A.J., Hatanaka, M., Petrova, G.P., Harris, T.V., Li, X., Ke, Z., Liu, F., Li, H.B., The ONIOM method and its applications. *Chem. Rev., 115,* 12, 5678–5796, 2015.
38. Bani-Yaseen, A.D., Al-Jaber, A.S. and Ali, H.M., Probing the solute-solvent interaction of an azo-bonded prodrug in neat and binary media: Combined experimental and computational study. *Sci. Rep.*, 9, 1, 3023, 2019.
39. Rivera, M., Dommett, M., Crespo-Otero, R., ONIOM (QM: QM') Electrostatic Embedding Schemes for Photochemistry in Molecular Crystals. *J. Chem. Theory Comput., 15,* 4, 2504–2516, 2019.

40. Crespo-Otero, R. and Blancafort, L., A Global Potential Energy Surface Approach to the Photophysics of AIEgens: The Role of Conical Intersections, in: *Handbook of Aggregation-Induced Emission*, pp. 411–454, 2022.
41. Quantum computers offer possibilities for cancer detection, in: *TNO*, 2022 Jul 2, Available from: https://www.tno.nl/en/focus-areas/information-communication-technology/roadmaps/trusted-ict/quantum/quantum-computer-cancer-detection/.
42. Roche taps into quantum computing software for Alzheimer's disease research | Fierce Biotech, 2022 Jul 2, Available from: https://www.fiercebiotech.com/medtech/roche-taps-into-quantum-computing-software-for-alzheimer-s-disease-research.
43. Dommett, M., Rivera, M., Smith, M.T., Crespo-Otero, R., Molecular and crystalline requirements for solid state fluorescence exploiting excited state intramolecular proton transfer. *J. Mater. Chem. C*, *8*, 7, 2558–2568, 2020.
44. Saunders, L.K., Pallipurath, A.R., Gutmann, M.J., Nowell, H., Zhang, N., Allan, D.R., A quantum crystallographic approach to short hydrogen bonds. *CrystEngComm*, *23*, 35, 6180–6190, 2021.
45. Li, Y., Zhao, T., Qing, L., Yu, H., Xu, X., Li, P. and Zhao, S., Solvation dynamics in simple fluids: effect of solute size and potential. *Chem. Eng. Sci.*, 232, 116371, 2021.
46. Bastos, M. and Camps, I., First-principles calculations of nickel, cadmium, and lead adsorption on a single-walled (10, 0) carbon nanotube. *J. Mol. Model.*, *20*, 2, 1–8, 2014.
47. Fidiani, E., Modeling of diatomic molecule using the Morse potential and the Verlet algorithm. *AIP Conf. Proc.*, 1719, 1, 030001, 2016, March.
48. Raghavachari, K. and Saha, A., Accurate composite and fragment-based quantum chemical models for large molecules. *Chem. Rev.*, *115*, 12, 5643–5677, 2015.
49. Römer, L. and Scheibel, T., The elaborate structure of spider silk: structure and function of a natural high performance fiber. *Prion*, *2*, 4, 154–161, 2008.
50. Exploring quantum computing use cases for manufacturing, in: *IBM*, 2022 Jul 2, Available from: https://www.ibm.com/thought-leadership/institute-business-value/report/quantum-manufacturing.
51. Quantum Computing Pharma R&D, in: *Biogen Case StudyAccenture*, 2022 Jul 2, Available from: https://www.accenture.com/gr-en/case-studies/life-sciences/quantum-computing-advanced-drug-discovery.
52. Advanced computing in pharma: 3 reasons why quantum computing could disrupt the pharma R&D, in: Alcimed, 2022 Jul 2, Available from: https://www.alcimed.com/en/alcim-articles/advanced-computing-in-pharma-3-reasons-why-quantum-computing-could-disrupt-the-pharma-rd/.
53. Ambainis, A., Variable time amplitude amplification and a faster quantum algorithm for solving systems of linear equations, in: *arXiv preprint*, 2010, arXiv:1010.4458.

54. Möller, M. and Vuik, C., On the impact of quantum computing technology on future developments in high-performance scientific computing. *Ethics Inf. Technol.*, *19*, 4, 253–269, 2017.
55. Feynman, R.P., Simulating physics with computers, in: *Feynman and computation*, pp. 133–153, CRC Press, Department of Physics, California Institute of Technology, 91107, Pasadena, California, 2018.
56. Hartman-Caballero, K., Quantum Computing and Drug Discovery, 2021.
57. Arute, F., Arya, K., Babbush, R., Bacon, D., Bardin, J.C., Barends, R., Biswas, R., Boixo, S., Brandao, F.G., Buell, D.A., Burkett, B., Quantum supremacy using a programmable superconducting processor. *Nature*, *574*, 7779, 505–510, 2019.
58. Azhagiya Singam, E.R., Zhang, Y., Magnin, G., Miranda-Carvajal, I., Coates, L., Thakkar, R., Poblete, H., Comer, J., Thermodynamics of adsorption on graphenic surfaces from aqueous solution. *J. Chem. Theory Comput.*, *15*, 2, 1302–1316, 2018.
59. A Note Quantum Computing in Drug Discovery, in: Analytics Steps, 2022 Jul 2, Available from: https://www.analyticssteps.com/blogs/note-quantum-computing-drug-discovery.
60. Blais, A., Gambetta, J., Wallraff, A., Schuster, D.I., Girvin, S.M., Devoret, M.H., Schoelkopf, R.J., Quantum-information processing with circuit quantum electrodynamics. *Phys. Rev. A*, *75*, 3, 032329, 2007.
61. Ghosh, D., Upmanyu, N., Shukla, T., Shrivastava, T.P., Cell and organ drug targeting, in: *Nanomaterials for Drug Delivery and Therapy*, pp. 1–31, William Andrew Publishing, Faculty of Applied Chemistry and Materials Science, University Politehnica of Bucharest, Romania, 2019.
62. Zhu, G., Subaşı, Y., Whitfield, J.D., Hafezi, M., Hardware-efficient fermionic simulation with a cavity–QED system. *NPJ Quantum Inf.*, *4*, 1, 1–10, 2018.
63. Biamonte, J., Wittek, P., Pancotti, N., Rebentrost, P., Wiebe, N., Lloyd, S., Quantum machine learning. *Nature*, *549*, 7671, 195–202, 2017.
64. Holevo, A., *Probabilistic and Statistical Aspects of Quantum Theory*, Edizioni della Normale, 1982.
65. Smith, A., Kim, M.S., Pollmann, F., Knolle, J., Simulating quantum many-body dynamics on a current digital quantum computer. *NPJ Quantum Inf.*, *5*, 1, 1–13, 2019.
66. Boixo, S., Isakov, S.V., Smelyanskiy, V.N., Babbush, R., Ding, N., Jiang, Z., Bremner, M.J., Martinis, J.M., Neven, H., Characterizing quantum supremacy in near-term devices. *Nat. Phys.*, *14*, 6, 595–600, 2018.
67. Sarma, S.D., Freedman, M., Nayak, C., Topological quantum computation Physics Today. *Quantum Comput. A Gentle Introd*, *593*, 32, 2006.
68. Savage, N., Quantum diamond sensors. *Nature*, *591*, 7851, 37–37, 2021.
69. Quantum Biosensing: Medicine at the Smallest Scales, in: Chicago Quantum Exchange, 2022 Jul 2, Available from: https://chicagoquantum.org/news/quantum-biosensing-medicine-smallest-scales.

70. Biosensors: Quantum technology promises better biosensors, in: Laser Focus World, 2022 Jul 2, Available from: https://www.laserfocusworld.com/test-measurement/research/article/14039524/quantum-technology-promises-better-biosensors.
71. Özkan, A., Atar, N., Yola, M.L., Enhanced surface plasmon resonance (SPR) signals based on immobilization of core-shell nanoparticles incorporated boron nitride nanosheets: Development of molecularly imprinted SPR nanosensor for anticancer drug, etoposide. *Biosens. Bioelectron.*, *130*, 293–298, 2019.
72. Rosch-Grace, D. and Straub, J., Analysis of the likelihood of quantum computing proliferation. *Technol. Soc*, *68*, 101880, 2022.
73. Song, L., Pan, M., Zhao, R., Deng, J., Wu, Y., Recent advances, challenges and perspectives in enantioselective release. *J. Control. Release*, *324*, 156–171, 2020.
74. Arif, W.M., Elsinga, P.H., Gasca-Salas, C., Versluis, M., Martínez-Fernández, R., Dierckx, R.A., Borra, R.J., Luurtsema, G., Focused ultrasound for opening blood-brain barrier and drug delivery monitored with positron emission tomography. *J. Control. Release*, *324*, 303–316, 2020.
75. Ekhteiari Salmas, R., Serhat Is, Y., Durdagi, S., Stein, M., Yurtsever, M., A QM protein–ligand investigation of antipsychotic drugs with the dopamine D2 Receptor (D2R). *J. Biomol. Struct. Dyn.*, *36*, 10, 2668–2677, 2018.
76. Solenov, D., Brieler, J., Scherrer, J.F., The potential of quantum computing and machine learning to advance clinical research and change the practice of medicine. *Mo. Med.*, *115*, 5, 463, 2018.
77. Kamat, S., Kumari, M., Jayabaskaran, C., Nano-engineered tools in the diagnosis, therapeutics, prevention, and mitigation of SARS-CoV-2. *J. Control. Release*, *338*, 813–836, 2021.
78. Hassanzadeh, P., The capabilities of nanoelectronic 2-D materials for bio-inspired computing and drug delivery indicate their significance in modern drug design. *Life Sci.*, *279*, 119272, 2021.
79. Zinner, M., Dahlhausen, F., Boehme, P., Ehlers, J., Bieske, L., and Fehring, L.,2021. Quantum computing's potential for drug discovery: Early stage industry dynamics. *Drug Discov. Today*, *26*, 7, 1680–1688, 2021.
80. Li, P., Xiao, H., Shang, F., Tong, X., Li, X., Cao, M., A hybrid quantum-inspired neural networks with sequence inputs. *Neurocomputing*, *117*, 81–90, 2013.
81. Türkpençe, D., Akıncı, T.Ç., Şeker, S., A steady state quantum classifier. *Phys. Lett. A*, *383*, 13, 1410–1418, 2019.
82. Hassanzadeh, P., Towards the quantum-enabled technologies for development of drugs or delivery systems. *J. Control. Release.*, 324, 260–79, 2020 Aug 10.
83. Zhang, J., Norinder, U., Svensson, F., Deep learning-based conformal prediction of toxicity. *J. Chem. Inf. Model.*, *61*, 6, 2648–2657, 2021.

84. Alam, M. and Ghosh, S., QNet: A Scalable and Noise-Resilient Quantum Neural Network Architecture for Noisy Intermediate-Scale Quantum Computers. *Front. Phys.*, 702, 2022.
85. Guo, Z., Li, B., Cheng, L.T., Zhou, S., McCammon, J.A., Che, J., Identification of protein–ligand binding sites by the level-set variational implicit-solvent approach. *J. Chem. Theory Comput.*, *11*, 2, 753–765, 2015.
86. Fukunishi, Y. and Nakamura, H., Prediction of ligand-binding sites of proteins by molecular docking calculation for a random ligand library. *Protein Sci.*, *20*, 1, 95–106, 2011.
87. Quantum Computing: Know the Impact on Drug Discovery and Development, 2022 Jul 2, Available from: https://www.analyticsinsight.net/quantum-computing-know-the-impact-on-drug-discovery-and-development/.
88. Colgan, K.J., Looking two steps ahead. *Am. J. Health Syst. Pharm.*, *79*, 7, 556–563, 2022.
89. Aggarwal, L., Sachdeva, S. and Goswami, P., Quantum healthcare computing using precision based granular approach. *Appl. Soft Comput.*, 144, 110458, 2023.

7

Role of Quantum Computing Simulations in Targeted Drug Delivery of Liposomes

Rupali Sharma[1]*, Suman Khurana[1,2], Arun Mittal[3], Parveen Kumar Goyal[2], Kavita Sangwan[2] and Satish Sardana[1]

[1]*Amity Institute of Pharmacy, Amity University Haryana Amity Education Valley, Manesar, Gurugram, Haryana, India*
[2]*Department of Pharmacy, Panipat Institute of Engineering and Technology (PIET), Samalkha, Panipat, Haryana, India*
[3]*SRM Modinagar College of Pharmacy, SRMIST, Delhi-NCR Campus, Ghaziabad, India*

Abstract

Nanotechnology-based drug delivery systems are promising domains of innovation as the healthcare system is continuously getting improved by employing targeted delivery of drugs which is chiefly designed with the integration of nanotechnology and computer simulations. Liposomes-based drug delivery system is used for delivering a variety of drugs (including hydrophilic, hydrophobic, and amphiphilic) in one of the highly impacted systems which can be further scaled up by employing quantum computing. In liposomes, the aqueous core is encircled with one or more concentric layers of phospholipid and helps in penetrating the cell membranes. The phospholipids, because of their distinct physio-chemical properties, significantly affect the stability of liposomes. The selection of the suitable phospholipid and its inclusion, at optimized concentration, in designing the liposomal system determines the efficacy (in terms of membrane penetration) as well as stability of liposomes. Computer simulation-based liposomal models can help in resolving the issues about the transport of drugs across the plasma membrane, ion channels, generation of vesicles, fusion with membrane, and mechanisms governing the activity of cell-membrane–associated proteins. Several computational simulation models, such as the molecular dynamics simulation model, dissipative particle dynamics model, coarse-grained (CG) MARTINI, and implicit solvent model, can be used for simulating liposome systems and helping fabricate the drug delivery system by studying the varying mechanisms involved in the formation of liposomes.

**Corresponding author*: rsharma9@ggn.amity.edu; rupali6kalra@gmail.com

Rishabha Malviya, Sonali Sundram and Dhanalekshmi Unnikrishnan Meenakshi (eds.)
Drug Delivery Systems Using Quantum Computing, (195–230) © 2024 Scrivener Publishing LLC

Computational models are also helpful in optimizing the varying experimental conditions such as temperature, duration of a specific experimental step, selection of water-lipid ratio, and interactions and linkage of nanoparticles with the cell membrane. The involvement of quantum computing techniques in creating liposomal systems can be quite beneficial in designing the most efficient and appropriate drug delivery system. This chapter is mainly focused on the mechanisms, methods, and applications of quantum computing simulation techniques for designing targeted drug delivery systems, especially for liposomes.

Keywords: Drug delivery, liposomes, targeted delivery, quantum computing

7.1 Introduction

Drug discovery simply means to identify a therapeutically effective molecule that can be adopted in the treatment strategies and/or diagnostic protocol of any disease. In the process of drug discovery and development which takes more than 10 years and costs billions of dollars, the suitable proposed drug candidate is identified, synthesized, characterized, optimized, screened, assayed, and validated for therapeutic efficacy [1]. Firstly, the molecule(s) exhibiting significant therapeutic activity, known as a hit, is identified. To identify the hit molecule from the generated pool of promising drug candidates, these candidates are screened for interactions and binding on the biological target molecule(s) that regulates and determines the aetiological aspects of the disease [2]. Now, the hit molecule, after evaluating for some specific pharmacokinetic parameters such as absorption, distribution, metabolism, excretion, and toxicity, which are considered for selecting a better drug candidate, is stated as the lead compound [3]. The process of screening a plethora of molecules against a huge library of larger molecules, especially biological targets is carried out by employing high-throughput screening (HTS), which simulates the *in vitro* and *in vivo* experiments. Despite the better efficiency of these automated techniques over conventional methodology, their demand is diminishing day by day since the need for inventing and developing new drug molecules is increasing steeply. The application of commercial computers in drug discovery and development can improve the quality as well as quantity of lead molecules by accelerating the process of HTS and other conventional computerized processes associated with computational chemistry, computer-aided drug design (CADD), etc. [4]. The inclusion of recent computer techniques, especially quantum-based, can improve the efficiency of statistical analysis and reduce the cost of computational modeling systems for both tiny compounds and biological targets. The changes can easily be adopted, and challenges in developing new drug molecules can be resolved with the help of quantum computers by employing the concepts of machine learning, artificial intelligence, sensor technology, augmented and virtual reality, etc., which are beyond the pace of classical computers. Classical computers use *bits* for information processing while quantum computers work in a combination of two-level quantum states, i.e., *qubits* and execute computations with extraordinary speed in exceptional ways [2]. Once a potential drug is developed, the next vital step is to administer the same in the biological system where it can bind with the active site of the target molecule and produce therapeutic activity. The consideration of administration/delivery of the drug molecule to the desired site is one of the most important steps in the formulation study because most of the pharmacological actions including side effects are determined by it. The suitable drug delivery systems should have some properties such as proper drug distribution, transport, prolonged localization of drugs at target sites, controlled rate of drug release without affecting therapeutic properties, and minimum leakage of the free drugs.

Nowadays, nanomedicines have a significant impact on the healthcare system, especially by improving the varying aspects of drug delivery. These have a diameter between 10 to 100 nm [5]. The varying nano molecules, with a diameter between 10 to 100 nm [5], are based on carrier systems, such as liposomes [6, 7], carbon nanotubes [8, 9], dendrimers [10–12], and polymeric micelles [13–15], and have been developed, with each has its advantages. The interactions and behavior of these nanocarriers with vasculature systems, cell membranes, and active sites and the physicochemical characteristics such as size, shape, surface charge, and chemical compositions significantly influence drug release [16], which is why physical characterization is a necessary step in the development of nanomedicines. The design and structural optimization of nanocarriers face several practical limitations and experimental challenges that could be time-consuming and expensive affairs. Despite having good *in vitro* potential, the lack of *in vivo* efficacy is another major challenge associated with nanocarrier-mediated drug delivery systems [17]. In order to resolve the abovementioned problems, computational simulation techniques that give molecular information and interaction with physicochemical properties of drugs, as well as biological targets, are used in the modern drug discovery and development era. These simulations will help in assisting the rational design of drug delivery systems and bridge the gap by simultaneous analysis of pharmacodynamics, pharmacokinetics, and *in vitro–in vivo* correlations. These simulation techniques help in the prediction of interactions between drug and carrier, drug delivery, and associated mechanisms before performing the wet laboratory experimental procedure, and they will reduce the cost of new drug products.

7.2 Liposomes

Liposomes are small synthetic vesicles and have a spherical shape composed of triglycerides and nontoxic phospholipids. Drug delivery with liposomes is promising due to their size and hydrophilic and hydrophobic properties along with biocompatibility [18]. Both hydrophilic pharmaceuticals (e.g., aqueous-cored doxorubicin medicine called Doxil) and hydrophobic chemicals (e.g., amphotericin B–entrapped, AmBisome), van der Waals (vdW) forces submerged in the lamellae, can be encased in liposomes [19–21]. Phospholipids A glycerol molecule joined to a phosphate group makes up [amphiphilic lipids] (PO4 2) and two saturated or unsaturated chains of fatty acids [22]. Additionally, an ester bonds the phosphate to an organic molecule like choline or ethanolamine [23]. Phospholipids are important components that give liposomes their distinct features, such as how they encapsulate substances and how they are functionalized in the body [24]. As the basic biological elements of cell membranes, phospholipids, cell membranes, and liposomes can coexist during the release mechanisms [25].

The number, size, and surface charge of lamellae of a liposome, as well as the bilayer's stiffness, surface alterations, and production process, significantly impact the liposome's characteristics [26]. For example, the ammonium sulfate approach, like the vincristine pH gradient technique, would yield a massive number of amphipathic medicines like doxorubicin [19]. The critical micelle concentration (CMC), or the very short range of concentration within which dispersions are polar and hydrophilic display and have a sudden shift in physical characteristics, is another crucial parameter for generating amphiphiles that self-assemble, like lipids, liposomes, and surfactants. Monomeric phospholipids are present at concentrations below the CMC; at the CMC, molecular aggregation produces micelles, and the dispersion's morphological attributes are altered. The CMC values are influenced by a variety of internal and external parameters, comprising the amphiphilic molecule's arrangement of the hydrophobic and hydrophilic domains, the medium's composition, ionic strength, dielectric constant, pH, and others [27].

Liposomes improve medication, therapeutic index, and effectiveness (actinomycin D). Encapsulation improves the stability of liposomes. For systemic and non-systemic applications, liposomes are flexible, completely nontoxic, and biodegradable. They are also non-immunogenic and biocompatible. The encapsulating agent's toxicity is reduced by liposomes (amphotericin B, Taxol). Liposomes aid in the reduction of hazardous medication exposure to sensitive tissues. Flexibility in ligand selection achieves active targeting with targeted locations [28, 29]. However, there are some disadvantages such as significant production costs, molecular leakage, fusing and encapsulated medicine, and short half-life. Phospholipids are more prone to oxidation and hydrolysis reactions [28, 29].

7.3 Liposome Classification

Liposomes are categorized depending on the manner of preparation [30], composition and application [31], and structural parameters [32].

7.3.1 Based on Preparation Methods

Based on preparatory methods, liposomes are classified as vesicle extrusion technique (VET), reverse phase evaporation method (REV), stable pluri lamellar vesicle (SPLV), dehydration-rehydration method (DRV), frozen and thawed multilamellar vesicle (FATMLV), and multilamellar vesicle made by reverse phase evaporation method (MLV-REV) [30].

7.3.2 Based on Compositional and Structural Characteristics

Based on composition and structural characteristics, liposomes are of different types such as conventional liposomes (CV), furogenic liposomes (RSVE), PH-sensitive liposomes, long circulatory liposomes (LCL), immuno-liposomes (IL), and cationic liposomes [31, 32].

7.4 Methods of Liposome Preparation

To achieve effective drug capture, narrow particle length of shelf life and size distribution is the primary objective of the optimal liposome formulation process. The lipid must first be hydrated for all liposome production techniques before the particles are size-sorted and the medication that is not encapsulated is removed. Active loading methods and passive loading mechanical dispersion methods are both employed to create liposomes. For lipid-soluble pharmaceuticals with a strong lipid membrane affinity, the passive loading approach can be used to obtain high drug encapsulation efficiency. To meet the demands of therapeutic efficacy, there are various methods for the production of liposomes with variable release rates [7, 32, 33]. By establishing gradients of ion diffusion or medications between the exterior and aqueous phases inside, the drugs can be loaded using the active loading strategy. There are numerous techniques for creating liposomes [34, 35].

There are four standard stages to all liposome preparation techniques:

- Lipids are dried down from organic solvents.
- The lipid dispersion in fluid mediums.
- Cleaning the final liposome.
- Examining the finished item.

7.5 Drug Loading Method

Two types of techniques are employed for loading the drugs into the liposomes passive loading and active loading.

7.5.1 Passive Loading Technique

These techniques include the mechanical dispersion method, detergent removal method, and solvent dispersion method. There are several diverse mechanical dispersion types techniques, such as extrusion, French pressure cell, sonication of thawed-from-freeze liposomes, and hand or non-hand shaking membrane extrusion reconstituted dried vesicles [34, 35]. The solvent dispersion technique includes ether injection, ethanol injection, and reverse phase evaporation methods. Detergent removal methods include dialysis, gel permeation chromatography, and dilution.

7.5.1.1 Sonication

Sonication is one of the most used methods for preparing an SUV. Here, MLVs are passively sonicated using a probe sonicator or a bath-type sonicator. Major drawbacks of this technique including its extremely low inner volume and encapsulation efficiency, phospholipid and chemical deterioration, the removal of big molecules, metal contamination of the probe tip, and the presence of MLV and SUV are all things to consider [34]. Sonication includes two types: probe sonication and bath sonication.

7.5.1.2 French Pressure Cell

Extrusion MLV is extruded through a tiny aperture in a French pressure cell [34]. The proteins do not appear to be as pretentious during the French press vesicle approach as they do during sonication, which is one of its key advantages [18]. An intriguing observation is that entrapped solutes created by sonication or detergent removal seem to be remembered by French press vesicles for a longer period than they are remembered by SUVs [36–38].

7.5.1.3 Freeze-Thawed Liposomes

SUVs freeze quickly and defrost slowly. Aggregated materials are dispersed to LUV by quick sonication. SUVs fuse during the freezing and thawing processes, resulting in the formation of unilamellar vesicles [39–41].

7.5.1.4 Solvent Evaporation with Ether Injection

An aqueous solution of the intended object is gradually mixed with a lipid solution dissolved in ether-methanol or diethyl ether at 55° to 65°C or under decreased pressure. Liposomes are produced because of the elimination of ether under a vacuum. The main drawbacks of the approach are the heterogeneity of the population exposure of the molecule to organic solvents at high temperatures and wavelengths (70 to 200 nm) [42].

7.5.1.5 Alcohol Infusion

A massive amount of buffer is quickly injected with an ethanol lipid solution. The MLVs instantly form. It is difficult to eliminate all of the ethanol. There is a strong possibility that different biologically active macromolecules will become inactive in the presence of ethanol, even in very small quantities, because it, with water, generates an azeotrope. The population is diverse (30 to 110 nm), and the concentration of liposomes is quite low [42].

7.5.1.6 Method of Reverse Phase Evaporation

This method enhanced liposome technology since it made it possible to create a large ratio of aqueous to lipid space liposomes and the ability to hold a significant amount of the supplied aqueous material for the first time. The formation of inverted micelles is the foundation of reverse-phase evaporation.

The gel state collapses at a crucial stage in this process, disrupting part of the inverted micelles. The aqueous volume-to-lipid ratios of reverse-phase evaporation–produced liposomes compared to multilamellar liposomes or hand-shaken liposomes are four times greater, and they may be created from a variety of lipid formulations [35, 43].

7.5.1.7 *Removal of Non-Encapsulated Material with Detergent*

Dialysis: Lipids have been solubilized by detergents when employed when micelle concentrations are critical (CMC). The micelles' phospholipid content improves as the detergent is removed, and they eventually converge to form LUVs. Dialysis was used to remove the detergents [44–46].

7.5.2 Active Loading Technique

7.5.2.1 *Pro-Liposome*

In a pro-liposome, lipids and medicine are coated onto a soluble carrier to create free-flowing granular material that, upon hydration, transforms into an isotonic liposomal solution. The pro-liposome strategy might make it possible to mass-produce liposomes containing lipophilic medicines, at a low cost.

7.5.2.2 *Lyophilization*

Lyophilization is the process of removing water from frozen items under very low pressure (freeze-drying). The method is typically used to dry thermolabile materials that could be ruined by heat-drying. This method has a tremendous deal of promise to address issues with liposomal stability over the long run. Leakage of components that have been trapped may occur both during the freeze-drying process and upon reconstitution.

7.6 Newer Approaches to Liposomes

Liposomes could be unable to enter the GI tract's mucosal barrier because of their frequently enormous size. This has led to the hypothesis that, before reaching the epithelial cells for absorption, drugs are either released for absorption or converted and enter the GI tract into mixed micelles [47]. However, macromolecular medications like insulin have also been linked to M cell–mediated uptake, which has generated interest in the absorption

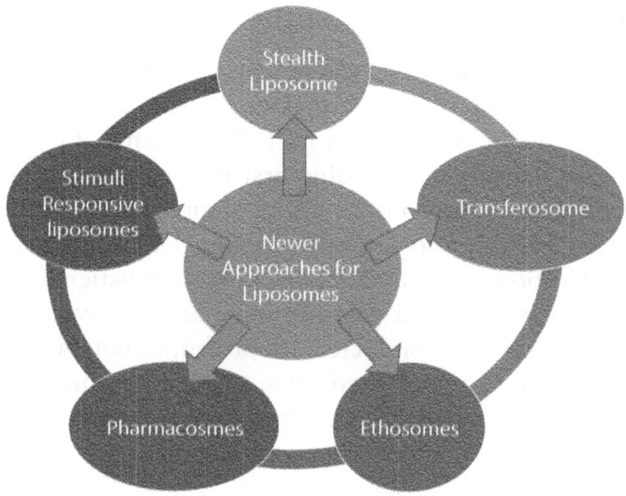

Figure 7.1 Newer approaches to liposome drug delivery.

pathway for liposomes and other nanoparticles [48]. However, because M cells only represent up to 1% of the total population of epithelial cells, it has not been thought that absorption of intact liposomes constitutes a substantial mechanism [49]. Liposomes were covered in mucoadhesive chitosan derivatives and acid-resistant polymers or had their surfaces modified with polyethylene glycol (PEG) 2000 to shield them from the GI tract's harsh environment [50]. Daidzein was released continuously from pegylated liposomes at pH 1.2 and pH 6.9, indicating the GI tract's continued stability. Hydrophilic PEG chains may operate as a steric barrier, preventing epithelial cells from absorbing pegylated liposomes in their undamaged state. Therefore, the increased BA may be caused by the pegylated liposomes' increased stability in the GI tract [47].

Newer approaches for liposomes are shown in Figure 7.1 and explained further.

7.6.1 Stealth Liposomes (Improving Circulation Time)

Longer-circulating liposomes were created by covering them with a hydrophilic polymer like PEG, which was proven to inhibit some blood proteins from being absorbed and so increase their time in circulation. Stealth liposomes were the name given to these. Long-circulating liposomes have better bioavailability, non-saturable, dose-independent kinetics, and log-linear behavior [52].

7.6.2 Improving Elasticity (Transferosomes)

The limited penetration effectiveness of treatment medicines over the skin hinders the development of transdermal medication administration, which is a critical therapeutic requirement [52]. The benefits of the use of cutaneous liposomal delivery are avoiding first-pass metabolism, minimizing inconvenient effects, improving the utility of medications with short half-lives, enhancing pharmacological and physiological responses, avoiding a change in medication levels, and, finally, providing patients with comfort and security [53, 54]. Many *ex vivo* and some *in vitro* investigations using transferosomes have been conducted. Ibuprofen, terbinafine, and emodin have all been shown to be delivered using elastic liposomes [55–57]. They have been found to increase skin penetration of a variety of pharmaceuticals [58].

7.6.3 Ethosomes (Improving Skin Penetration)

A unique liposome formulation known as ethosomes has been created by leveraging the penetrating capabilities of ethanol to increase the effectiveness of liposomes' penetration over the skin [59]. It has been suggested that ethanol, vesicles, and skin lipids work together synergistically to allow ethosomes to penetrate the skin because ethosomes exhibit increased skin penetration to ethanol alone [60]. According to "the ethanol effect" [61], which occurs when the polar head group of lipid molecules is where ethanol interacts with them, the fluidity of the membrane increases while the lipid multilayer's density decreases. The "ethosomes effect" is hypothesized to occur after this initial encounter, in which lipid penetration and permeation occur because of the ethosomes and skin lipid fusion, releasing the medication into the skin layers [62].

7.6.4 Pharmacosomes (Improvement in Medication Delivery for Poorly Soluble Medicines)

Drugs bind to phospholipids in pharmacosomes, which are amphiphilic phospholipid complexes, via hydrogen, covalent, or electrostatic bonding [63]. They are believed to increase the bioavailability of medicines that are poorly lipophilic as well as poorly water-soluble. Pharmacosomes can exist as hexagonal aggregates or ultrafine micelles depending on their chemical makeup [64]. Ketoprofen [65] is one drug that has been lately enclosed in pharmacosomes with outstanding findings of laboratory *in vitro* research. Based on analytical data, compared to free ketoprofen

(49.77%), the solubility of the drug was 93.3%. In further *in vitro* investigations, it was discovered that furosemide-bound pharmacosomes boosted the drug's permeability by more than 28% when compared to free furosemide [66].

7.6.5 Nebulized Liposomes and Stimuli-Responsive Liposomes

Numerous malignancies, especially initial liver cancer, breast cancer that has returned through the chest wall, colorectal cancer, pancreatic cancer, and metastatic liver cancer, are handled with ThermoDox (Celsion Corporation, Lawrenceville, NJ). This formulation contains lysophosphatidylcholine. Additionally, nebulized liposomes have made it into medical settings. One such solution for inhaling amikacin into the lungs is Insmed's liposomal amikacin (Arikace) [67].

7.7 Quantum Computing

Quantum computing is a brand-new, emerging technology that works effectively and is beyond complex for traditional computers by applying the principles of quantum mechanics when the supercomputers (massive, traditional computers with many CPU and GPU cores) are not able to solve some problems and get stuck in problems with high complexity. For example, modeling the behavior of individual atoms in a molecule due to all the different electrons interacting with one another is an example of a problem that is complex. Another example of a complex problem is sorting through a big database of protein sequences and proteins' behavior is determined by the data's fine patterns and fold in complexity for becoming useful in biological machines. One hundred amino acids in a protein can fold in many ways and these combinations of folds cannot be handled by conventional computers. Nowadays, complicated issues are solved in a different way using quantum computing by generating multidimensional spaces with the aid of quantum algorithms. In the event of a protein folding issue, that pattern might be the collection of folds that requires the least amount of energy to assemble. Compared to supercomputers, quantum computers are beautiful, energy-efficient gadgets. While a quantum computer employs qubits (CUE-bits) to execute multidimensional quantum algorithms, a conventional computer uses bits to carry out its operations. The differences between classical computing and quantum computing are given in Table 7.1.

Table 7.1 Difference between classical computing and quantum computing.

Classical computing	Quantum computing
Transistors are involved in calculation which can represent either 0 or 1.	Qubits are involved in calculations that can represent 0 or 1.
The number of transistors has a 1:1 relationship with power.	Power grows exponentially as qubits increase in number.
These work at room temperature and have minimal mistake rates.	These have a high rate of inaccuracy and must be kept extremely cold.
Daily processing is handled by classical computers.	Suitable for complex problems such as optimization problems, data analysis, and simulations.

Quantum computers work with the help of superfluids, superconductors, superposition, and entanglement to solve complex problems. About one 100th of a degree above absolute zero is the temperature required for this quantum computing technology. Superfluids are utilized to produce superconductors and meet this temperature requirement. The electrons

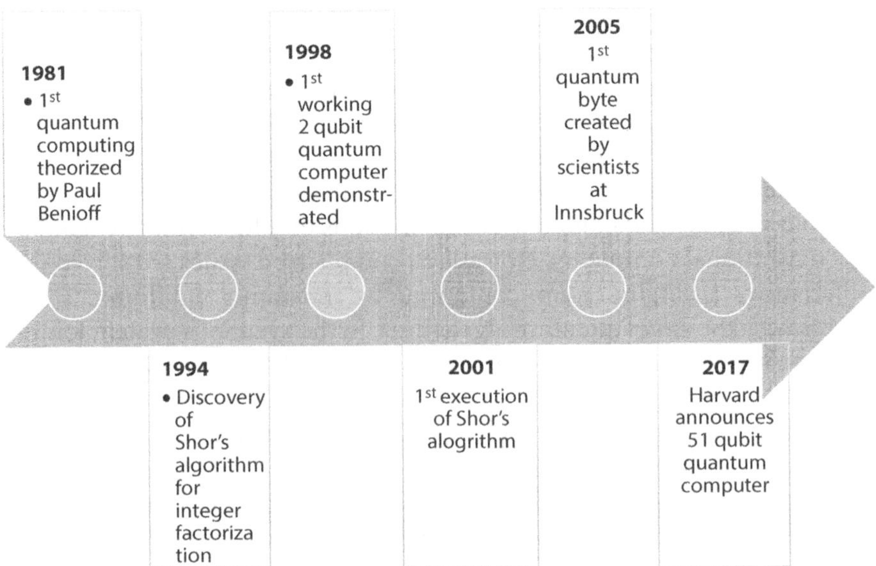

Figure 7.2 History of quantum computing.

produce "Cooper pairs" when they go through superconductors. These pairs carry a charge beyond insulators or barriers through quantum tunneling. As superconducting qubits, two superconductors positioned on either side of an insulator create a Josephson junction. By putting the quantum data it carries into a state of superposition, which is a synthesis of all potential qubit configurations, it can pull off a vital trick. The superposition of qubit groups can produce intricate, multidimensional computational landscapes. The acts of two separate entities are connected through entanglement, a feature of quantum mechanics. When two qubits are entangled, changes to one qubit have an immediate impact on the other. These links are used by quantum algorithms to address complex problems. Quantum computers are used to speed up the process of computational modeling used in drug delivery [68]. The history of quantum computing is shown in Figure 7.2.

7.8 Computational Modeling for Drug Delivery Process

Quantum mechanics speeds up the process of computational modeling used for the rational design of nanocarriers. Computational modeling includes molecular dynamics and multiscale computational techniques. The Newtonian equation of motion is utilized in molecular dynamic simulation to construct a movie of a group of atoms giggling across time, from which statistical mechanics can be used to ascertain the system's typical structural and physical characteristics [69]. The study of biological and soft matter systems has frequently utilized molecular dynamics modeling. One of the pharmaceutical industry's most vibrant and quickly expanding segments is computational drug delivery. The distribution ingredient, having therapeutic efficacy in a safe and optimum dosage form, forms the basis of any medication delivery system. The analysis of nanoscale particles like liposomes is a very important endeavor that necessitates a substantial amount of experimental data but also leaves some gaps that cannot be filled using conventional methods. The study of interactions that are happening at nanoscales is now possible thanks to recent advances in molecular dynamics. The free energy of a molecule to transit through a lipid layer can be investigated using techniques like replica exchange and umbrella sampling. CG MARTINI, dissipative particle dynamics model, and implicit solvent mode are a few examples of computational simulations that can be used to describe liposome systems. Computational simulations have produced several novels targeted delivery systems such as the lipid membrane and nanoparticle tethered together as performed by a complete modernistic drug delivery vehicle. Computational models are the foundation for formulating new medications and creating novel dosage forms [70].

Computational techniques to predict medication delivery before trials are becoming more and more tempting to lower the cost of drug development and research. The medication delivery system has seen significant advancements in molecular dynamics simulation techniques. Such technologies can be used to create and optimize drug delivery systems. Molecular dynamic simulations are very helpful in tackling problems that are challenging to examine in drug delivery laboratory studies. For the strategy and enhancement of drug delivery systems, where rates of medication release are controlled like how liposomal delivery works through membrane transfer, computational approaches to assess the permeability of lipid bilayers can be used. The most effective nanoparticle delivery method, liposomal delivery, is especially interesting for the therapeutic administration of substances to inflammatory and solid tumor areas. By limiting medication obtaining normal tissue while enabling access to malignant tissue due to the increased effect of permeability and retention seen in tumor tissue, drug encapsulation in a liposome can significantly modify the distribution of drugs [71].

7.9 Correlation Between Quantum Computing Simulation and Targeted Delivery of Liposomes

We have studied the computational simulation techniques used in the drug delivery process. Quantum computing will increase the speed in comparison to classical computing; but with these simulation techniques, some nanocarriers are completely simulated such as cyclodextrins [72] calixarenes [73], and low-number dendrimers [74]. In the case of nanomedicines, different approaches are used for getting information such as small micelles as models of the complete system [75], the involvement of drugs with the particular components of the system, and the bilayers as a model of liposomes. By utilizing the computational molecular modeling method for bilayer membranes on the liposomal delivery system's synthetic membranes, like experiments, computational modeling is used for investigating the major and cellular levels of structure of liposome; and to comprehend how the liposome's overall behavior interacts with its environment in the bloodstream, computational modeling might be applied [76].

In multiscale computational techniques, multiple levels of physical detail are included in descriptions of the systems. Quantum mechanics is used to characterize a molecular system at the highest level. Quantum mechanics calculations are directly applied to various drug delivery systems, for instance, nanoparticles, liposomes, quantum dots, carbon nanotubes, and indirectly, molecular dynamics simulation models. However, in real-world applications, quantum mechanics calculations are restricted to tiny systems with few degrees of freedom. It is possible to employ QM/MM, or hybrid quantum mechanics and molecular mechanical models if only a portion of the system needs QM [77–79]. Taking the degrees of freedom into account for the electrons and representing them as partial charges on atoms, one can simplify the systems to replicate larger systems. vdW-charged spheres are used to describe how atoms interact and Lennard-Jones (LJ) forces in accordance with the molecular mechanical framework. Potentials for harmonic restraint, including bond, angle, and dihedral terms, are used to hold all the atoms together, introducing the concept of molecules, a force field in which large-scale calculations of nanocarriers are performed with the involvement of interactions in simulated systems. The interaction set contains all values of parameters for the simulations. The force-field parameters are adjusted with experimental and quantum data to match the spectroscopic, thermodynamic, and crystallographic data of the molecule. CHARMM (version 36m for amino acids at this time) [80], 36 for lipids, DNA, and RNA, as well as 36 for lipids [81–82], and AMBER are a few examples of force fields that are frequently employed. For molecular dynamics simulations incorporating lipids, most applications have been for CHARMM force fields. A variety of software programs, including CHARMM [84], AMBER [85], NAMD [86], GROMACS [87], Open MM [88], and LAMMPS [89], have included molecular dynamics as a mature approach over the years. Using the available computational power, a molecular dynamic simulation typically produces a trajectory (or movie) for each atom in the system, with the spatial and temporal scales set to 10 to 102 nanometers and 0.1 to 10 microseconds, respectively.

Coarse graining in molecular dynamics refers to the construction of models in which particles represent groups of atoms rather than individual atoms. The two components of a common coarse-graining approach are the formation of CG beads out of groups of atoms and the parameterization of the CG bead's practical potential (or interactions). Chemical or biomolecular system molecular dynamics simulations, two general techniques have been thoroughly investigated. The MARTINI model represents one, which is based on chemistry, while the other is represented by the physics-based approach [90]. In MARTINI models, representations of lipids are done by head group beads that have two or more tail beads to simulate some of the essential characteristics of the molecules. A straightforward and empirical LJ potential depends on the sorts of interacting beads to represent the interaction between the beads. A physics-based coarse-graining approach, in contrast, produces beads by selecting atoms with comparable dynamics and parameterizing them via force matching [91, 92] or atomic fluctuation replication [93]. In its early stages of research, the force-matching method has made encouraging progress in modeling multicomponent lipid bilayers, which has helped to illuminate protein self-assembly [94], but it has not been applied to lipid nanoparticle simulations as frequently as the MARTINI models.

7.10 Computational Simulation of Lipid Membranes

Three different simulation levels can be used to model molecules: the first one is fully atomistic, where each atom is explicitly included as a site for interaction; the second is the united atom, where excluding non-polar hydrogen atoms, all atoms are explicitly included in the simulation; and the third one is CG, where three to five atoms in the group are considered as a single site for interaction. Phospholipids and dipalmitoyl phosphatidylcholine (DPPC) are the two lipids that are most modeled. Time scales of 100 ns are possible with completely atomistic or united atom representation; however, coarse-graining speeds up molecular dynamic simulations, allowing for longer times up to microseconds and length scales (10 nm). This is because the potential energy landscape is smoother and there are fewer particles overall. Problems that require capturing the system's basic mechanics rather than its chemical details typically use coarse graining. Various force fields are utilized in coarse graining for lipids [95–98], but MARTINI, created by Marrink and colleagues [99–101], is the force that is most frequently used. The parameters for phospholipids, sphingolipids, water, and other solvent molecules and ions, as well as cholesterol, are scattered throughout the MARTINI force field.

7.11 Properties Measured by Simulation

7.11.1 Mechanical and Structural Properties

The structural properties of lipid membranes, which can be studied by computational modeling, include the area covered by lipids, the density profile of the membrane, and the thickness of the membrane. The above-mentioned properties are used for the characterization of lipid membranes and how the structural properties can be constructed. By dividing the average size of each lipid layer's surface area by the average area of the simulated cell's bilayer, it is possible to compute the area per lipid. For a full description of other lipid components, extra molecules, or grid-based techniques, Voronoi tessellation or grid-based techniques may be used [102–104]. The lipid membrane thickness and hydrophobic thickness can be calculated by the density profiles. The density profile is measured by taking the average of the system components such as water, lipid head group, and tail group involved in the lipid layer.

The lipid membranes' mechanical characteristics can be calculated via compressibility modulus through computational simulation. Compressibility modulus denoted by KA is a measurement of energy needed for compressing and stretching the bilayer and can be calculated by taking the average square fluctuation in the lipid membrane area over time [105]. By measuring the height of the bilayer over the membrane's surface and examining the undulatory modes, one may simulate the energy necessary to adjust the membrane's curvature from its optimal state [106, 107].

7.11.2 Dynamic Properties

The dynamic properties include the ability of lipids to move in the bilayer, and this mobility is calculated by the lateral diffusion constant (D) of different lipids present in the bilayer. Dynamic properties can be calculated by using the Einstein relation to calculate the slope of the lipid molecules' average mean squared displacement over lengthy periods (in the plane of the bilayer). Molecular dynamics can also be used to show lipid dynamics, such as rotational relaxation and lipid flip-flopping between the leaflets of the bilayer [108, 109].

7.11.3 Molecule Permeation

The characterization of lipid membrane transport for the compounds is done using the membrane permeability, P. It can be calculated experimentally using membrane permeability experiments that measure flux, or the amount of permeant passing through a unit area in each length of time, divided by the difference in concentration of the permanent across the membrane. A molecule's permeability depends on its local partition coefficient and its diffusion coefficient at that depth. The molecular inhomogeneity of a lipid membrane varies with depth along the bilayer's normal direction.

A sequence of biased simulations in which the molecule is introduced and simulated at discrete points along the membrane normal can be used to calculate the local partition coefficient and Diffusion coefficient in a molecular simulation [110–112].

7.12 Liposomal Drug Delivery System

To study the role of computational simulation in the liposomal drug delivery system, first, we need to study the liposomal drug delivery system and the factors that affect it. Generally, the liposomal drug delivery system can be divided into four systems, as shown in Figure 7.3: ligand-targeted, sterically stabilized, conventional liposomes, and combinations of all the aforementioned, known as theranostic liposomes. These initial liposomes are created from neutral, cationic, or anionic phospholipids, and cholesterol, all of which are encapsulated within an aqueous layer. Sterically stabilized liposomes were introduced due to problems in conventional such as rapid elimination from the bloodstream as it affects the therapeutic efficacy. To deliver medications to specific cell types or organs that express receptors or biomolecules as ligands, ligand-targeted liposomes are utilized. Ligands can be antibodies, peptides, proteins, and carbohydrates.

Wc have studied these liposomal delivery systems that are used for better efficacy of therapeutics. In addition to these, there are various ways that liposome formulation can be altered in terms of selection of components, size, surface charge, and lamellarity for optimization of their effect. The size and lamellarity of the liposome during storage and blood flow, as well as the bilayer membrane's structure and its interactions with the circulation, are essential for the rational design of the liposome delivery system. The research on liposome membrane structure and the interaction between liposome and bloodstream are done using computational modeling to address this issue. For the structure of the membrane and the larger scale components of the liposome structure, computational models with all-atom resolution are used.

The experimental techniques will be discussed in further sections, providing information about the structure and behavior of the liposomal drug delivery system in terms of the length scale of the lipid membrane and the entire liposome. From the experiments, a multiscale picture of the liposome is not available regarding the mechanism of liposomal properties that affects the complete liposomal system. This can be resolved with the help of computational modeling using computational simulation techniques that will be discussed in a further section.

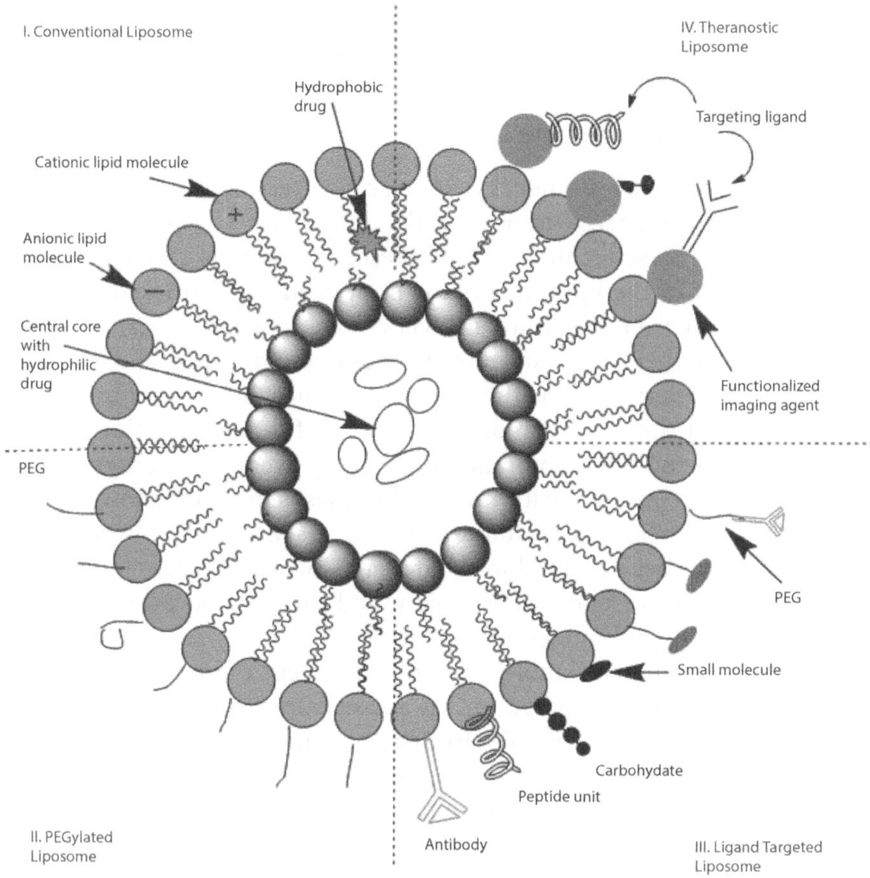

Figure 7.3 Different types of liposomal drug delivery systems. (I) Conventional liposome, (II) PEGylated liposome, (III) ligand-targeted liposome, (IV) theranostic liposome.

7.13 Experimental Techniques and Role of Computational Simulation in Liposomal Drug Delivery System

The main properties such as liposome size, surface charge, and lamellarity affect the liposome behavior in storage and bloodstream. In some cases, it depends on the components of the liposome, i.e., lipids used in membranes, and in other cases, the assembly of membranes in large-scale liposomes will affect the size and lamellarity. We will study the lipids used in the formulations and experimental techniques for characterizing the complete liposome as well as studying lipid membranes.

7.13.1 Lipids Membrane

Lipid membranes are studied by fluorescence techniques such as fluorescence resonant energy transfer and fluorescence recovery after photobleaching. In these techniques, a fluorescent moiety is conjugated with the component of the lipid membrane of liposome and tracked easily by microscopy *in vitro* cell culture and via optical imaging *in vivo* environment. However, there are chances of changes in the structure and the physicochemical properties of the functional molecules due to the conjugation process used in this methodology. Changes in functional molecules will lead to changes in lipid membranes followed by overall properties of liposomes. The main problem with these techniques is that they show promising results *in vitro* but are not followed by success *in vivo* [51]. In analyzing this problem, we need to develop and use analytical techniques that are not based on this type of fluorescent moiety, and this may be solved with the help of computational modeling.

The choice of lipids is another main component of liposome formulation. Lipids are selected based on the phase transition temperature. The lipid membrane's transition point from gel to liquid or crystalline depends on temperature. At body temperature, for example, liposomes composed of lipids that have a phase transition temperature below 37°C quickly release their substance, DPPC, or 1,2-dipalmitoyl sn-glycero-3-phosphocholine [113]. Since 1,2 distearoyl-sn-glycero-3 phosphatidylcholine (DSPC) does not phase transition at body temperature, it is employed to create stable liposomes [51]. The phase transition temperature is affected by the addition of drugs with modified surfaces like PEG and identifying ligands. Differential scanning calorimetry is used to measure the phase transition of the lipid membrane and drug liposome interactions [51]. Cationic lipids, for example, myristyl trimethyl ammonium propene [DMTAP], are used for the formation of liposomes that are used for gene therapy as they will possess a positive charge and will bind with anionic DNA [114].

For the rational development of liposomal drug delivery system, there is a need to understand the interaction of lipid membrane and small molecules at the molecular level about how the small molecules assemble in terms of transversal positioning, orientation, and partitioning, distribute in the membrane, and the manner lipid bilayers are affected by small molecules in terms of structure, i.e., order, curvature and packing, electrostatic potential, mechanical properties such as rigidity and pressure, partitioning in the membrane, and, most importantly, permeability. The partitioning of drugs and small molecules in lipid membranes is affected by both the composition, structure, and phase characteristics of membranes as well as the chemical and structural characteristics of molecules. The permeability means the ability of drug molecules to cross through the lipid membrane, and this parameter is calculated by molecular diffusivity and partitioning. Experimentally, the overall permeability coefficient can be calculated but unable to calculate the separate contribution from diffusion and partitioning. Computational simulation can be used for investigating both portioning and diffusivity of drug molecules in lipid membranes, which will be discussed in the further section.

The first simulation of the lipid membrane was done at the beginning of the 90s [115]. The molecular dynamic simulation with the all-atom resolution was performed by Wilson and Pohorille and demonstrated the mechanism for crossing the lipid membranes by the charged particles. By measuring the free energy barrier via the lipid membrane in this simulation, permeability may be determined with ease [116]. Calculations of permeability were completed using algorithms like replica exchange and umbrella sampling [117–120].

Selected b-blockers were simulated using molecular dynamics with a bilayer of 1-palmitoyl-2-oleoyl-sn-glycero-3-phosphocholine. According to the morphologies of the free energy profiles and the internal lipid membrane structure, they can be separated into two groups. Electrostatic contact is crucial for the first category of medications, according to energy analysis. The results of the examination of hydrogen bonds also confirm that the first group of b-blockers form more hydrogen bonds than the others. Sequences of computed permeability are comparable to those of experiments. The calculations' findings indicate that the energy difference between the free energy profile's maximum and minimum impacts how difficult it is for a medication to traverse a membrane. Drug-membrane interactions are greatly influenced by a drug's polarity, and drug permeability is also greatly influenced by the number of hydrogen bonds that are created between a drug and a membrane [121].

For permeability enhancement, a CG simulation was performed in a phospholipid membrane bilayer with the addition of lysolipid in terms of their lipid phase transition temperature. This temperature is calculated by properties of lipids such as the area covered per radial distribution of lipid-lipid interactions and the lateral diffusion constant. When lysolipid is present, the phase shift from gel to liquid crystalline state occurs at the same time as the peak in permeability, and the greater oscillations in the lipid bilayer are correlated with this permeability peak. There is a correlation between the phase transition temperature and lysolipid because, above Tm, there is no dependence of permeability on the amount of lysolipid. It was also found that a small amount of lysolipid, i.e., less than 15 mol %, maintains the gel phase and raises the phase transition temperature, whereas more lysolipid, or 20 mol %, lowers the Tm to that of pure DPPC and prevents the gel phase from forming in bilayers with 30 mol % lysolipid [122]. From the various explained simulation models, we can observe the structure of the lipid bilayer and the factors affecting the permeability of the membrane rather than performing the different experiments for characterization of lipid membrane and permeability as it will indirectly affect the liposomal drug delivery system.

7.13.2 Size, Surface Charge, and Zeta Potential

For the study of entire liposomes, we need to investigate experimental techniques that are used for the characterization of liposomes that can be used in the design of liposomal delivery systems. The liposome characterization includes size, surface charge, and lamellarity. The size of the liposome is important because it has an impact on the pharmacokinetics, tissue distribution, and elimination of the liposome. It is the first and most important assessment of liposome stability. Dynamic light scattering (DLS) and nanoparticle tracking analysis (NTA) are used to calculate the size. The DLS principle asserts that the intensity of scattered light is proportional to the sixth power of the diameter based on the Stokes-Einstein equation. DLS techniques that quickly measure the size and size distribution are most widely used [126], but this process is unable to differentiate between individual liposomes and clusters of liposomes and gives false readings, and it is also sensitive to impurities present in very low quantities [123, 124]. NTA is also used for determining the particle size, and the principle is the same as DLS but has high resolution. It can also track individual particles, do independent analyses, and measure sizes between 10 and 2000 nm [123, 125].

DLS and NTA are used for measuring the zeta potential of liposomes. Zeta potential is a measurement of a particle's surface charge. In the case of liposomes, it is the difference between a stationary hydration layer and a dispersion medium. Interaction of liposome with cell surface, i.e., biological barrier, cellular uptake, and intracellular distribution, is affected by surface charge.

The molecular dynamic simulation was performed to demonstrate the mechanism of small molecules through a lipid membrane. In this simulation, it was investigated how the size, hydrophobicity, and shape of the penetrants affect the permeability through the lipid membrane. For this purpose, the solubility-diffusion model is used for the calculation of permeability coefficients of oxygen and ammonia. For several LJ particles with various sizes and shapes, solubility and diffusion data are computed. According to the findings, the shape of the permeation resistance profile is determined by the free energy of solubility, and in the case of hydrophobic particles, the membrane interior will function as a trap barrier. In contrast, the hydrophilic penetrant tail region of lipids will function as a barrier for permeation resistance. The simulations have demonstrated that the lipid membrane functions as a permeation barrier because the bilayer is inhomogeneous and anisotropic, and the permeation properties vary greatly for various penetrants. It is possible to provide a better description using the four-region model. We have shown that region 3, the dense area of the lipid tails, typically exhibits the strongest resistance to the total permeation process for a range of penetrants. This is especially true for hydrophilic penetrants, for which experimental access to penetration rates is available. Thus, in contrast to the permeation process of bulk alkane phases, this region is also in charge of most of the anomalous effects that are experimentally detected in the lipid membrane conformational changes of micro penetrants [126].

7.13.3 Morphology and Lamellarity

In comparison to DLS and NTA, cryo-TEM techniques depict direct images for size, shape, morphology, and lamellarity. This technique is used for the determination of lamellarity, i.e., the number of the bilayer of the liposome. Drug release, storage qualities, surface-bound ligand addition, drug entrapment, and other properties are all impacted by the liposome's lamellarity [127, 128]. One-dimensional 31 P NMR [129], small angle x-ray scattering [130], and trapped volume measurement [131] are used for the evaluation of the lamellarity of liposomes.

7.14 Computational Simulation Study of Liposomes

The drug release from liposomes depends on the bilayer permeability [132, 133]. We have studied the various simulation models applied for lipid bilayer in the previous section. Self-assembly is another feature in the development of nanocarriers, including nanoparticles and liposomes, for desirable properties [134]. Computational simulations are performed for lipid-based aggregates for drug delivery. MD simulations are used to comprehensively explain the large-scale structure and dynamics of liposomes, including their morphological traits and self-assembly processes, which may be directly related to the systems' effectiveness as delivery vehicles.

Molecular dynamics studies were performed with liposomes for photosensitizing drugs. For 2500 lipids, a CG liposome model is created with various drug-loading compositions. The chemical hypericin was chosen because it is a natural substance that has been suggested as a component of photodynamic treatment. For this drug, a CG model was developed and standardized against equivalent simulations of an atomistic membrane bilayer model for 10 microseconds with liposomes containing 21 to 84 hypericin molecules. The distribution of the hypericin, their orientations inside the lipid bilayer, and the mean force required to transport a hypericin molecule from the internal water "droplet" across the liposome bilayer were calculated using the simulation models. Simulations of liposomes with varying hypericin concentrations on a microsecond time scale revealed that the liposome expands as the hypericin concentration rises, which is consistent with experimental studies. At the concentrations utilized here, hypericin had only little impact on the liposome other than this slight enlargement, and no significant structural changes could be identified. The hydrophobic part is toward the lipid interior and hydrophilic ends of the hypericin are positioned to enhance interactions with the polar headgroup area of the outer lipid layer, i.e., toward the water phase. The findings show that the hypericin current CG model can be used in large-scale investigations to examine the precise process of drug transport from a liposome to a membrane utilizing CG molecular dynamics at an almost atomistic resolution [135].

A molecular dynamics simulation study of vesicle fusion at the atomic level was carried out because, when a membrane-bounded vesicle fuses with a target membrane, it is essential for drug delivery, exocytosis, and intracellular trafficking. With the help of molecular dynamics simulations, the fusion of small unilamellar vesicles made of a DPPC/palmitic acid 1:2 combination has been analyzed thoroughly. The simulations, which were carried out at 350 to 370 K, confirmed that temperature and pH had an impact on the fusing of DPPC/palmitic acid vesicles in earlier research. A vesicle that was simulated under periodic boundary conditions was merged with its periodic picture to make the computations computationally possible on a nanosecond timeline, forcing the lipid into the hemifusion diaphragm's interior causing a fusion pore to grow and enlarge. This result opens the door to atomic-level research on a wide range of mesoscopic biological phenomena. In conclusion, atomistically precise models can be used to simulate the observed vesicle fusion behavior. At the same time, the findings support that the CG models, which appear to capture the key details, such as the possibility of a transitory pore forming next to the hemifusion diaphragm, could cause exchange and leakage of phospholipid between the inner and outer leaflets. This research also identifies characteristics unique to lipid composition. Fatty acid molecules can easily cross a lipid bilayer's hydrophobic core in their protonated state at low pH and interchange across the leaflets even in the absence of a pore. This may make it easier to modify the lipid distribution to account for variations in leaflet curvature brought on by membrane fusion. Due to kinetic entrapment, the hemifusion state was metastable on the nanosecond time frame of simulations at equilibrium, but as soon as a local defect was introduced into the hemifusion diaphragm, the state swiftly changed into a completely fused one. With this discovery, it is now possible to examine in greater depth how lipid composition and (fusion) peptides affect vesicle fusion [136].

A CG simulation of local anesthetic, i.e., prilocaine, an amino amide local anesthetic often used in dentistry, for the encapsulation encased within a liposome was carried out. Key insights into the prilocaine-vesicle interactions were obtained from molecular dynamics simulations of prilocaine at various protonation phases. All neutral prilocaine molecules were found to hop between the interior and outer vesicle monolayers and instantly diffuse into the vesicle's hydrophobic area, adopting an asymmetric bimodal density distribution.

Instead, originally dissolved in water, protonated prilocaine molecules were only found on the liposome's outer monolayer, where they had a high rate of exchange with the water phase and no access to the liposome's interior. Hydrophobicity is what drives prilocaine drug absorption within the liposome, despite evidence of organized electrostatic contact between protonated prilocaine tails and dipolar lipid head groups. The simulations also demonstrate that beginning with a configuration with protonated prilocaine dispersed inside the lipid membrane, a significant portion of protonated prilocaine is still trapped within the interior water phase of the liposome. This demonstrates that the method of production can significantly affect the effectiveness of medications at low pH and liposome-prilocaine complexes. The MARTINI framework was used to develop the parameters for the prilocaine force field, and this expanded force field was validated by comparing the results of CG simulations to those of simulations of prilocaine contained in a flat lipid bilayer that was atomistic. The simulations also demonstrate that, when starting from a configuration with protonated prilocaine dispersed throughout the lipid membrane, a significant portion of protonated prilocaine is retained within the liposome's internal water phase. This shows that the preparation process can have a significant impact on the efficacy of drugs at low pH and liposome-prilocaine complexes. A closer look reveals an organized interaction between protonated prilocaine tail groups and the vesicle membrane, which are positioned close to the lipid heads with opposing charges. For absorbed neutral prilocaine, however, no preferred orientation or structure was discovered. Therefore, whereas protonation occurs in a structured interaction between the drug and the host, our simulated results indicate that hydrophobicity is the primary factor influencing drug encapsulation. The findings further demonstrate that the protonated prilocaine distribution is influenced by the preparation conditions, just as the starting conditions have a significant impact on the number of prilocaine molecules trapped in the liposome. This finding shows that various liposome-drug complex production methods that result in prilocaine confined inside vesicles may boost overall drug potency [137].

From the various studied computer simulation models, the various properties of liposomes can be calculated via computers rather than performed experimentally. This will reduce the expenses and the time on the work performed experimentally.

7.15 Conclusion

In this chapter, we have studied the challenges in the development of drug delivery systems, especially in the context of nanomedicine. The design and the structural optimization of nanocarriers face practical limitations through experiments, and it could be time-consuming and expensive. Another challenge is that these drug carrier systems show activity *in vitro* but not *in vivo* comparatively. Integrated studies combining computational and experimental methods are urgently required. The physicochemical characteristics of medications and carriers, as well as their molecular interactions, are revealed using computational simulation techniques. These simulations will help in assisting the rational design with improved efficacies. These simulations fill the gap between experimentally measured attributes of nanocarrier and molecular mechanisms. They help in the evaluation of biological systems and pathomechanisms of disease, rational drug design, drug-body interactions, and pharmacodynamics and pharmacokinetics of various therapeutics. Before carrying out the experiment procedure, these simulation techniques aid in the prediction of drug carrier interactions, drug distribution, and associated mechanisms, which saves a lot of money.

In order to gather crucial data for the creation of novel liposome-based pharmacological therapies, we have given an overview of the state of the art in the investigation of lipid membrane-based systems using a method that combines experimental and computational approaches. We have shared the results of the investigations done on the molecular dynamics of lipid membranes and liposomes. In this chapter, the following major points are emphasized such as a valuable tool for the development of therapeutics based on liposomal drug delivery is the synergistic application of computational approaches, such as molecular dynamic modeling, with experimental methodologies and an emphasis on mechanistic knowledge. In this study, MD simulation can serve two functions: as a tool to gain the mechanistic understanding that contextualizes experimental findings and as a source of fresh ideas for developing nano vectors. Additionally, a multiscale approach is necessary for both experimental and computational techniques. The speed of computational modeling employed in the creation of liposomes will be accelerated by quantum computing. It will provide a rationally optimized drug carrier with promised results.

References

1. DiMasi, J.A., Hansen, R.W., Grabowski, H.G., The price of innovation: new estimates of drug development costs. *J. Health Econ.*, 22, 151, 2003.
2. Cao, Y., Romero, J., Aspuru-Guzik, A., Potential of quantum computing for drug discovery. *IBM J. Res. Dev.*, 21, 62, 2018.
3. Sliwoski, G., Kothiwale, S., Meiler, J., Lowe, E.W., Computational methods in drug discovery. *Pharmacol. Rev.*, 66, 334, 2014.
4. Van Drie, J.H., Computer-aided drug design: the next 20 years. *J. Comput. Aided Mol. Des.*, 2, 591, 2007.
5. Parveen, S., Misra, R., Sahoo, S.K., Nanoparticles: a boon to drug delivery, therapeutics, diagnostics and imaging. *Nanomed. Nanotechnol. Biol. Med.*, 8, 147, 2012.
6. Pattni, B.S., Chupin, V.V., Torchilin, V.P., New developments in liposomal drug delivery. *Chem. Rev.*, 115, 10938, 2015.
7. Allen, T.M. and Cullis, P.R., Liposomal drug delivery systems: from concept to clinical applications. *Adv. Drug Deliv. Rev.*, 65, 36, 2013.
8. Bianco, A., Kostarelos, K., Prato, M., Applications of carbon nanotubes in drug delivery. *Curr. Opin. Chem. Biol.*, 9, 674, 2005.
9. Lacerda, L., Bianco, A., Prato, M., Kostarelos, K., Carbon nanotubes as nanomedicines: from toxicology to pharmacology. *Adv. Drug Deliv. Rev.*, 58, 1460, 2006.

10. Gillies, E.R. and Frechet, J.M., Dendrimers and dendritic polymers in drug delivery. *Drug Discov. Today*, 10, 35, 2005.
11. Lee, C.C., MacKay, J.A., Fréchet, J.M., Szoka, F.C., Designing dendrimers for biological applications. *Nat. Biotechnol.*, 23, 1517, 2005.
12. Svenson, S. and Tomalia, D.A., Dendrimers in biomedical applications—reflections on the field. *Adv. Drug Deliv. Rev.*, 64, 102, 2012.
13. Sanna, V., Pala, N., Sechi, M., Targeted therapy using nanotechnology: focus on cancer. *Int. J. Nanomed.*, 9, 467, 2014.
14. Yokoyama, M., Polymeric micelles as drug carriers: their lights and shadows. *J. Drug Targeting*, 22, 576, 2014.
15. Kataoka, K., Harada, A., Nagasaki, Y., Block copolymer micelles for drug delivery: design, characterization and biological significance. *Adv. Drug Deliv. Rev.*, 1, 64, 2012.
16. Blanco, E., Shen, H., Ferrari, M., Principles of nanoparticle design for overcoming biological barriers to drug delivery. *Nat. Biotechnol.*, 33, 941, 2015.
17. Zhang, Y., Chan, H.F., Leong, K.W., Advanced materials and processing for drug delivery: The past and the future. *Adv. Drug Deliv. Rev.*, 65, 104, 2013.
18. Akbarzadeh, A., Rezaei-Sadabady, R., Davaran, S., Joo, S.W., Zarghami, N., Hanifehpour, Y., Samiei, M., Kouhi, M., Nejati-Koshki, K., Liposome: classification, preparation, and applications. *Nanoscale Res. Lett.*, 8, 1, 2013.
19. Senapati, S., Mahanta, A.K., Kumar, S., Maiti, P., Controlled drug delivery vehicles for cancer treatment and their performance. *Signal Transduction Targeted Ther.*, 16, 1, 2018.
20. Gonda, A., Zhao, N., Shah, J.V., Calvelli, H.R., Kantamneni, H., Francis, N.L., Ganapathy, V., Engineering tumor-targeting nanoparticles as vehicles for precision nanomedicine. *Med. One*, 4, e190021, 2019.
21. Gao, Y., Wijewardhana, C., Mann, J.F., Virus-like particle, liposome, and polymeric particle-based vaccines against HIV-1. *Front. Immunol.*, 28, 345, 2018.
22. Pinot, M., Vanni, S., Pagnotta, S., Lacas-Gervais, S., Payet, L.A., Ferreira, T., Gautier, R., Goud, B., Antonny, B., Barelli, H., Polyunsaturated phospholipids facilitate membrane deformation and fission by endocytic proteins. *Science*, 345, 69, 2014.
23. Monteiro, N., Martins, A., Reis, R.L., Neves, N.M., Liposomes in tissue engineering and regenerative medicine. *JR Soc. Interface.*, 11, 20140459, 2014.
24. Hussain, A., Singh, S., Sharma, D., Webster, T.J., Shafaat, K., Faruk, A., Elastic liposomes as novel carriers: recent advances in drug delivery. *Int. J. Nanomed.*, 12, 5087, 2017.
25. Rothfeld, L.I., *Structure and function of biological membranes*, 1st ed, Cambridge: Academic Press, 1971.
26. Olusanya, T.O., Haj Ahmad, R.R., Ibegbu, D.M., Smith, J.R., Elkordy, A.A., Liposomal drug delivery systems and anticancer drugs. *Molecules*, 23, 907, 2018.

27. Priev, A., Zalipsky, S., Cohen, R., Barenholz, Y., Determination of critical micelle concentration of lipid polymers and other amphiphiles: comparison of sound velocity and fluorescent measurements. *Langmuir*, 218, 612, 2002.
28. Anwekar, H., Patel, S., Singhai, A.K., Liposome-as drug carriers. *Int. J. Pharm. Life Sci.*, 2, 945–951, 2011.
29. Dua, J.S., Rana, A.C., Bhandari, A.K., Liposome: methods of preparation and applications. *Int. J. Pharm. Stud. Res.*, 3, 14, 2012.
30. Moghimi Pour, E. and Handali, S., Liposomes as Drug Delivery Systems: Properties and Applications. *Res. J. Pharm. Biol. Chem. Sci.*, 4, 1, 170, January–March 2013. https:// www.researchgate.net/publication/261774146.
31. Sercombe, L., Veerati, T., Moheimani, F., Wu, S.Y., Sood, A.K., Hua, S., Advances and challenges of liposome assisted drug delivery. *Front. Pharmacol.*, 6, 286, 2015.
32. Olga Popovska, J.S., Kavrakovski, Z., Rafajlovska, V., An Overview: Methods for Preparation and Characterization of Liposomes as Drug Delivery Systems. *Int. J. Pharm. Phytopharmacon.*, 3, 13, 2013.
33. Shashi, K., Satinder, K., Bharat, P., A Complete review on: Liposomes. *IRJP*, 3, 10–16, 2012.
34. Riaz, M., Liposome preparation method. *Pak. J. Pharm. Sci.*, 9, 65, 1996.
35. Himanshu, A., Sitasharan, P., Singhai, A.K., Liposomes as drug carriers. *IJPLS*, 2, 945, 2011.
36. Song, H., Geng, H.Q., Ruan, J., Wang, K., Bao, C.C., Wang, J., Peng, X., Zhang, X.Q., Cui, D.X., Development of polysorbate 80/phospholipid mixed micellar formation for docetaxel and assessment of its *in vivo* distribution in animal models. *Nanoscale Res. Lett.*, 6, 354, 2011.
37. Zhang, Y., Relations between size and function of substance particles. *Nano. Biomed. Eng.*, 3, 1, 2011.
38. Mozafari, M.R., Liposomes: an overview of manufacturing techniques. *Cell. Mol. Biol. Lett.*, 10, 711, 2005.
39. Pick, U., Liposomes with a large trapping capacity prepared by freezing and thawing of sonicated phospholipid mixtures. *Arch. Biochem. Biophys.*, 212, 186, 1981.
40. Ohsawa, T., Miura, H., Harada, K., Improvement of encapsulation efficiency of water-soluble drugs in liposomes formed by the freeze-thawing method. *Chem. Pharm. Bull.*, 33, 3945, 1985.
41. Liu, L. and Yonetaini, T., Preparation and characterization of liposome encapsulated hemoglobin by a freeze-thaw method. *J. Microencapsul.*, 11, 409, 1994.
42. Bhupathy Raj, M., Liposomal Drug Delivery Systems – A Review. *Int. J. Pharm. Sci. Rev. Res.*, 55, 295, 2013.
43. Kataria, S., Sandhu, P., Bilandi, A., Akanksha, M., Kapoor, B., Seth, G.L., Bihani, S.D., Stealth liposomes: a review. *IJRAP*, 2, 1534, 2011.
44. Daemen, T., Hofstede, G., Ten Kate, M.T., Bakker-Woudenberg, I.A.J.M., Scherphof, G.L., Liposomal doxorubicin induced toxicity: depletion and

impairment of phagocytic activity of liver macrophages. *Int. Cancer*, 61, 761, 1995.
45. Kirby, C.J. and Gregoriadis, G., A simple procedure for preparing liposomes capable of high encapsulation efficiency under mild conditions, in: *Liposome Technology*, vol. 19, 1984.
46. Alpes, H., Allmann, K., Plattner, H., Reichert, J., Rick, R., Schulz, S., Formation of large unilamellar vesicles using alkyl maltoside detergents. *Biochim. Biophys. Acta*, 862, 294, 1986.
47. Lee, M.K., Liposomes for enhanced bioavailability of water-insoluble drugs: *in vivo* evidence and recent approaches. *Pharmaceutics*, 12, 264, 2020.
48. Niu, M., Lu, Y., Hovgaard, L., Guan, P., Tan, Y., Lian, R., Qi, J., Wu, W., Hypoglycemic activity and oral bioavailability of insulin-loaded liposomes containing bile salts in rats: the effect of cholate type, particle size and administered dose. *Eur. J. Pharm. Biopharm.*, 81, 265, 2012.
49. Lopes, M.A., Abrahim, B.A., Cabral, L.M., Rodrigues, C.R., Seiça, R.M., de Baptista Veiga, F.J., Ribeiro, A.J., Intestinal absorption of insulin nanoparticles: contribution of M cells. *Nanomed. Nanotechnol. Biol. Med.*, 10, 1139, 2014.
50. Nguyen, T.X., Huang, L., Gauthier, M., Yang, G., Wang, Q., Recent advances in liposome surface modification for oral drug delivery. *Nanomedicine*, 11, 1169, 2016.
51. Van Slooten, M.L., Boerman, O., Romøren, K. *et al.*, Liposomes as a sustained release system for human interferon-gamma: biopharmaceutical aspects. *Biochim. Biophys. Acta*, 1530, 134, 2001.
52. Langer, R., Transdermal drug delivery: past progress, current status, and future prospects. *Adv. Drug Deliv. Rev.*, 56, 557, 2004.
53. Bouwstra, J.A., Honeywell-Nguyen, P.L., Gooris, G.S., Ponec, M., Structure of the skin barrier and its modulation by vesicular formulations. *Prog. Lipid Res.*, 42, 1, 2003.
54. Lee, E.H., Kim, A., Oh, Y.K., Kim, C.K., Effect of edge activators on the formation and transfection efficiency of ultradeformable liposomes. *Biomaterials*, 26, 205, 2005.
55. Irfan, M., Verma, S., Ram, A., Preparation and characterization of ibuprofen loaded transferosome as a novel carrier for transdermal drug delivery system. *Asian J. Pharm. Clin. Res.*, 5, 162, 2012.
56. Ghannoum, M., Isham, N., Henry, W., Kroon, H.A., Yurdakul, S., Evaluation of the morphological effects of TDT 067 (terbinafine in Transfersome) and conventional terbinafine on dermatophyte hyphae *in vitro* and *in vivo*. *Antimicrob. Agents Chemother.*, 56, 2530, 2012.
57. Lu, K., Xie, S., Han, S., Zhang, J., Chang, X., Chao, J., Huang, Q., Yuan, Q., Lin, H., Xu, L., Shen, C., Preparation of a nano emodin transfersome and study on its anti-obesity mechanism in adipose tissue of diet-induced obese rats. *J. Transl. Med.*, 12, 1, 2014.

58. Mohammed, A.R., Weston, N., Coombes, A.G., Fitzgerald, M., Perrie, Y., Liposome formulation of poorly water soluble drugs: optimisation of drug loading and ESEM analysis of stability. *Int. J. Pharm.*, 285, 23, 2004.
59. Zylberberg, C. and Matosevic, S., Pharmaceutical liposomal drug delivery: a review of new delivery systems and a look at the regulatory landscape. *Drug Deliv.*, 23, 3319, 2016.
60. Touitou, E., Dayan, N., Bergelson, L., Godin, B., Eliaz, M., Ethosomes—novel vesicular carriers for enhanced delivery: characterization and skin penetration properties. *J. Control. Release.*, 65, 403, 2000.
61. Elsayed, M.M., Abdallah, O.Y., Naggar, V.F., Khalafallah, N.M., Lipid vesicles for skin delivery of drugs: reviewing three decades of research. *Int. J. Pharm.*, 332, 1, 2007.
62. Fan, C., Li, X., Zhou, Y., Zhao, Y., Ma, S., Li, W., Liu, Y., Li, G., Enhanced topical delivery of tetrandrine by ethosomes for treatment of arthritis. *Biomed. Res. Int.*, 2013, 161943, 2013.
63. Semalty, A., Semalty, M., Rawat, B.S., Singh, D., Rawat, M.S., Pharmacosomes:the lipid-based new drug delivery system. *Expert Opin. Drug Deliv.*, 6, 599, 2009.
64. Shivhare, R., Pathak, A., Shrivastava, N., Singh, C., Tiwari, G., Goyal, R., An update review on novel advanced ocular drug delivery system. *World J. Pharm. Pharm. Sci.*, 1, 545, 2012.
65. Kamalesh, M., Baviskar, D., Wagh, K., Baviskar, K., Formulation and evaluation of pharmacosomes of Ketoprofen. *Indo Am. J. Pharm. Res.*, 4, 1363, 2014.
66. Chatap, V.K., Pati, P.L., Patil, S.D., *In-vitro, ex-vivo* characterization of Furosemide bounded pharmacosomes for improvement of solubility and permeability. *Adv. Pharmacol. Pharm.*, 2, 67, 2014.
67. Li, Z., Zhang, Y., Wurtz, W., Lee, J.K., Malinin, V.S., Durwas-Krishnan, S., Meers, P., Perkins, W.R., Characterization of nebulized liposomal amikacin (Arikace™) as a function of droplet size. *J. Aerosol Med. Pulm. Drug Deliv.*, 21, 245, 2008.
68. https://www.ibm.com/topics/quantum-computing#:~:text=Quantum%20computing%20is%20a%20rapidly,available%20to%20thousands%20of%20developers.
69. Frenkel, D. and Smit, B., *Understanding Molecular Simulation; From Algorithms to Applications*, Academic Press, San Diego, San Francisco, New York, 2001.
70. Jiwani, M., Pathak, H., Rohit, M., Solanki, H.K., Sarkar, D., Recent Approaches in Computational Drug Delivery System. *MAT J. Res. Rev. Drugs Drugs Dev.*, 2, 1–10.
71. Li, Y. and Hou, T., Computational Simulation of Drug Delivery at Molecular Level. *Curr. Med. Chem.*, 17, 36, 2010.
72. de Jesus, M.B., Fraceto, L.F., Martini, M.F. *et al.*, Non-inclusion complexes between riboflavin and cyclodextrins. *J. Pharm. Pharmacol.*, 64, 832, 2012.

73. Braga, M.A., Martini, M.F., Pickholz, M. *et al.*, Clonidine complexation with hydroxypropyl-beta-cyclodextrin: From physico-chemical characterization to *in vivo* adjuvant effect in local anesthesia. *J. Pharm. Biomed. Anal.*, 119, 27, 2016.
74. Korchowiec, B., Korchowiec, J., Gorczyca, M. *et al.*, Molecular organization of nalidixate conjugated calixarenes in bacterial model membranes probed by molecular dynamics simulation and Langmuir monolayer studies. *J. Phys. Chem. B*, 119, 2990, 2015.
75. Paulo, P.M.R., Lopes, J.N.C., Costa, S.M.B., Molecular dynamics simulations of charged dendrimers: Low-to-intermediate half-generation P AMAMs. *J. Phys. Chem. B*, 111, 10651, 2007.
76. Bunker, A., Magarkar, A., Viitala, T., Rational design of liposomal drug delivery systems, a review: combined experimental and computational studies of lipid membranes, liposomes and their PEGylation. *Biochim. Biophys. Acta, Biomembr.*, 1858, 2334, 2016.
77. Gooneie, A., Schuschnigg, S., Holzer, C., A review of multiscale computational methods in polymeric materials. *Polymers*, 9, 16, 2017.
78. Sherwood, P., Brooks, B.R., Sansom, M.S., Multiscale methods for macromolecular simulations. *Curr. Opin. Struct. Biol.*, 18, 630, 2008.
79. Meier, K., Choutko, A., Dolenc, J., Eichenberger, A.P., Riniker, S., Van Gunsteren, W.F., Multi-resolution simulation of biomolecular systems: a review of methodological issues. *Angew. Chem. Int. Ed.*, 52, 2820, 2013.
80. Huang, J., Rauscher, S., Nawrocki, G., Ran, T., Feig, M., de Groot, B.L., Grubmüller, H., MacKerell, A.D., CHARMM36m: an improved force field for folded and intrinsically disordered proteins. *Nat. Methods*, 14, 71, 2017.
81. Klauda, J.B., Venable, R.M., Freites, J.A., O'Connor, J.W., Tobias, D.J., Mondragon-Ramirez, C., Vorobyov, I., MacKerell Jr., A.D., Pastor, R.W., Update of the CHARMM all-atom additive force field for lipids: validation on six lipid types. *J. Phys. Chem. B*, 114, 7830, 2010.
82. Hart, K., Foloppe, N., Baker, C.M., Denning, E.J., Nilsson, L., MacKerell Jr., A.D., Optimization of the CHARMM additive force field for DNA: Improved treatment of the BI/BII conformational equilibrium. *J. Chem. Theory Comput.*, 8, 348, 2012.
83. Denning, E.J., Priyakumar, U.D., Nilsson, L., Mackerell Jr., A.D., Impact of 2′- hydroxyl sampling on the conformational properties of RNA: update of the CHARMM all-atom additive force field for RNA. *J. Comput. Chem.*, 32, 1929, 2011.
84. Brooks, B.R., Brooks III, C.L., Mackerell Jr., A.D., Nilsson, L., Petrella, R.J., Roux, B., Won, Y., Archontis, G., Bartels, C., Boresch, S., Caflisch, A., CHARMM: the biomolecular simulation program. *J. Comput. Chem.*, 30, 154, 52009.
85. Case, D.A., Cheatham 3rd, T.E., Darden, T., Gohlke, H., Luo, R., Merz Jr., K.M., Onufriev, Simmerling, C., Wang, B., Woods, R.J., The Amber biomolecular simulation programs. *J. Comput. Chem.*, 26, 1668, 2005.

86. Phillips, J.C., Braun, R., Wang, W., Gumbart, J., Tajkhorshid, E., Villa, E., Schulten, K., Scalable molecular dynamics with NAMD. *J. Comput. Chem.*, 26, 1781, 2005.
87. Abraham, M.J., Murtola, T., Schulz, R., Pá ll, S., Smith, J.C., Hess, B. *et al.*, GROMACS: High performance molecular simulations through multi-level parallelism from laptops to supercomputers. *SoftwareX*, 1–2, 19, 2015.
88. Eastman, P., Swails, J., Chodera, J.D., McGibbon, R.T., Zhao, Y., Beauchamp, K.A., Wang, L.P., Simmonett, A.C., Harrigan, M.P., Stern, C.D., Wiewiora, R.P., OpenMM 7: Rapid development of high performance algorithms for molecular dynamics. *PLoS Comput. Biol.*, 13, 1–17, 2017.
89. Prieve, D.C. and Russel, W.B., Simplified predictions of Hamaker constants fromLifshitz theory. *J. Colloid Interface Sci.*, 125, 1, 1988.
90. Pak, A.J. and Voth, G.A., Advances in coarse-grained modeling of macromolecular complexes. *Curr. Opin. Struct. Biol.*, 52, 119, 2018.
91. Lu, L., Dama, J.F., Voth, G.A., Fitting coarse-grained distribution functions through an iterative force-matching method. *J. Chem. Phys.*, 139, 12013, 2013.
92. Izvekov, S. and Voth, G.A., Effective force field for liquid hydrogen fluoride from ab initio molecular dynamics simulation using the force-matching method. *J. Phys. Chem. B*, 109, 573, 2005.
93. Lyman, E., Pfaendtner, J., Voth, G.A., Systematic multiscale parameterization of heterogeneous elastic network models of proteins. *Biophys. J.*, 95, 4183, 2008.
94. Grime, J., Dama, J.F., Ganser-Pornillos, B.K., Woodward, C.L., Jensen, G.J., Yeager, M., Voth, G.A., Coarse-grained simulation reveals key features of HIV-1 capsid self-assembly. *Nat. Commun.*, 7, 1, 2016.
95. Hadley, K.R. and McCabe, C., A simulation study of the self-assembly of coarse-grained skin lipids. *Soft Matter.*, 8, 4802, 2012.
96. Izvekov, S. and Voth, G.A., Multiscale coarse-graining of mixed phospholipid/cholesterol bilayers. *J. Chem. Theory Comput.*, 2, 637, 2006.
97. Orsi, M., Haubertin, D.Y., Sanderson, W.E., Essex, J.W., A quantitative coarse-grain model for lipid bilayers. *J. Phys. Chem. B*, 112, 802, 2008.
98. Shelley, J.C., Shelley, M.Y., Reeder, R.C., Bandyopadhyay, S., Klein, M.L., A coarse grain model for phospholipid simulations. *J. Phys. Chem. B*, 105, 4464, 2001.
99. Marrink, S.J., de Vries, A.H., Mark, A.E., Coarse grained model for semi-quantitative lipid simulations. *J. Phys. Chem. B*, 108, 750, 2004.
100. Marrink, S.J., Risselada, H.J., Yefimov, S., Tieleman, D.P., de Vries, A.H., The MARTINI force field: coarse grained model for biomolecular simulations. *J. Phys. Chem. B*, 111, 7812, 2007.
101. Marrink, S.J. and Tieleman, D.P., Perspective on the Martini model. *Chem. Soc. Rev.*, 42, 6801, 2013.

102. Falck, E., Patra, M., Karttunen, M., Hyvonen, M.T., Vattulainen, I., Lessons of slicing membranes: interplay of packing, free area, and lateral diffusion in phospholipid/cholesterol bilayers. *Biophys. J.*, 87, 1076, 2004.
103. Lopez, C.F., Nielsen, S.O., Ensing, B., Moore, P.B., Klein, M.L., Structure and dynamics of model pore insertion into a membrane. *Biophys. J.*, 88, 3083, 2005.
104. Allen, W.J., Lemkul, J.A., Bevan, D.R., Grid MAT-MD: a grid-based membrane analysis tool for use with molecular dynamics. *J. Comput. Chem.*, 30, 1952, 2009.
105. Feller, S.E. and Pastor, R.W., Constant surface tension simulations of lipid bilayers: the sensitivity of surface areas and compressibilities. *J. Chem. Phys.*, 111, 1281, 1999.
106. Marrink, S.J. and Mark, A.E., Effect of undulations on surface tension in simulated bilayers. *J. Chem. Phys. B*, 105, 6122, 2001.
107. Lindahl, E. and Edholm, O., Mesoscopic undulations and thickness fluctuations in lipid bilayers from molecular dynamics simulations. *Biophys. J.*, 9, 426, 2000.
108. Moore, P.B., Lopez, C.F., Klein, M.L., Dynamical properties of a hydrated lipid bilayer from a multi nanosecond molecular dynamics simulation. *Biophys. J.*, 81, 2484, 2001.
109. Gurtovenko, A.A., Onike, O.I., Anwar, J., Chemically induced phospholipid translocation across biological membranes. *Langmuir*, 24, 9656, 2008.
110. Bemporad, D., Essex, J.W., Luttmann, C., Permeation of small molecules through a lipid bilayer: A computer simulation study. *J. Phys. Chem. B*, 108, 4875, 2004.
111. Notman, R. and Anwar, J., Breaching the skin barrier — Insights from molecular simulation of model membranes. *Adv. Drug Deliv. Rev.*, 65, 237, 2013.
112. Trzesniak, D., Kunz, A.P.E., van Gunsteren, W.F., A comparison of methods to compute the potential of mean force. *ChemPhysChem.*, 8, 162, 2007.
113. Paasonen, L., Romberg, B., Storm, G., Yliperttula, M., Urtti, A., Hennink, W.E., Temperature-sensitive poly (N-(2-hydroxypropyl) methacrylamide mono/dilactate)-coated liposomes for triggered contents release. *Bioconjug. Chem.*, 18, 2131, 2007.
114. Gurtovenko, A.A., Patra, M., Karttunen, M., Vattulainen, I., Cationic DMPC/DMTAP lipid bilayers: molecular dynamics study. *Biophys. J.*, 86, 3461, 2004.
115. Drouffe, J.M., Maggs, A.C., Leibler, S., Computer simulations of self-assembled membranes. *Science*, 254, 1353, 1991.
116. Bemporad, D., Luttmann, C., Essex, J.W., Computer simulation of small molecule permeation across a lipid bilayer: dependence on bilayer properties and solute volume, size, and cross-sectional area. *Biophys. J.*, 87, 1, 2004.
117. Bunker, A. and Dünweg, B., Parallel excluded volume tempering for polymer melts. *Phys. Rev. E*, 63, 016701, 2000.

118. St-Pierre, J.F., Karttunen, M., Mousseau, N., Rog, T., Bunke,r A., Use of umbrella sampling to calculate the entrance/exit pathway for Z-Pro-Prolinal inhibitor in prolyl oligopeptidase. *J. Chem. Theory Comput.*, 7, 1583, 2011.
119. Nandy, B. and Maiti, P.K., Bunker, A., Force biased molecular dynamics simulation study of effect of dendrimer generation on interaction with DNA. *J. Chem. Theory Comput.*, 9, 722, 2013.
120. Mori, T., Miyashita, N., Im, W., Feig, M., Sugita, Y., Molecular dynamics simulations of biological membranes and membrane proteins using enhanced conformational sampling algorithms. *Biochim. Biophys. Acta, Biomembr.*, 1858, 1635, 2016.
121. Wang, H., Ren, X., Meng, F., Molecular dynamics simulation of six β-blocker drugs passing across POPC bilayer. *Mol. Simul.*, 42, 56, 2016.
122. Winter, N.D. and Schatz, G.C., Coarse-grained molecular dynamics study of permeability enhancement in DPPC bilayers by incorporation of lysolipid. *J. Phys. Chem. B*, 114, 5053, 2010.
123. Berger, N., Sachse, A., Bender, J., Schubert, R., Brand, M., Filter extrusion of liposomes using different devices: comparison of liposome size, encapsulation efficiency, and process characteristics. *Int. J. Pharm.*, 223, 55, 2001.
124. Provder, T., Challenges in particle size distribution measurement past, present and for the 21st century. *Prog. Org. Coat.*, 32, 143, 1997.
125. Ohlsson, G., Tabaei, S.R., Beech, J., Kvassman, J., Johanson, U., Kjellbom, P., Tegenfeldt, J.O., Höök, F., Solute transport on the sub 100 ms scale across the lipid bilayer membrane of individual proteoliposomes. *Lab. Chip.*, 12, 4635, 2012.
126. Marrink, S.J. and Berendsen, H.J., Permeation process of small molecules across lipid membranes studied by molecular dynamics simulations. *J. Phys. Chem.*, 100, 16729, 1996.
127. Betageri, G.V. and Parsons, D.L., Drug encapsulation and release from multilamellar and unilamellar liposomes. *Int. J. Pharm.*, 81, 235–241, 1992.
128. Du Plessis, J., Ramachandran, C., Weiner, N., Müller, D.G., The influence of lipid composition and lamellarity of liposomes on the physical stability of liposomes upon storage. *Int. J. Pharm.*, 127, 273, 1996.
129. Stampoulis, P., Ueda, T., Matsumoto, M., Terasawa, H., Miyano, K., Sumimoto, H., Shimada, I., Atypical membrane-embedded phosphatidylinositol 3, 4-bisphosphate (PI (3, 4) P2)-binding site on p47phox phox homology (PX) domain revealed by NMR. *J. Biol. Chem.*, 287, 17848, 2012.
130. Bouwstra, J.A., Gooris, G.S., Bras, W., Talsma, H., Small angle X-ray scattering: possibilities and limitations in characterization of vesicles. *Chem. Phys. Lipids*, 64, 83, 1993.
131. Magarkar, A., Róg, T., Bunker, A., Molecular dynamics simulation of inverse-phosphocholine lipids. *J. Phys. Chem. C*, 118, 19444, 2014.
132. Winter, N.D., Murphy, R.K.J., O' Halloran, T.V., G.C., Development and modeling of arsenic-trioxide-loaded thermosensitive liposomes for anticancer drug delivery. *J. Liposome Res.*, 21, 106, Schatz2011.

133. Messina, P.V., Miguel Besada-Porto,J., Ruso, J.M.Self-assembly drugs: from micelles to nanomedicine. *Curr. Top. Med. Chem.*, 14, 555, 2014.
134. Müller, M., Mackeben, S., Müller-Goymann, C.C., Physicochemical characterisation of liposomes with encapsulated local anaesthetics. *Int. J. Pharm.*, 274, 139, 2004.
135. Jambeck, J.P., Eriksson, E.S., Laaksonen, A., Lyubartsev, A.P., Eriksson, L.A., Molecular dynamics studies of liposomes as carriers for photosensitizing drugs: development, validation, and simulations with a coarse-grained model. *J. Chem. Theory Comput.*, 10, 5, 2014.
136. Knecht, V. and Marrink, S.J., Molecular dynamics simulations of lipid vesicle fusion in atomic detail. *Biophys. J.*, 92, 4254, 2007.
137. Pickholz, M. and Giupponi, G., Coarse grained simulations of local anesthetics encapsulated into a liposome. *J. Phys. Chem. B*, 114, 7009–7015, 2010.

8

Quantum Computational Methods and Computer-Aided Drug Design in Transdermal Drug Delivery of Nanoparticles

Jahasultana Mohammed[1], Haritha Lanka[1], Roja Rani Budha[2], Rajasekhar Reddy Alavala[3], G.S.N. Koteswara Rao[3]* and Surya Kovvasu[4]

[1]*Department of Pharmaceutics, K L College of Pharmacy, Koneru Lakshmaiah Education Foundation, Guntur, Andhra Pradesh, India*
[2]*Department of Pharmacology, H.K. College of Pharmacy, Oshiwara, Jogeshwari (W), Mumbai, India*
[3]*Shobhaben Pratapbhai Patel School of Pharmacy & Technology Management, SVKM's NMIMS, Vile Parle (W), Mumbai, Maharashtra, India*
[4]*Department of Pharmaceutical Sciences, College of Pharmacy, Western University of Health Sciences, Pomona, CA, USA*

Abstract

The pharmaceutical industry revolves around the development of molecular formulations to treat/cure diseases. Their main activity is to find and develop molecules that could help cure diseases and disorders. The use of computational approaches, like quantum mechanics (QM), molecular dynamics simulation, mathematical modeling, physiologically based pharmacokinetic (PBPK) modeling, process simulation, artificial intelligence (AI), and machine algorithm learning in pharmaceutical research has expanded dramatically in recent decades. The primary principle involved in quantum computing (QC) is quantum physics, which is more efficient, accurate, and faster especially when dealing with large molecules. QC could improve the effectiveness of current CADD (computer-aided drug design) tools that impact the development process including designing proteins. The main purpose of the article is to offer an update on nanoparticulate

**Corresponding author*: drgsnkrao@gmail.com

Rishabha Malviya, Sonali Sundram and Dhanalekshmi Unnikrishnan Meenakshi (eds.)
Drug Delivery Systems Using Quantum Computing, (231–246) © 2024 Scrivener Publishing LLC

transdermal drug delivery systems (TDDSs) using quantum computational approaches, which involve the delivery of a drug into or through a skin membrane. Current transdermal drug delivery technologies improve skin permeability for low-dose drugs, which are lipophilic with low molecular weight molecules. Nanoparticles have been discovered to have advantages in medication delivery to target cells or tissues as their structural and chemical compositions play a major role in their transport. Designing the best nanoparticle TDDS is a bit challenging nowadays and leads to an increase in the number of experimental experiments to investigate the properties of the nanoparticle. This chapter discusses the most recent developments in nanoparticles as DDSs. Complementary experimental and computational studies have improved the understanding of nanocarrier mechanism of action and underlying interactions with drugs. The chapter focuses on computer modeling for DDSs. TDDS nanoparticles play an essential role in NDDS for successful therapy with fewer or no side effects than conventional techniques.

Keywords: Quantum computation methods, CADD, transdermal, nanoparticles, mathematical modeling, physiologically based pharmacokinetic (PBPK) modeling, artificial intelligence (AI)

8.1 Introduction

Skin, the most visible and largest organ in the body, protects against environmental stresses and germs, prevents the loss of water and other essential body fluids, and provides protection between the inside of the body and the outside world [1, 2]. In addition, several drugs applied topically are effectively prevented by the skin [3]. Medical professionals and pharmacologists interested in using the skin as a route of drug entry, as well as toxicologists and environmentalists worried about the risks of topical exposure to significant amounts of air, water, and industrial chemicals, would all benefit from looking at nature and understanding the origin of the skin's properties. They would also gain knowledge of the physicochemical properties of substances that ascertain their capacity to enter the circulatory system through the skin [4]. A popular form of administration, transdermal drug delivery (TDD), commonly referred to as "Patches," involves delivering the medication through the skin for local or systemic distribution [5]. TDDs provide several benefits, including the ability to escape first-pass hepatic metabolism that medications taken orally experience, higher patient compliance compared to more painful/invasive alternative methods requiring needle injection, a reduction in dose frequency, and target delivery [6, 7]. Transdermal systems can also be self-administered and are non-invasive. These are typically affordable [6]. Aside from these advantages, it has been challenging to transport only a small number of medications via the skin barrier efficiently using this method [8, 9]. Stratum corneum (SC) is a main selective barrier that only allows tiny and moderately lipophilic molecules to reach the inner layers of the skin. Following this SC layer are the living epidermis, dermis, and subcutaneous connective tissue, all of which provide a lower degree of further resistance [10]. To restrict the medications that can get through the bilayer, the SC analyzes two variables. The first is its pharmacological efficacy, which necessitates blood concentrations in the ng/ml range. The second criterion entails meeting the physicochemical requirements, which include lipophilicity, mobility, and diffusion [11, 12]. There are consequently not many commercially accessible transdermal delivery formulations. Therefore, it is necessary to compromise the SC's structural integrity to ensure the penetration of medicines. There are two recognized approaches to get beyond the obstacle. Iontophoresis, sonophoresis, microneedles, and electroporation are examples of active procedures that use external energy. Passive methods use nanoparticles (NPs) and chemical penetration enhancers [13–15]. The earlier procedure, while somewhat quicker, irritates the skin and creates rashes; therefore, it is not the best option. Certain substances have been demonstrated to cause the bilayer structure's persistent breakdown even when using passive approaches [16–19].

To reduce the consequences of these issues, NPs, which display specific physiochemical size-dependent features, have been used in place of the enhancers [20]. Solid colloidal particles called nanoparticles have a range from 10 to 100 nm [21]. This is the main cause of the recent increase in the number of NP-containing goods on the market. They also ensure controlled release over a prolonged period and improve active ingredient absorption via the skin. These NPs may be created from a variety of substances, including triglycerides, sugars, polymers, metal alloys, and organic and inorganic components [22, 23]. Because NPs may hold or carry a broad range of therapeutic and diagnostic agents, including functional molecules, peptides, and nucleic acids, and then release them in a controlled manner, they are increasingly being used as TDD carriers [24].

The typical approach to drug development of NPs as TDD carriers is a time-consuming, imprecise, and expensive process. Therefore, to get around the problems with old approaches, modern technologies are applied, including computer-aided quantum computational methods. The term "computer-aided molecular design" (CAMD) or "computer-aided drug design" (CADD) describes the practice of using computers in drug research [25, 26]. CADD is a specialized field that computationally models drug-receptor interactions to determine whether or how strongly a specific molecule will attach to a target. This method has become the most popular strategy to drastically limit the number of potential pharmaceutical compounds from a vast library by foretelling which compounds will be active and inactive. This method makes high-throughput screening significantly quicker and less expensive without sacrificing the precision of lead identification. Hydrogen bonding interactions, electrostatic, and hydrophobic interactions are the three types of binding mechanisms that ligands might use to attach to receptors. Because partial to total desolvation must occur, the solvation energies of the ligand and the receptor site are also critical in this process [25, 27]. The discipline of pharmaceutics relies heavily on computational techniques such as quantum mechanics (QM), computational analysis, physiologically based pharmacokinetic (PBPK) simulation, molecular dynamics (MD) simulation, machine learning (ML), and machine algorithms [16].

The spatial positions of electrons and other things at the atomic and molecular scale are precisely described by QM using the Schrödinger equation. The structure and physicochemical characteristics of a molecule can be predicted. The MD simulation simulates the real physical motion of molecules and atoms in accordance with Newtonian physics rules. The physical, kinetic, and energetic characteristics of drugs and excipients, as well as the biological processes of preparations, may be studied using molecular modeling. Its foundation is molecular mechanics and the empirical force field. The numerical depiction of a physical phenomenon, such as one that takes place in a production line, is called process modeling. PBPK simulation can be used to forecast how the formulation will behave in terms of PK/PD in people. By employing an algorithm to create the quantitative formulation prediction model from substantial volumes of acquired experimental data, AI and machine learning algorithms can produce data-driven predictions. An intelligent AI system with proper design may significantly speed up development, optimize formulations, reduce costs, maintain product consistency, and collect and retain the specialized knowledge and skills of formulation specialists [28].

8.2 Skin Lipid Membranes for TDD

Transdermal medication distribution to or through the skin has many benefits over "stratum corneum (SC)," but it is still exceedingly difficult. The SC is typically depicted as a brick-and-mortar frame made of skin cells with pancake patterns (corneocytes) that are enclosed in a lipid matrix (Figure 8.1) [8, 29, 30]. SC lipids are in charge of the skin's most crucial

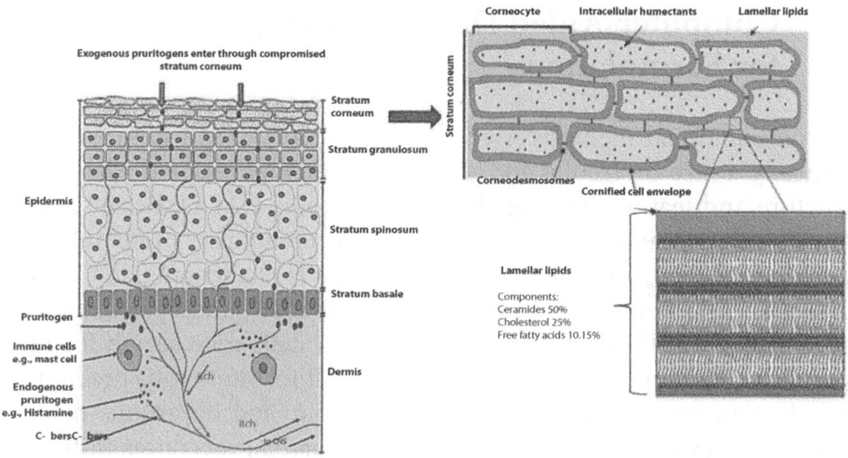

Figure 8.1 Structure of skin in detail.

barrier function. These lipids combine to form lamellar structures, which are typically believed to be sandwich-like bilayer or trilayer structures that exist in a gel-like phase [31]. The SC lipids' molecular configuration is still a mystery. Therefore, it's crucial to comprehend the molecular organization and structure of the SC skin barrier as well as the mechanisms by which molecules enter the SC while creating new dermal and transdermal administration techniques [32, 33].

The primary elements of the lipid mixture that create the lamellar layers that encircle the corneocytes are cholesterol (CHOL), ceramides, and free fatty acids. These components all work together to provide the SC with strong barrier properties while keeping its softness and flexibility. Out of the three components, the ceramide lipid class is widely recognized as being essential to the skin's membrane permeability [34]. The visible ceramide hydrogen bonding network between the polar head groups of neighboring ceramides and other lamellar layers is one of the important features that allow this SC to operate as a formidable barrier [31].

The SC's structural integrity is provided by this network. It is believed that the function of CHOL is to impart fluidity to a hard structure [31]. On the other hand, free fatty acids are thought to arrange lipids because they improve the packing density of hydrocarbon chains. The lack of a conclusive model, either experimentally or computationally, necessitates the development of a technique to create a model that serves as a conduit for a skin lipid replacement while preserving the skin's distinctive barrier qualities [35].

8.3 Computer-Aided Drug Designing for Formulation

All computational methods used to identify, create, and improve physiologically active molecules for potential medicinal use with the desired structure and features are together referred to as CADD. Because of the tremendous demand for medications and the complexity of many biological systems, the clinical development process is challenging and complicated [36, 37]. New medications are occasionally found by accident, but more frequently, they are created as part of a deliberate search for novel approaches to treating diseases. Since they call for numerous predictions, including pharmacokinetic, pharmacodynamic, and hazardous qualities before synthesis, the conventional trial-and-error procedures utilized in drug development are extremely uneconomical [38–40].

CADD is a specialized field that includes computational algorithms and graphical approaches that provide data on drug-receptor interactions. Applications and databases used in bioinformatics are closely related to CADD techniques.

8.4 Advantages of CADD

Drug discovery programs can greatly benefit from CADD tools.

- It can speed up the process of biological and synthetic tests while reducing the amount of work required.
- Additionally, it provides the most desirable drug candidate and filters silica to remove undesirable qualities.
- Researchers can determine how medicines and receptors interact. It has a high success rate and, in contrast to conventional screening techniques, can *in silico* search enormous libraries of biological molecules.
- The entire strategy reduces the possibility of failure in the latter stage of drug research [27, 34, 41].

8.5 System and Model

8.5.1 Force Field

In the computation of molecular docking (MD), the mobility of a collision occurring (compounds, electrons, globules, and so on) is calculated, and the orientation of the particles is regulated by the pressure applied by other atoms and molecules [42, 43]. The interatomic potential, often known as the force field, is used to calculate particular interactions or energy derivation. The force energy and particle acceleration are linked by the equation below:

$$F = -\frac{d\mu(r)}{dr} = ma = m\frac{d^2r}{dt} \tag{8.1}$$

while

F = the force applied to the particles,
μ = the association potential of the particle,
m = the particle's mass,
a = acceleration, and
r = particle's vector position.

$$\mu(r) = \frac{1}{2}\sum_{i=1}^{i=n}\sum_{j=1}^{j=n}\mu_{ij(r_{ij})} \tag{8.2}$$

while

i and j are the particles,
n represents the total particles in the system,
r_{ij} = distance between the two particles, and
μ_{ij} = force between the two particles.

There are several force fields, including GROMOS, OPLS, AMBER, CHARMM, and Berger, that can imitate biological molecules [36, 42–45]. The kind of system that will be mimicked will dictate the force field that will be used. When simulating a given system, it is best not to merge two force fields. These force fields could be classified into two categories: united-atomic force fields (CHARMM, AMBER, and OPLS) and all-atomic force fields (OPLS, GROMOS, and Berger). Each atom in the first may be seen with clarity, however, in the second, hydrogen atoms are bound to carbon by extensive alkane chains and aromatic rings. These force fields may simulate highly complex atomistic phenomena including hydrogen bonding networks, structural components, and interactions between drugs and proteins.

The timescales that these force fields can achieve with adequate processing power are constrained, though. On-time ranges of many microseconds, denaturation of proteins and drug and NP penetration through the epidermis or plasma membrane both take place; all atomic (AA) and unified atomic (UA) force fields are unable to accurately estimate these processes. It is simple to explore the phenomenon using a higher-order force field like MARTINI [46, 47]. In this force field, a single bead is formed by the combination of three or four groups (CH_3, CH_2, etc.). The contact between these spheres is mediated by the Lennard-Jones (LJ) potential, which is calibrated using experimental dispersion energies. Bonds and angles are often used to connect the beads. By altering the settings for characteristics like density, binding, and angular distribution, the same parameters can be obtained via atomistic simulation. Compared to the AA and UA force fields, the MARTINI force field computes far more quickly.

8.5.2 Software Supporting MD and Visualization

For the simulation process, there are both paid programs (such as AMBER and Material Studio) and open-source codes (like GROMACS, LAMMPS, NAMD, and CHARMM) [37, 48]. The program is utilized according to the nature of the problem; for example, AMBER, GROMACS, and CHARMM are frequently used for biomolecular simulations, but LAMMPS is frequently used for materials that are organic and polymeric. Analyzing MD trajectories often involves the use of the VMD application.

8.6 Computational Components

The simulations are developed using parameters like ensemble (Initially utilizing NVT for symmetry, and later switching to NPT at once), pressure, temperature, LJ potential, time constant, and sampling. For every move, the paired list would be recorded along with the configuration of samples tested.

8.7 Procedure

8.7.1 Framework for *In Silico* NP Design

The framework for *in silico* NP design has been outlined in Figure 8.2.

1. **Skin Model:** The self-assembling models of ceramide, long-chain fatty acids, and saturated fat in water are employed in

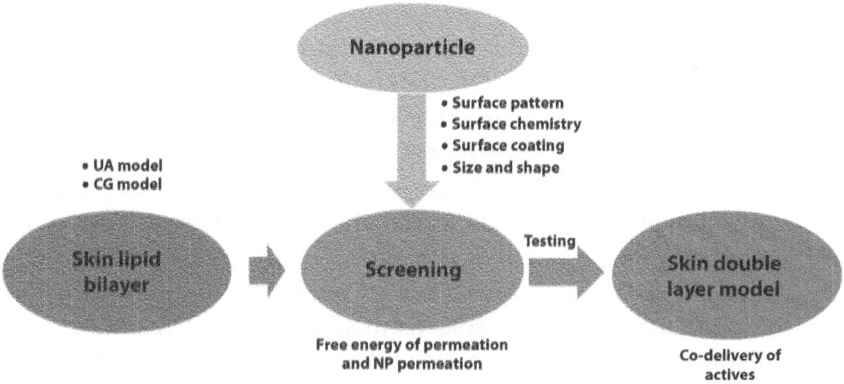

Figure 8.2 Framework for *in silico* nanoparticle design, its outline of the process for creating and analyzing nanoparticles using computational methods.

simulations with the coarse-grained skin model. There are two models created, one with one layer and the other with two layers.

2. **NPs:** Different-sized, -shaped, chemical, and surface-structured NPs are produced.
3. **Screening:** Regular MD simulations of NP utilizing the skin bilayer model are carried out to evaluate their validity. The energy released by permeation via the skin monolayer model is used to screen the NPs.
4. **Testing:** Using the skin bilayer model, the screened NPs serve as carriers for the active ingredients (protein and medication) in co-delivery applications.

First, using Monte Carlo simulation, NPs with various surface patterns and chemistries are generated. These NPs flow via a skin monolayer model, and constrained simulation is utilized to determine their free energy. To validate the energy statistics, selected NPs are assessed on a skin bilayer modeling in accordance with the free energy profile. Ultimately, the potential for co-delivery applications of a protein including screened NPs is assessed.

8.7.2 Integumentary Model

The SC's lipid layer regulates the skin's capacity to function as a barrier. Three major molecular groups—ceramide, triglycerides, and free fatty acids—compose this lipid layer. There are over 300 different kinds of sphingolipids and long-chain fatty acids in the skin [37, 49]. A simple model with a fixed ratio (1:1:1) of CHOL, monounsaturated fatty acids, and ceramide NS was employed to reproduce SC lipids (CER). We will now show you how to simulate the lipid bilayer of the skin. As previously mentioned, selecting the appropriate field force is essential for molecular dynamic simulations. Using three separate force fields, three different types of skin lipid models may be created: AA, UA, and CG [14, 50]. GROMOS, Berger (UA), and CHARMM (AA) force fields are often used to mimic skin lipids [51]. The calculation time is the sole distinction between UA and AA models' simulation setup methods (AA simulations run about four times at a slower rate than UA simulations).

8.7.3 NP Model

NPs' lipid bilayer penetration in the skin is influenced by their size. Different crystals can be used to create NPs of various sizes [4, 24, 52]. Using 1:1 mapping, every component of the NP is put into the beads. The beads' LJ interaction parameter can be employed to change the NP's hydrophobicity and hydrophilicity properties. We consider the charged forms of interaction beads (Q). Additionally, each bead is broken down into several subtypes, making it possible to depict the underlying atomic structure's chemical makeup more precisely. Subtypes of a specific bead type can be identified by two factors: (1) the degree of polarity and (2) hydrogen bonding. The MARTINI force field has a total of 18 unique variations that offer these mapping techniques. The MARTINI force field contains the hydrophobic and very hydrophilic beads P5 and C1, respectively. Here, it is possible to build roughly two different types of particles, each with a unique surface pattern and surface chemistry [53, 54]. While the latter can be accomplished by adjusting the beads' interaction characteristics, the former could be managed by adjusting where the beads are placed on the NP's surface.

8.7.4 Free Permeability Energy

The lipid bilayer's barrier can be identified by measuring the energy state of molecules that break through it. To determine the energy state, the heterogeneous dissolution systemic absorption model is utilized. For a symmetric phospholipid configuration with a standard in the z-direction, the mean force potential is provided by

$$\Delta G(z) = -\int_{outside}^{z} \langle F(z',t) \rangle dz' \tag{8.3}$$

while

F'(z, t) = restrained stress on the NPs for a specific z value,

ΔG = possibility of pressure ratio, and

$F(z',t)$ = equal to force as it has been computed on average.

The typical bilayer, whose centroid is denoted by z = 0 nm, serves as the reaction coordinate in limited simulations. First, an NP is generated for at least 200 ns while being pinned at a location 6 nm in z-distance from the COM bilayer. The simulations of forced pull then use the balanced structure [50, 55]. With a velocity of 0.002 nm/ps, the NPs are progressively being pulled in the direction of the center of the bilayer.

8.8 Testing, Co-Delivery of Actives, and Active Screening

Based on the kinetic energy profile produced using the modeling NPs approach, the NPs are screened. As they are stuck there, NPs with local dips and spikes near the functional group or inside the bilayer are poor candidates for carriers. Once the NPs have been screened, more testing on the skin bilayer model must be done to verify the results from the free energy simulation [56, 57]. To do this, NPs are introduced to the bilayer membrane of the skin, and moisture content (mc) is increased until the NPs are completely immersed in the water layer. The calculation was first carried out in the NVT cluster. The system is then placed through a production run utilizing the variables from the NP modeling, followed by an NPT equilibration [58]. The capacity of the NPs to deliver a payload is further explored after being screened using free energy computations and conventional permeation simulations (drug, protein, etc.).

8.9 Future Challenges

Unquestionably expensive traditional pharmaceutical development forces contemporary pharmaceutical businesses to find a more effective strategy to create competitive goods. Some researchers advance their work by fusing pharmaceutical and digital technologies. These studies speed up progress. As a result, recognizing the crucial benefit that computational pharmaceutics offers is the first step toward comprehending it. In addition to saving time and money, computer procedures also expose more pharmacy knowledge. Traditional pharmaceutical research and development rely on empirical evidence, trial and error, and decision-making, leaving certain fundamental concerns unsolved. But if mankind is to live and flourish, we must adopt a new perspective since we cannot address issues at the same level at which they were generated.

8.10 Conclusion

Transdermal medication administration can address the poor bioavailability of many oral pharmaceuticals, the discomfort and inconvenience of injections, and the restricted controlled-release choices of both. The SC helps the skin's barrier function. NPs must be able to overcome the SC barrier and enter cells for therapeutic application. Many approaches have been developed to disturb the SC to help in NP administration, including gene guns, microneedles, ultrasound, electroporation, and tape stripping. QC approaches are more effective than these. Several QM-based approaches now play a key part in CADD and will have a bigger influence in the future due to increased processing power and efficient algorithms. QM applications must balance accuracy and efficiency. Each step of drug development should use the most suitable methodology, and QM methods should only be used if they offer a meaningful benefit over conventional procedures.

References

1. Ruan, R., Chen, M., Zou, L., Wei, P., Liu, J., Ding, W. *et al.*, Recent advances in peptides for enhancing transdermal macromolecular drug delivery. *Ther. Deliv.*, 7, 89–100, 2016, https://doi.org/10.4155/tde.15.94.
2. Subedi, R.K., Oh, S.Y., Chun, M.K., Choi, H.K., Recent advances in transdermal drug delivery. *Arch. Pharm. Res.*, 33, 339–51, 2010, https://doi.org/10.1007/s12272-010-0301-7.
3. Ghasemiyeh, P. and Mohammadi-Samani, S., Potential of nanoparticles as permeation enhancers and targeted delivery options for skin: Advantages and disadvantages. *Drug Des. Devel. Ther.*, 14, 3271–89, 2020, https://doi.org/10.2147/ DDDT.S264648.
4. Experimental, V., Drug Permeation Through Human Skin: Theory and *in Vitro*. *Exp. Meas.*, 21, 985–96, 1975.

5. Delivery, N.T.D. and Targeting, T., Nanoparticle-Enabled Transdermal Drug Delivery Tissue Targeting. *Molecules*, 21, 12, 7–9, 2016, https://doi.org/10.3390/molecules21121719.
6. Prausnitz, M.R. and Langer, R., Transdermal drug delivery. *Nat. Biotechnol.*, 26, 11, 1261–8, 2008. https://doi.org/10.1038/nbt.1504.
7. Tanner, T. and Marks, R., Delivering drugs by the transdermal route: review and comment. *Skin Res. Technol.*, 14, 3, 249–60, 2008, https://doi.org/10.1111/j.1600-0846.2008.00316.x.
8. Elias, P.M., Epidermal lipids, barrier function, and desquamation. *J. Invest. Dermatol.*, 80, 44s–49s, 1983, https://doi.org/10.1111/1523-1747.ep12537108.
9. Drug, A. and Reviews, D., Breaking the skin barrier. *Adv. Drug Delivery Rev.*, 56, 555–6, 2004. https://doi. org/10.1016/j.addr.2003.10.022.
10. Gupta, R. and Rai, B., *In-silico* design of nanoparticles for transdermal drug delivery application. *Nanoscale*, 10, 4940–51, 2018, https://doi.org/10.1039/c7nr07898f.
11. Benson, H.A.E. and Namjoshi, S., Proteins and peptides: Strategies for delivery to and across the skin. *J. Pharm. Sci.*, 97, 3591–610, 2008, https://doi.org/10.1002/ jps.21277.
12. Naik, A., Kalia, Y.N., Guy, R.H., Guy, R.H., Transdermal drug delivery: overcoming the skin' s barrier function. *Pharm. Sci. Technol. Today*, 3, 9, 318–326, 2000.
13. Mathur, V., Satrawala, Y., Rajput, M.S., Physical and chemical penetration enhancers in transdermal drug delivery system. *Asian J. Pharm.*, 4, 3, 173–84, 2010, https://doi. org/10.4103/0973-8398.72115.
14. Gupta, R. and Rai, B., Penetration of Gold Nanoparticles through Human Skin: Unraveling Its Mechanisms at the Molecular Scale. *J. Phys. Chem. B*, 120, 7133–42, 2016, https://doi.org/10.1021/acs.jpcb.6b03212.
15. Vitorino, C., Almeida, A., Sousa, J., Lamarche, I., Gobin, P., Marchand, S. *et al.*, European Journal of Pharmaceutics and Biopharmaceutics Passive and active strategies for transdermal delivery using co-encapsulating nanostructured lipid carriers : *In vitro* vs. *in vivo* studies. *Eur. J. Pharm. Biopharm.*, 86, 133–44, 2014, https://doi.org/10.1016/j.ejpb.2013.12.004.
16. Wang, W., Ye, Z., Gao, H., Ouyang, D., Computational pharmaceutics - A new paradigm of drug delivery. *J. Control. Release*, 338, 119–36, 2021, https://doi. org/10.1016/j.jconrel.2021.08.030.
17. Delouise, L.A., Applications of Nanotechnology in Dermatology. *J. Invest. Dermatol.*, 132, 964–75, 2012, https://doi.org/10.1038/jid.2011.425.
18. Cappel, J., Effect of nanoparticles on transdermal drug delivery. *J. Microencapsulation*, 8, 3, 369–74, 1991.
19. Desai, P., Patlolla, R.A.M.R., Singh, M., Interaction of nanoparticles and cellpenetrating peptides with skin for transdermal drug delivery. *Mol. Membr. Biol.*, 27, 7, 247–59, 2010, https://doi.org/10.3109/09687688.2010.522203.

20. Marren, K., Dimethyl Sulfoxide: An Effective Penetration Enhancer for Topical Administration of NSAIDs. *Phys. Sportsmed.*, 39, 3, 75–82, 2011. https://doi.org/10.3810/ psm.2011.09.1923.
21. Ealias, A.M., A review on the classification, characterisation, synthesis of nanoparticles and their application A review on the classification, characterisation, synthesis of nanoparticles and their application. *Mater. Sci. Eng.*, 263, 3, 032019, 1–14, 2017, https://doi.org/10.1088/1757-899X/263/3/032019.
22. Akash, B. and Ravindranath, S., A Review on Novel Drug Delivery System: A Recent Trend. *J. Drug Deliv. Ther.*, 9, 3, 517–21, 2019. https://doi.org/10.1016/S0169-409X(02)00113-8.
24. Lohcharoenkal, W., Wang, L., Chen, Y.C., Rojanasakul, Y., Protein Nanoparticles as Drug Delivery Carriers for Cancer Therapy. *BioMed Res. Int.*, 2014, 1–12, 2014.
25. Song, C.M., Lim, S.J., Chuantong, J., Recent advances in computer-aided drug design. *Briefings Bioinf.*, 10, 5, 579–591, 2009, https://doi.org/10.1093/bib/bbp023.
26. Veselovsky, A.V., Zharkova, M.S., Poroikov, V.V., Nicklaus, M.C., SAR and QSAR in Environmental Computer-aided design and discovery of protein – protein interaction inhibitors as agents for anti-HIV therapy n.d. *J. Chem. Inf. Model.*, 54, 1, 37–41, https://doi. org/10.1080/1062936X.2014.898689.
27. Nolte, R.T., Wisely, G.B., Westin, S., Cobb, J.E., Lambert, M.H., Kurokawa, R. *et al.*, Ligand binding and co-activator assembly of the peroxisome. *Nature*, 395, 6698, 137–43, 1998.
28. Rowe, R.C., Roberts, R.J., Rowe, R.C., Artificial intelligence in pharmaceutical product formulation: knowledge-based and expert systems. *Pharm. Sci. Technol. Today*, 1, 4, 153–159, 1998.
29. Karande, P., Jain, A., Ergun, K., Kispersky, V., Mitragotri, S., Design principles of chemical penetration enhancers for transdermal drug delivery. *Biochim. Biophys. Acta*, 1788, 2362–2373, 2005.
30. Bouwstra, J.A. and Ponec, M., The skin barrier in healthy and diseased state. *Biochim. Biophys. Acta, Biomembr.*, 1758, 2080–95, 2006, https://doi.org/10.1016/j. bbamem.2006.06.021.
31. Harrison, I.P. and Spada, F., Breaking the Itch – Scratch Cycle : Topical Options for the Management of Chronic Cutaneous Itch in Atopic Dermatitis, *Medicines*, 6, 76, 1–14, 2019.
32. Alkilani, A.Z., Mccrudden, M.T.C., Donnelly, R.F., Transdermal Drug Delivery: Innovative Pharmaceutical Developments Based on Disruption of the Barrier Properties of the stratum corneum. *Pharmaceutics*, 7, 4, 438–70, 2015, https://doi.org/10.3390/ pharmaceutics7040438.
33. Wokovich, A.M., Prodduturi, S., Doub, W.H., Hussain, A.S., Buhse, L.F., Transdermal drug delivery system (TDDS) adhesion as a critical safety, efficacy and quality attribute. *Eur. J. Pharm.*

Biopharm., 64, 1, 1–8, 2006, https://doi.org/10.1016/j.ejpb.2006.03.009.

34. Wegener, M., Neubert, R., Rettig, W., Wartewig, S., Structure of stratum corneum lipids characterized by FT-Raman spectroscopy and DSC. III. *Mixtures of ceramides and cholesterol. Biophys. J.*, 72, 5, 88, 73–82, 1997.
35. Manikkath, J., Manikkath, A., Mutalik, S., Delving Deeper into Dermal and Transdermal Drug Delivery: Factors and Mechanisms Associated with Nanocarrier-mediated Strategies. *Curr. Pharm. Des.*, 24, 27, 3210–22, 2018, https://doi.org/10.2174/138 1612824666180924122640.
36. Jorgensen, W.L., Maxwell, D.S., Tirado-rives, J., Development and Testing of the OPLS All-Atom Force Field on Conformational Energetics and Properties of Organic Liquids. *J. Am. Chem. Soc.*, 118, 45, 7863, 11225–36, 1996, https://doi.org/10.1021/ja9621760.
37. Hess, B., Uppsala, S-, Lindahl, E., GROMACS 4: Algorithms for Highly Efficient, Load-Balanced, and Scalable Molecular Simulation. *Journal of the American Chemical Society*, 118, 45, 435–47, 2008, https://doi.org/10.1021/ct700301q.
38. Kathe, K. and Kathpalia, H., Film forming systems for topical and transdermal drug delivery. *Asian J. Pharm. Sci.*, 12, 487–97, 2017, https://doi.org/10.1016/j. ajps.2017.07.004.
39. Patel, D., Chaudhary, S.A., Parmar, B., Bhura, N., The Pharma Innovation Transdermal Drug Delivery System: A Review. *Pharma. Innov.*, 1, 66–75, 2012.
40. Wiedersberg, S. and Guy, R.H., Transdermal drug delivery: 30+ years of war and still fighting. *J. Control. Release*, 190, 150–6, 2014, https://doi.org/10.1016/j. jconrel.2014.05.022.
41. Pârvu, L., QSAR - a piece of drug design. *J. Cell. Mol. Med.*, 7, 3, 333–5, 2003.
42. Oostenbrink, C., Villa, A., Mark, A.E., Van Gunsteren, W.F., A Biomolecular Force Field Based on the Free Enthalpy of Hydration and Solvation: The GROMOS Force-Field Parameter Sets 53A5 and 53A6. *J. Comput. Chem.*, 25, 13, 1656–1676, 2004, https://doi. org/10.1002/jcc.20090.
43. Yasmeen, E., Riaz, M., Ashfaq, U.A., Computer aided drug designing: A safari to drug formulation. *Adv. Med. Biol.*, 115, 18–20, 2017.
44. Wang, J., Wolf, R.M., Caldwell, J.W., Kollman, P.A., Case, D.A., Development and Testing of a General Amber Force Field. *J. Comput. Chem.*, 25, 9, 56531, 2004.
45. Berger, O., Edholm, O., Jahnig, F., Molecular Dynamics Simulations of a Fluid Bilayer of Dipalmitoylphosphatidylcholine at Full Hydration, Constant Pressure. *Biophys. J.*, 72, 5, 2002–13, 2013, https://doi.org/10.1016/S0006-3495(97)78845-3.
46. De Vries, A.H., The MARTINI Force Field: Coarse Grained Model for Biomolecular Simulations. *J. Phys. Chem. B*, 111, 27, 7812–24, 2007, https://doi.org/10.1021/jp071097f.

47. Marrink, S.J., De, Vries A.H., Mark, A.E., Coarse Grained Model for Semiquantitative Lipid Simulations. *J. Phys. Chem. B*, 108, 2, 750–60, 2004.
48. Case, D.A., Cheatham, T.E., Darden, T.O.M., Gohlke, H., Luo, R.A.Y., Merz, K.M. *et al.*, The Amber Biomolecular Simulation Programs. *J. Comput. Chem.*, 26, 16, 12255, 2005. https://doi. org/10.1002/jcc.20290.
49. Humphrey, W., Dalke, A., Schulten, K., VMD : Visual Molecular Dynamics. *J. Mol. Graphics*, 14, 1, 7855, 33–8, 1996.
50. Jepps, O.G., Dancik, Y., Anissimov, Y.G., Roberts, M.S., Modeling the human skin barrier — Towards a better understanding of dermal absorption. *Adv. Drug Deliv. Rev.*, 65, 152–68, 2013, https://doi.org/10.1016/j.addr.2012.04.003.
51. Guo, S., Moore, T.C., Iacovella, C.R., Strickland, L.A., Mccabe, C., Simulation Study of the Structure and Phase Behavior of Ceramide Bilayers and the Role of Lipid Headgroup Chemistry. *J. Chem. Theory Comput.*, 9, 5116–5126, 2013.
52. Caflisch, A., Quantum Mechanical Methods for Drug Design. *Curr. Top. Med. Chem.*, 10, 1, 33–45, 2009, https://doi.org/10.2174/156802610790232242.
53. Karande, P. and Mitragotri, S., Biochimica et Biophysica Acta Enhancement of transdermal drug delivery via synergistic action of chemicals. *BBA Biomembr.*, 1788, 2362–73, 2009, https://doi.org/10.1016/j.bbamem.2009.08.015.
54. Labouta, H.I., El-khordagui, L.K., Schneider, M., Nanoscale Mechanism and determinants of nanoparticle penetration through human skin. *Nanoscale*, 3, 12, 4989–99, 2011, https://doi.org/10.1039/c1nr11109d.
55. Filipovic, N., Saveljic, I., Rac, V., Graells, B.O., Bijelic, G., Computational and experimental model of transdermal iontophorethic drug delivery system. *Int. J. Pharm.*, 533, 383–8, 2017, https://doi.org/10.1016/j.ijpharm.2017.05.066.
56. O'Neill, D.W., Noh, S.Y., Notman, R., Computer Simulations of Lipid Membranes and Liposomes for Drug Delivery. *Comput. Pharm. Appl. Mol. Model Drug Deliv.*, 2015, 101–22, 2015, https://doi.org/10.1002/9781118573983.ch6.
57. Alexander, A., Dwivedi, S., Ajazuddin, Giri, T.K., Saraf, S., Saraf, S. *et al.*, Approaches for breaking the barriers of drug permeation through transdermal drug delivery. *J. Control. Release*, 164, 26–40, 2012, https://doi.org/10.1016/j. jconrel.2012.09.017.
58. Durand, C., Alhammad, A., Willett, K. C., Practical considerations for optimal transdermal drug delivery. *Am. J. Health Syst. Pharm.*, 69, 2, 116–124, 2012. https://doi.org/10.2146/ajhp110158

9

Computational Approaches for Drug Delivery of Nanoparticles

Urvashi Sharma[1]*, Hemant Khambete[1], Nitu Singh[2], Sanjay Jain[1], Neelam Jain[3] and Kamal Dua[4]

[1]*Faculty of Pharmacy, Medi-Caps University, Indore, Madhya Pradesh, India*
[2]*RGS College of Pharmacy, Lucknow, U.P., India*
[3]*Faculty of Pharmacy, Oriental University, Indore, Madhya Pradesh, India*
[4]*The Discipline of Pharmacy, Graduate School of Health, University of Technology, Sydney, NSW, Australia*

Abstract

Nanoparticulate drug delivery is a novel and rapidly emerging concept for targeted delivery of bioactive to the predefined site in a controlled way and diagnosis and treatment of various diseases. Despite numerous advantages, drug delivery through nanocarrier-mediated systems is complex and expensive due to their smaller size. To resolve all these complications, computational modeling is being incorporated into designing the new drug delivery systems and understanding the problems of existing nanoparticles. Computer simulation optimizes the designing of drug carrier systems, hence allowing the prediction of drug delivery efficiency to specific targets. Simulations not only can screen the lead compound and carrier substance but also can provide deep insights into features of fabricated nanoparticles, thus allowing the formation of better and more effective nanoparticles suitable for patients. In the drug delivery process, computational modeling is also envisioned to provide solutions to problems regarding surface chemistry, target properties, carrier specifications, formulation approach, etc. The commonly used computational approaches for targeting nanoparticles are the continuum approach, stochastical modeling, etc. These approaches provide knowledge about drug transport, dissolution process, cell interaction, and much other crucial information that can be acquired using them. The present chapter summarizes the background of drug delivery,

**Corresponding author*: urvashi.ekta12@gmail.com; ORCID: https://orcid.org/0000-0001-7819-2688

Rishabha Malviya, Sonali Sundram and Dhanalekshmi Unnikrishnan Meenakshi (eds.)
Drug Delivery Systems Using Quantum Computing, (247–282) © 2024 Scrivener Publishing LLC

concepts of nanoparticles, and approaches involved in the computational modeling of nanoparticulate drug delivery systems.

Keywords: Nanoparticulate drug delivery, computational modeling, simulations, continuum approach, stochastical modeling

9.1 Drug Delivery System

Drugs have been used for a long time to help people live longer and healthier lives. In recent decades, drug delivery practice has drastically altered with even more changes on the horizon. Biomedical engineering has made significant contributions to understanding the physiological barriers, to effectively delivering drugs without hindrance to various body parts like drug movement to the blood circulation and transport of drugs *via* cells and tissues, as well as the development of various novel drug delivery modes that are now in clinical use. Drug delivery systems are developed devices that deliver drug medicinal substances with precision and control. These delivery systems regulate the rate of drug release as well as the site where it is released in the body. Some systems are capable of controlling both, hence release kinetics and targeted delivery can be obtained from the same delivery device [1].

Drug delivery techniques have made a significant contribution to the conversion of prospective medicines into successful therapies. The strategies of drug delivery swiftly evolved to accommodate changing drug delivery needs as the drug therapeutic landscape changed. Small molecular medicines were the major therapeutic class of concern a few decades ago, as their bioavailability is significantly influenced by their physicochemical properties, so the delivery efforts initially concentrated on enhancing the solubility of the medications, regulating drug release, altering their pharmacokinetics (PK) and enhancing drug action. Moreover, entirely new therapeutic functionalities have been offered by new generations of treatments, incorporating proteins and peptides, monoclonal antibodies (mAbs), and nucleic acids, though additional hurdles also arose as a result of the increased roles, including protein and peptide stability, as well as intracellular delivery needs [2].

According to the National Institute of Health in the United States, a drug delivery system (DDS) refers to the formulation that facilitates the introduction of a drug substance into the body to produce a therapeutic response by managing the rate and releasing the drug at the target site with proper safety and efficacy. This procedure includes the delivery of the active agent, the release of therapeutic constituents from the product, and its transportation *via* biological membranes to the site of action [3].

Numerous disadvantages are associated with traditional DDSs such as poor bioavailability, adverse effects, insufficient drug loading capacity, inadequate variable plasma drug levels, therapeutic efficiency, and lack of targeted drug delivery. These conventional DDSs do not selectively deliver the pharmaceuticals to body cells which may result in significant side effects, lower drug concentration at the target region, and multiple drug resistance (MDR). For instance, in cancer treatment, drug delivery to cancerous cells in traditional ways can cause nephrotoxicity, neurotoxicity, and cardiotoxicity in normal tissues. These disadvantages have prompted scientists to learn more about novel DDS [4]. For proper drug administration, several elements must be considered; some of them are as follows:

- Physicochemical characteristics of drugs, excipients, and other carriers.
- Drug interactions and effects on the body.
- Enhancement of drug action.
- The patient's comfort and well-being [5].

Drug delivery systems can increase the administration and efficacy of pharmaceutical molecules such as antibodies, peptides, vaccines, medicines, and enzymes and hence have the potential to enhance millions of patients' life quality. The capacity to construct controlled-release versions of current medications has altered the entire drug development economics. By generating a new product, DDS can improve performance and efficacy while also extending the patent life of an existing substance [6]. In comparison to the new lead molecules (approximately $500 million in 10 years), novel drug delivery systems can be created considerably faster at a lesser cost (roughly $20 to 50 million and 3 to 4 years). Liposomal amphotericin B was the first DDS, approved by the FDA in 1990. Until now, more than 10 DDSs are already commercially accessible for the treatment of disorders ranging from cancer to fungal infections to muscular atrophy [7].

9.1.1 Conventional vs. Controlled Drug Delivery Systems

Apart from ensuring satisfactory patient compliance, the improvement of therapeutic efficacy of incorporated medication(s) has long been the aim of drug delivery systems through increasing the bioavailability or reducing the side effects. The traditional delivery systems like tablets, capsules, and syrups are nevertheless effective for most pharmacological agents where minor changes in plasma profile have little effect on the therapeutic outcome. However, these are quickly excreted from the body, so the drug level is not well maintained within the therapeutic index. Traditional dosage forms offer a single, brief bursting of the drug which metabolizes faster after a single standard dose with the rise in drug level, followed by an exponential drop. Therefore, the duration may be insufficient to have a meaningful therapeutic impact, resulting in a subtherapeutic response [8]. However, in the case of potent medicines, delivery methods must manage drug release in addition to delivering the medication at the absorption site, to keep the plasma drug level within the therapeutic index for the entire course of treatment as seen in Figure 9.1 [9].

The function of this kind of drug delivery system is especially important in pathological circumstances such as CNS disorders and cancer where the excessive release of potent medicines might produce significant side effects [10].

Any drug therapy must be able to quickly acquire therapeutic plasma levels and keep them there during treatment to achieve its therapeutic goals. After considering the PK properties of the therapeutic agent, and integrating the correct dose of the active substance with proper dosing frequency, traditional delivery systems were created to fulfill these goals. However, one must consider the compound's biological half-life and therapeutic index when choosing the most suitable dose frequency for optimum therapeutic efficacy [11].

Multiple ways have been developed to keep drug concentrations in plasma above minimal effective concentration (MEC) but somewhat less than hazardous value. Such as multiple doses given at regular intervals may appear to be a better option than a single dose, but multiple doses also cause fluctuations in drug levels, which frequently fall below the therapeutic range or may rise above dangerous ones. In addition, patient compliance is also poor when many dosages are taken in one day. Another method

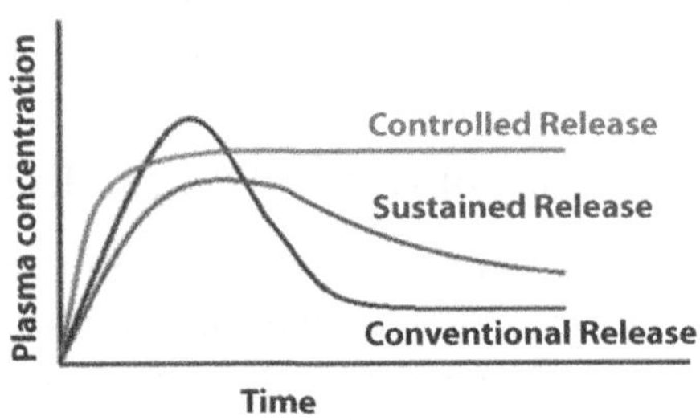

Figure 9.1 Plasma concentration vs. time profile in various formulations [8].

Table 9.1 Pros and cons of conventional and controlled drug delivery systems [12].

Pros of conventional DDS	Pros of controlled DDS
Convenient administration	Predetermined or controlled release of medication
Better and nonintrusive IVIVC	Specific or targeted drug delivery with longer drug residence
Higher shelf-life, low cost	Low dosing frequency and improved bioavailability
Accurate and measured unit dosage form	Protection from metabolism by enzymes/chemicals
Cons of conventional DDS	**Cons of controlled DDS**
Poor absorption from the site of administration	Dose dumping
No target specificity	Possible toxicity of materials used
Premature metabolism and excretion from the body	Poorer IVIVC with invasive implantation or removal procedure
Poor bioavailability and repeated dosing	Higher manufacturing cost

is to give a single dose that is higher than the required dose, resulting in side effects that are unrelated to the drug's intended effects. To maintain consistent plasma drug levels within the therapeutic window with better drug safety, controlled-release drug delivery methods are required, and an additional benefit is targeted therapeutic impact for a longer period [12]. Various pros and cons of conventional and controlled drug delivery systems are presented in Table 9.1.

9.1.2 History of DDS Development

In the 1920s, the need for routine and well-controlled insulin administration to manage diabetes was first observed by Frederick Banting, Charles Best, and John Macleod. Before the major advancements in DDS in the 20th century, these discoveries provided an understanding of the traditional "peak and trough" drug PK curve in advance. In the early 1950s, pharmaceutical formulations that could extend pharmacological action while reducing dose frequency hit the market [13]. These findings provided an early understanding of the traditional "peak and bottom" PK curve of drugs.

Before 1950, all medications were being produced as pills or capsule formulations which released the loaded drug as soon as it came into contact with water, with no control over the release kinetics of the drug. They are usually made up of a drug and a variety of excipients like sugars and starches, which provide protection and stability to the formulation. The basic idea behind this is as the tablet is swallowed, it dissolves in the stomach and releases a drug that gets absorbed into circulation. This strategy works for many medications, but not for those that can be broken down in the stomach. Insulin is frequently used as an example of a drug that must be injected because it fails to survive in the acidic environment of the stomach; therefore, it is administered into fatty tissues from where it gets absorbed into the bloodstream [14].

This led to the evolution of the controlled-release drug delivery system. CRDDS refers to a device that regulates the release of drugs in the body at a predetermined rate. This technology assures to sustain drug levels within the body while controlling drug levels at the target site [15]. As seen in Table 9.2, the history of CRDDS is divided into three generations, which have been discussed in this section.

- **First Generation**

The first-generation (1G) drug delivery can be seen from 1950 to 1980. Smith Kline & French inaugurated Dexedrine Spansules, the first ever commercially available controlled-release product, in 1952. This comprised macrospheres covered with fatty substances like a mixture of glyceryl monostearate and beeswax, which resembled a system commonly recognized as a reservoir employable for controlled drug release and dissolution [17]. The first pulmonary drug delivery system that received FDA approval was a pressurized metered dosage inhaler, this revolutionized the therapeutic aerosol market and was launched in 1956 by Riker Laboratories, Inc. [18]. These events marked the start of the 1G of drug delivery. Some others also contributed to this era, Alejandro Zaffaroni envisioned DDSs with repeatable and predictable drug release kinetics in the mid-1960s. In 1968, Zaffaroni also founded Alza Corp., which was the first of its kind and established on the notion of controlled drug delivery [19].

During the 1G, the majority of the fundamental understanding of drug release mechanisms particularly for oral and transdermal dosage forms was gained. These 1G produced devices with zero-order release

Table 9.2 Evolution of modified release drug delivery systems [16, 21].

1st generation	2nd generation	3rd generation
1950–1980	1980–2010	2010 onwards
Fundamentals of controlled release	Smart delivery system	Modulated delivery system
Oral delivery Once a day, twice a day	Zero-order release Zero- vs. first-order release	Poorly soluble drug - Nontoxic excipients On-off insulin release
Transdermal delivery Once a day, once a week	Smart polymers and hydrogels - Environment sensitive - Self-regulated release	Smart polymers and hydrogels - Signal specificity and sensitivity - Rapid response kinetics
Drug release mechanisms - Dissolution - Diffusion - Osmosis - Ion exchange	Peptide and protein - Durable depot with biodegradable polymer - Delivery to pulmonary sites	Peptide and protein - Delivery for >6 months - Control of release kinetics - Non-invasive delivery
	Nanoparticles - Tumor-specific deliver	Targeted delivery - Anticancer drugs - Nontoxic to nontarget cells - Overcoming the BBB
Successful control of physicochemical properties of delivery	Cannot overcome the biological barriers	- Should surmount both biological and physiochemical barriers - Long-term delivery systems of 6 to 12 months with a minimum burst effect

kinetics-maintained plasma drug concentration by utilizing a range of release mechanisms, such as dissolution, diffusion, and ion exchange. Because of this, various types of once-daily or twice-daily delivery systems emerged during this time, which also characterized the creation of a macroscopic size scale for DDSs [20].

- **Second Generation**

The next era of drug delivery (from 1980 to 2010) featured both micro and nano drug delivery systems along with "smart" DDS technology. Depot DDSs developed as novel technologies of that time were based on polymeric microparticles, phase-separated formulations, and hydrogels. At this point, all of the research was focused on developing release systems with zero-order kinetics, believing that these systems are superior and aid in keeping a consistent drug plasma concentration. But with passing years, it was shown that for an oral delivery system, zero-order drug release does not ensure consistent concentration of drug within the blood. The reason for this is decreased drug concentration after a peak as the drug travels through the small intestine leading to a decline in absorption. Second, most medications such as nitroglycerine and insulin do not need to keep a consistent drug concentration in plasma as the drug throughout remains above the MEC. It took a long time to realize the intuitive insight that permitted the ability to build future medication delivery systems with greater flexibility [21].

The matrix/monolithic depot DDSs that were first authorized in 1989 by the FDA were Zoladex and Lupron, built on polyesters like poly(lactic acid) (PLA) and poly(lactic-co-glycolic acid) PLGA [22]. These degradable DDSs were micron-sized particles with first-order release kinetics. During the second generation, "smart" polymers and hydrogels or environmentally triggered or responsive polymers were created to facilitate drug distribution easier in response to environmental factors like pH or temperature [21, 23].

In the field of research, activities toward the creation of injectable nanocarriers flourished in the late 1980s and 1990s. Two developments sparked this enthusiasm, Enzon Pharmaceuticals, Inc. pioneered the idea of conjugating polyethylene glycol to proteins for extending the circulation period of therapeutics and minimizing immunogenicity came in 1981 [24]. Second, the creation of passively targeted solid tumor nanocarriers was greatly aided and encouraged by the enhanced permeability and retention (EPR) effect. This was first identified in 1984, resulting in the approval of a formulation having PEGylated liposomal doxorubicin named Doxil in 1995. These developments helped the creation of sophisticated, site-specific DDSs in this generation [25].

- **Third Generation**

To speed up the advancement, third-generation drug delivery that began in 2010 focused on regulating drug delivery systems and understanding how they function *in vivo.* Active targeting became a significant topic of interest throughout this time [26]. Although many early advancements in DDS concentrated on delivering small molecules, drug delivery systems for macromolecular therapeutics like peptides, proteins, and nucleic acids have recently received attention (e.g., DNA and siRNA). In 2018, Alnylam Pharmaceuticals, Inc. developed Onpattro (e.g., patisiran), a lipoidal nanoparticle (NP), which was the first RNAi drug to receive FDA approval [27]. The third generation was eventually determined by how well these new modified and tailored DDSs perform in clinical applications. The drug delivery systems of this generation promised to solve challenges associated with the formulation and biological obstacles that were observed while developing current drug delivery systems [28].

9.1.3 Drug Delivery Routes

Different anatomical pathways can be employed to introduce drugs into the human body. They may be designed to have systemic effects or to target specific organs or illnesses. In addition, each path has its benefits and

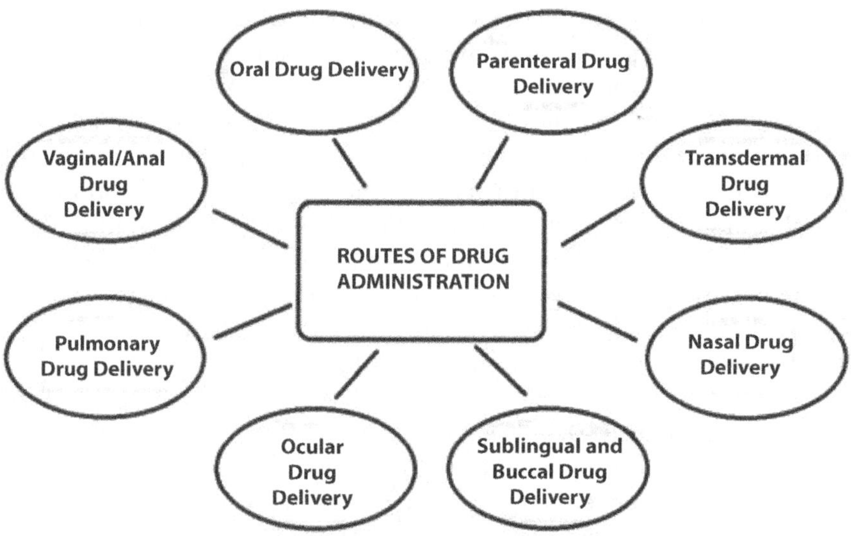

Figure 9.2 Routes of drug administration.

drawbacks. Various anatomical routes for drug administration are shown in Figure 9.2.

a. Oral Drug Delivery

The oral route, being the most popular and traditional route of medication administration, is used equally for conventional and innovative drug delivery. Medications are available mostly in the form of liquids, capsules, and tablets that can be taken orally. As it offers convenient administration, it is widely accepted universally by patients and its preference is understandable. The oral method has some more advantages of being safe, painless, and easy to use over time. However, the drug's solubility in water and its permeability determine whether it may be taken orally or not. Additionally, oral protein or peptide frequently suffers from the acidic environment and enzymatic system of the gastrointestinal tract, as it causes breakage of protein and thereby reduces the therapeutic effectiveness. As a result, several crucial strategies have been tried to enhance the stability of protein and peptide medications, including site-specific delivery systems, chemical peptide modification, and increased absorption. However, to enhance absorption and drug stability in oral drug delivery, significant advancements in new materials and technologies are still needed [29].

b. Parenteral Drug Delivery

Parenteral refers to injecting the drug substances in the body bypassing the gastrointestinal tract, which can be through intravenous (IV), intramuscular, subcutaneous, etc. Under several therapeutic conditions such as for targeted medication, injections are specifically administered into organs of interest. Glass or plastic syringes with needles are commonly used (disposable). Retractable needles or auto-destruct syringes, that lock following the injection, are available in nonreusable forms [30]. The major difficulty in the pharmaceutical business is with the parenteral administration of poorly soluble drugs, as it influences the efficiency of medication delivery to the target area. However, microemulsion formulations, when taken *via* the parenteral route, propose several significant advantages over traditional emulsion delivery systems, as clearance of microemulsions is not that fast, resulting in longer circulation durations [31].

c. Transdermal Drug Delivery

A method of delivering medications through the skin is transdermal drug delivery, which is being exploited nowadays for therapeutic purposes. This approach provides an option of introducing drugs through the skin and is different from those taken through the mouth, intravascular, subcutaneous, or transmucosal channels. The application of a formulation onto intact skin to deliver a drug systemically is known as transdermal drug delivery. In this route, medicine first enters the stratum corneum, followed by its passage through the deeper epidermis and dermis, before being absorbed in blood circulation by the dermal microcirculation. Two significant advantages are its noninvasive nature and suitability for patients who are bedridden or vomiting [32].

d. Nasal Drug Delivery

For several years, drugs have been nasally delivered for all kinds of therapeutic effects be they topical or systemic. Topical medications include those used to treat sinusitis, rhinitis, nasal congestion, allergies, or chronic disorders. Nasal drug delivery refers to the process of administering medication through the nasal cavity. Drugs as nasal spray are typically prescribed for the treatment of upper respiratory tract infections such as allergic rhinitis. However, in rare conditions, this administration strategy can be utilized to systemically give small molecular drugs, where quick onset is required, such as in the case of zolmitriptan, a prescription for the treatment of migraines (Zomig®). Due to the vascularized nasal mucosa, first-pass metabolism can be avoided and drugs are quickly transported to the systemic circulation. Nasal medication can also be administered using liquid or less frequently with powder forms [33, 34]. For several years, drugs have been nasally delivered for topical and systemic effects.

e. Sublingual and Buccal Drug Delivery

For dissolution and direct absorption into the circulatory system despite consuming the drug by swallowing, few medications are taken sublingually, i.e., by placing them under the tongue or buccally by keeping in the space between the gums and teeth. The best way to administer nitroglycerin, which is used to treat angina, is sublingual because it quickly reaches the bloodstream and does not pass through the intestinal wall or liver. However, the majority of drugs cannot be taken in this manner due to the possibility of improper or insufficient absorption [35].

f. Ocular Drug Delivery

Drug delivery researchers have had difficulty in delivering drugs through the eye because of the particular structure and physiology of the eye. Ocular barriers whether static, dynamic, or metabolic prevent drug absorption. Several alternative methods are there for delivering drugs to certain regions of the eye. By identifying and modifying medicines to target certain efflux and inflow transporters in the eye, scientists being able to solve several obstacles to getting drugs into ocular tissues has increased the scope of ocular drug delivery [36].

g. Pulmonary Drug Delivery

The process of drug administration through inhalation *via* the mouth and into the airways is known as pulmonary drug delivery. Diseases related to the lungs can be effectively treated by inhaling the drugs. Recent research has looked into pulmonary medicine delivery as a possible administration method for systemic illnesses because of the large surface area for absorption and higher permeability of the alveolar region. Dietary difficulties and interpatient metabolic variance do not affect pulmonary delivery, which is an added benefit. Local lung illness can also be effectively treated with inhaled medicines [37].

h. Vaginal/Anal Drug Delivery

In comparison to oral administration, vaginal/anal drug delivery provides a quicker drug response and better bioavailability. Likewise, rectal drugs also have local, i.e., laxative, and systemic actions such as analgesics, when alternative routes are unsuitable. Moreover, GIT disorders are unaffected by drug delivery through the vaginal canal, which skips presystemic metabolism. Hence, when it comes to issues with women's health and hormone delivery, the vaginal route is frequently explored. Hydrogels, suppositories, pills, and pessaries are just a few of the vaginal formulation possibilities available [38, 39].

9.2 Nanoparticles

Michael Faraday created gold particles on a nanoscale more than 150 years ago and achieved the first breakthrough that paved the way for huge changes in the field of medicine. Later, researchers coupled these colloidal gold particles with antibodies to create immune-gold staining, which is a target-specific staining technique. Initially, this was viewed as a forerunner to contemporary nanotechnology-based drug delivery applications [40]. Nanotechnology is used in drug delivery to deliver and target pharmacological, therapeutic, and diagnostic substances to cells using NPs. Two requisites from NPs are as follows:

a. First, the drug-NP conjugate must be capable of delivering the medications to the target area excluding the degradation in the gastrointestinal tract or lowering their effectiveness.

b. Second, it should strike target cells while minimizing side effects and causing no harm to other cells [41].

The term nano now refers to one billionth (10^{-9}) of a meter (SI length unit). It is a prefix derived from Nanos, a Greek word meaning dwarf. On the other hand, nanotechnology involves creating novel nanoscale materials through designing, developing, and using materials at the atomic, molecular, and macromolecular levels. These new nano-related terminologies like nanoscience, nanotechnology, nanomaterials, and nanochemistry are few that frequently appear in scientific publications [42].

NPs can be defined as particulate dispersions or solid particles having a size between 1 and 100 nanometers. In NP preparation, the matrix is used to dissolve, entrap, encapsulate, or bind the medication. All these nanoformulations such as NPs, nanospheres, and nanocapsules can be made as per the preparation method. Nanocapsules are matrix systems having uniformly dispersed drugs; in contrast, nanospheres are systems where the drug is contained within a hollow polymeric membrane [43].

9.2.1 Need for Developing NPs

Nanotechnology is the science of extremely small or minute particles. It involves the utilization and modification of materials on a very small scale. Because atoms and molecules act differently at this scale, it allows for several unexpected and intriguing applications. In recent years, interest in nanotechnology and nanoscience research has risen across a wide product category. It allows the production of new materials, notably those for medical applications, when traditional procedures may be limited [44].

In order to achieve site-specific drug activity at a suitable rate and dose regimen, it is important to manage the size of particles, the surface properties, and the release of therapeutically active compounds while designing nanoparticular delivery systems. Polymeric NPs have a few key advantages over liposomes. For example, they have significant controlled-release capabilities and can aid in improving the stability of medicines and proteins [45].

9.2.2 Advantages

Some of the advantages of nanoparticulate drug delivery systems are discussed here:

- The system can be utilized for a variety of administration routes, including oral, intraocular, parenteral, and nasal.
- NP's size and surface characteristics can be simply adjusted for parenteral delivery to accomplish passive as well as active targeting.
- They alter the distribution and subsequent clearance of drugs to ensure high therapeutic efficacy and minimal side effects, controlling and maintaining drug release during transit and at the site of localization.
- Targeted drug delivery can be performed by adding targeting ligands to the surface of the particle or *via* magnetic guidance.
- Nanoengineering allows researchers to exert precise control over biomaterials and polymer physical properties on a smaller scale than previously possible.
- Drug distribution to small locations within the body might be improved by NPs.
- By overcoming the resistance imposed by physiological barriers in the body, which is directly affected by particle size, NPs enable effective medication delivery to diverse areas of the body.
- NPs can help with efficient medication delivery by increasing bioavailability and boosting the aqueous solubility of poorly soluble medicines, allowing for structured drug release and precise drug targeting.
- The characteristics of NP surface can be modulated for proteins, small molecules, peptides, and nucleic acids for targeted medicine delivery. The immune system does not detect the loaded NPs, and they are effectively targeted to certain tissue types.
- Drug toxicity can be decreased and drug distribution can be made more efficient by targeting nano drug carriers.
- Nanocarriers have the potential to deliver biotechnological medications across many anatomical extremities of the body, such as the blood-brain barrier (BBB) [46, 47].

9.2.3 Classification of NPs

1. Organic NPs:

Organic or polymeric NPs, also known as dendrimers, micelles, liposomes, etc., are safe and biodegradable and a few have a hollow core, generally referred to as nanocapsules. These NPs are vulnerable to electromagnetic radiations and temperature and, hence, easily get affected by heat and light. Because of their distinct properties, they are an excellent alternative for drug delivery. Apart from their usual features like size, composition, surface, and shape, their drug-carrying capacity, stability, and type of delivery systems, whether an entrapped drug or adsorbed drug system, define their efficiency and area of application. These NPs are frequently utilized in the biomedical field because they are effective and may be introduced to a particular body part for a process referred to as targeted drug delivery [48].

2. Inorganic NPs:

Particles that do not contain carbon are known as inorganic NPs. Some examples are metal and metal oxide-based NPs [49].

a. Metallic NPs: These have a higher preference for medication and biosensor delivery as carriers. Many metals have been utilized for the creation of NPs, but silver, gold, and iron NPs are the most important for biomedical applications. A vast range of ligands, including sugar, peptides, proteins, and DNA, have been coupled to NPs. These NPs feature surface functionalization, making it simpler to coat them with ligands. These are employed for bioassays, active distribution of bioactive, detection and imaging, drug discovery, and several others because of their functionalization ability [50].

Metal precursors are the only materials used to make metal NPs. These NPs have distinctive optoelectrical features because of popular localized surface plasmon resonance (LSPR) characteristics. NPs of alkali and noble metals like Ag, Au, and Cu provide wide absorption bands in the visible zone of the EMR spectrum. Modern high-tech materials make use of metal NPs with controlled size, facet, and shape. Because of their remarkable optical characteristics, metal NPs are utilized in different scientific domains. To enhance the electronic stream and facilitate the capture of high-quality SEM images, NP layering with gold is frequently utilized in SEM sampling [51].

b. Ceramic NPs: These NPs are inorganic nonmetallic solids formed by heating and cooling. They can take on polycrystalline, amorphous, porous, dense, or hollow morphologies. These NPs are chemically inert and have a high thermal resistance. These are beneficial in the delivery of drugs for a variety of conditions, including bacterial infections, glaucoma, and cancer. Moreover, such NPs are gaining interest from scientists due to their numerous uses in catalysis, photocatalysis, dye photodegradation, and imaging [51, 52].

c. Polymeric NPs: Because of their varying particle size from 10 to 1000 nm and extended circulation in the blood, these polymeric NPs are continuously being investigated for targeting therapeutic agents. When the carrier size is less than 1 μm, an IV injection is possible, reducing the risk of irritating reactions. The NPs produced using polymers of natural and synthetic origin are getting interested because of their increased stability and higher potential for surface nanoengineering. By adjusting the polymeric properties and surface chemistry, these NPs can be customized to accomplish both controlled drug release and drug localization for a specific disease. Bio-safety and biocompatibility are two crucial properties to look for when using polymers in pharmaceutical formulations. Biodegradable polymers are extensively utilized for drug delivery because they can be broken down inside the body into harmless monomers. Natural polymers like alginate, collagen, or chitosan or synthetic polymers like PLA and PLGA are among the polymeric materials used to make NPs [53].

d. Lipid-Based NPs: They are spherical, having a diameter that ranges between 10 to 100 nanometers. It contains a lipid-based strong core and a network of lipophilic particles dissoluble in nature. Surface active agents such as emulsifiers stabilize the outer layer of NPs. These NPs have applications as drug carriers for the delivery of potential candidates, as well as in RNA release for cancer therapy [54].

e. Semiconductor NPs: Semiconductor NPs share the same properties as metals and nonmetals. They belong to the groups II-VI, III-V, or IV-VI of the periodic table. These particles have large bandgaps, which can be tuned to reveal a variety of features and are applicable in photo optics, electronics, photocatalysis, and water shedding. GaN, GaP, InP, and InAs are semiconductor NPs from bunch III-V; ZnO, ZnS, CdS, CdSe, and CdTe are semiconductor NPs from bunch II-VI; and silicon and germanium are semiconductor NPs from lot IV [54].

3. Carbon-Based NPs:

The two main types of carbon-based NPs are carbon nanotubes (CNTs) and fullerenes. CNTs were identified in 1991, available as a sheet of graphite that has been rolled up into a cylindrical shape, with a hexagonal network of carbon atoms that range in size from 1 to 100 nm and about 1 nm in diameter. Because these materials are 100 times stronger than steel, they are primarily employed for structural reinforcement. Single-walled CNTs (SWCNTs) and multi-walled CNTs (MWCNTs) are two types of CNTs [55].

Fullerenes are carbon allotropes with spherical cages holding about 28 to over 100 carbon atoms. Buckminsterfullerene is the name for C-60's structure, which resembles a hollow football. These structures have a pentagonal and hexagonal arrangement of carbon units. They are commercially useful because of their structure, electrical conductivity, electron affinity, and higher strength [55].

9.3 Problems in Conventional Manufacturing of NPs

Nanomedicines will certainly be 3D structures having many components arranged in optimum spatial patterns to be functional. As a result, minor variations in the approach or composition of these constructions can have a negative impact on the component superposition. Knowing the components' physicochemical characteristics, understanding the important components and their mutual interactions, and defining the primary features and their impact on performance can all help to create highly repeatable production methods for nanomedicines [56].

Before a nanomedicine reaches the clinic, it must pass a series of tests, beginning with precise characterization and successful manufacturing of these sophisticated constructions.

a. Characterization of NP's Component

The particle size and size distribution of NPs, as well as their drug-loading efficiency, surface morphology, charge on the particle surface, chemistry, surface adhesion, steric stabilization, drug release kinetics, and hemodynamic properties, all influence how they behave *in vitro* and *in vivo* [56]. NPs have been utilized for delivering small-molecular medicines, peptides, etc., through liposomes, nanoshells, gold NPs, quantum dots, dendrimers, CNTs, and nanocrystals [57]. Each system has a distinct architecture and unique critical qualities that may aid in improving NP performance in specific indications [56].

However, determining the right NP characteristics for certain indications is critical. Going one route may solve one problem, but it may cause another [56]. Doxil® is a fantastic illustration of how NP physicochemical characteristics can influence PK. It is a PEGylated doxorubicin in the form of liposomes with higher concentrations of drug at the tumor site, longer half-life, improved antitumor efficacy, and fewer adverse effects than non-PEGylated doxorubicins [58]. In contrast to PEGylated liposomes, which can greatly limit liposome uptake by macrophages, doxorubicin-loaded non-PEGylated liposomes are extremely affine toward the reticuloendothelial system (RES) and are rapidly eliminated from circulation through an action known as "opsonization". While Doxil's PEG coating reduces cardiotoxicity, it also causes an increase in skin concentrations, causing palmar-plantar erythrodysesthesia, a side effect sometimes also known as hand-foot syndrome [58]. Because there are no any reliable *in vivo* simulations of the varied behaviors of the numerous types of NPs being studied right now, the fabrication of NPs must be supported by empirical data and exhaustive preclinical research on animals.

b. Analysis and Characterization of NP Formulation

One of the most difficult aspects of nanomedicine development is determining the appropriate analytical tests for the complete characterization of NPs. Because nanomedicines are more complicated than traditional pharmaceuticals and every component has a distinct function, a higher level of testing is needed for a thorough evaluation of every component, and assessment of interactions and stoichiometry of these components, including spatial orientation, would be crucial. Being therapeutically effective, whether nanomedicines possess the required properties or not, can be guaranteed by numerous orthogonal characterization procedures [56].

Numerous proposed nanomedicines contain biological elements like nucleic acids or protein acids that, depending on the circumstances of production, may be susceptible to compositional changes [59, 60]. Although these bio elements might not serve as therapeutically active substances in NPs, they might help active ingredients to reach particular cells or distribute them throughout the body. Due to their crucial role in the efficacy or protection of the medicine, they cannot be regarded as inert excipients, and hence proper analytical studies should be carried out on such substances to obtain thorough results of characterization. The earlier these components can be understood during the initial drug developmental phase, the higher the chance there is of achieving an efficient, repeatable production method [56].

c. Scale-Up and Manufacturing

Pharmaceutical development has challenges and obstacles with the successful scaling up and manufacturing of nanomedicines. The production of three-dimensional (3D), nanometer-scale multicomponent systems by conventional drug manufacturing presents numerous difficulties for the scaling up of nanomedicine. Determining the product's key features requires a thorough understanding of the vital components and their interactions because the majority of pharmaceutical NPs have complicated structures with several components arranged in optimal spatial configurations. Early detection of these primary features during nanomedicine designing will aid in the selection of techniques for bulk production to create the critical process steps and decide the analytical standards, for ensuring product repeatability [56].

Sonication, milling, centrifugation, lyophilization, high-speed homogenization, emulsification, crosslinking, organic solvents, evaporation, and filtration are commonly used in the formulation of NPs. It is advantageous to evaluate which scaling-up technique would be most successful at the initial developmental stage for the production on a "small scale" in the laboratory. A sterile technique for producing nanomedicines that employ a novel administration route may encounter difficulties because of particle size and composition. Among these problems is the potential for NP destruction during sterilization processes like gamma irradiation or autoclaving, particularly when biomaterials are present [61–63]. If the NPs have a flexible structure, like liposomes, and their particle size is noticeably below 220 nm, sterilizing them with conventional sterile filters would not be a problem. For rigid-structural particles like polymeric, silica, and metallic-based NPs, sterilization employing filters is the sole alternative. However, the small pore size of typical filters may cause serious filtering problems, if the particle size distribution or size is near about 220 nm. A significant quantity of active compounds may be lost or degraded during filtration in this situation when the particle size is not far below 220 nm [56]. Although aseptic production is frequently a viable option, it might be challenging, particularly for multistage processes involving material handling and transfer in a sterile environment. In NP production, environmental safety is another issue. NPs can diffuse into the air while working with dry materials, resulting in NP deposition in the lungs and causing pulmonary toxicity [64, 65]. As a result, extra vigilance is essential when producing nanomedicines. Furthermore, adequate personal safety equipment is required during manufacturing because some NPs can cross the skin's protective layer, posing a danger of dermal exposure [65].

d. Regulatory Challenges in the Development of Nanomedicine

Pharmaceutical NPs with diverse structures and multiple components usually exhibit complexity where numerous components may have an impact on the therapeutic effectiveness of lead molecules, in contrast to traditional pharmaceuticals, which normally contain a single active ingredient. Nanomedicine regulation may encounter many challenges as a result of this complexity [56, 66].

The Food and Drug Administration (FDA), the European Therapies Agency (EMA), and other authorities are currently evaluating new NP-based medicines on a case-by-case basis using the usual benefit/risk analysis paradigm [56, 67, 68]. Because nanomedicines are a distinct group of therapeutic substances or formulations, there are few standards for evaluating them [56]. The FDA regulates pharmaceutical products in general under two statutes: the Federal Food, Drug, and Cosmetic Act (FDCA), which applies to every chemically synthesized medication and device; and for therapeutic products that are obtained from biological sources, the Public Health Service Act (PHSA) is applicable [66, 67]. These regulations have various definitions and policies depending on whether the product has a biological source, a mechanical action mode in the form of a device, or a chemical action mode *via* a drug [67]. Because nanomedicines fall under the category of complex products with many components, the FDA has yet to issue any specific advice. In other words, the FDA does not make a categorical judgment on whether nanomedicines are hazardous or safe and will keep on analyzing the unique features of each product. For identifying and evaluating combination products, FDA standards for transparency and uniformity were established in December 2016 [67].

A combination product combines three sorts of products: a drug and a device, a drug and a biologic, a device and a biologic, or all kinds of products. The FDA classifies nanomedicines as combination products, which are subject to the usual regulatory pathway but with additional safety and efficacy standards. The FDA, for example, has approved paclitaxel and doxorubicin nano-formulations as combination products for novel treatments of cancer. Wider concerns regarding the risks associated with nanotechnology and products containing NPs have been raised over the suitability of the present regulatory frameworks and practices, including the toxicity of NPs, the unintended consequences of NPs' ability to pass the BBB, and the continual activity of the NPs. As a result, the FDA has produced one draft and five final guideline documents on the use of nanotechnologies in FDA-regulated goods, such as nanomedicines. Before promoting their products, all six criteria encourage producers to interact with the company [67, 69, 70].

The European Medications Agency (EMA) divides nanomedicines into biological and nonbiological medicines [71]. For biological and nonbiological nanomedicines, more research is needed beyond plasma concentration measurements. A detailed evaluation of the nanomedicine's bioequivalence, quality safety, and efficacy with a benchmark medication is necessary to achieve therapeutic equivalency [68, 72]. Nonbiological complex medications (NBCDs) can be regulated by the biological system because they share some characteristics with biological nanomedicines [68, 73]. For example, the structure is not completely described and the *in vivo* effect depends on the production procedure, so the comparison is required during the life span. The EMA has already issued many reflection papers in the field of nanomedicine. These publications are used to guide developers through the process of obtaining marketing clearance for new nanomedicines and nanosimilars [68].

Recently, entering the period of generic nanomedicines, significant issues and challenges are being faced by generic drug manufacturers and regulatory bodies in defining a framework to assess generic nanomedicines and show appropriate bio equivalency comparable to branded nanomedicines. They also have the equivalent physicochemical characteristics with the same efficacy and safety [56, 68, 74]. The relationship between physicochemical properties, clinical PK, and safety of NPs is unclear, and typical animal models might not be sufficient to conclude the biodistribution of NPs along with toxicity in humans. This is equally significant while comparing a revolutionary NP-based drug to conventional formulations as well as when comparing a generically approved nanomedicine to an innovative product. The bioequivalence between generic and innovative nanomedicines may be impossible to predict based on similarities in results of conventional PK and toxicity investigations or simply by comparing the composition of medicinal products [74].

9.4 Computational Drug Delivery

Computational drug delivery is one of the most dynamic and fast-growing sectors of the pharmaceutical industry. Delivery of an active pharmaceutical ingredient (API) in a safe and optimal dosage form is the ground zero of any drug delivery system. Systematizing state-of-the-art drug delivery systems is a very tedious and premium process that may take years and is based on hit-and-miss analysis. Thus, it becomes a prerequisite to streamlining the formulation development process of drug delivery. To design a sophisticated and novel drug delivery system. computer-aided drug delivery techniques can be exploited [75, 76].

With all the advances in software and hardware in the field of virtual graphics and information technology, the use of 3D modeling of the carrier as well as the target site is possible. The multiscale modeling in drug delivery has three stages that are correlated with each other, as shown in Figure 9.3.

In the last decade, the field of drug design using computational pharmaceutics has also evolved to a very great amplitude. The computer-aided drug design such as QSAR and ligand docking helps to screen out thousands of molecules with time-saving and economic efficiencies [77, 78].

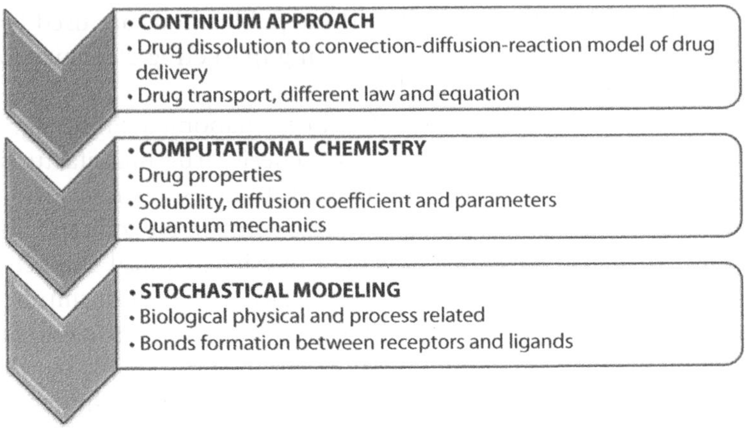

Figure 9.3 Multiscale modeling in drug delivery.

Computer-aided drug delivery focuses on the tactics to devise mathematical-based computer models to simulate the rationale of a drug delivery system in the human body. The first step in computer-aided drug delivery is to constitute a mathematical model, which is then rationalized based on the governing equations of various phenomena such as diffusion and absorption. The data obtained from mathematical models and equations then forms the outline of the computer model, which then can be used to simulate the effectiveness of a drug delivery process. The steps for accomplishing an optimized drug delivery system are outlined in Figure 9.4.

9.4.1 Computational Approach for Nanoparticulate Drug Delivery

The nanoparticulate drug delivery system is a state-of-the-art solution for drug delivery as it deciphers all the obstacles of the traditional drug delivery systems such as low solubility, limited half-life, no target specificity, and scarce translocation across barriers [79]. Computational techniques can be utilized to design optimized nanoparticulate drug delivery systems by employing mathematical and simulated models to analyze interactions between drugs and carriers, carriers and the target cells, biological macromolecules, and biological barriers [80]. Computational chemistry can be used to study quantum mechanics, molecular dynamics (MD), drug-loading, and physical attributes of the nanoparticulate drug delivery system [81, 82].

Ligand-based targeting is the most prominent approach employed for nanoparticulate systems. Nonetheless, elevated ligand affinity may lead to off-site binding of NPs. The off-site binding in the treatment of cancer may

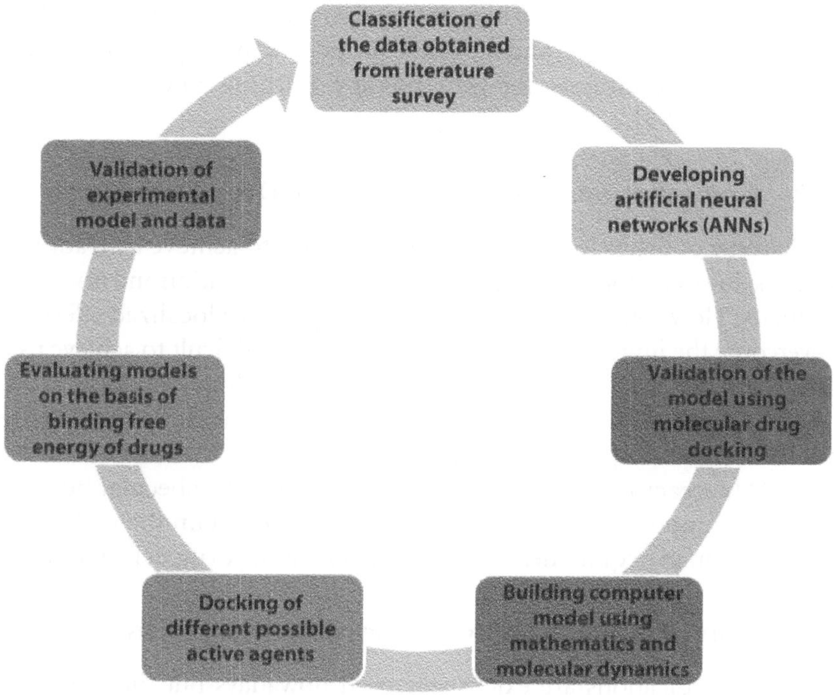

Figure 9.4 Steps for optimizing drug delivery.

vandalize healthy cells [83]. Hence, computational simulations are useful to quantify the net surface interactions, binding valency, selectivity, and different constants like partition and dissolution to increase the specificity of the delivery systems [84, 85].

Solid lipid NPs (SLN) are used widely as drug delivery carriers for particles ranging from macro to microscale. Computational simulations can be used to foresee the interactions and compatibility of drugs with SLN [86]. Initially, the modeling of the SLN is done using mathematical models and MD, which are simulated to observe the drug-SLN interactions [87, 88].

Gene/drug delivery through nanoparticulate systems is an emerging paradigm that is predominantly based on computer simulation to evaluate conformational and deformation changes upon the interaction of nanoparticulate systems with membranes, genetic material, etc. [89]. The computational techniques can also help to optimize parameters such as drug loading, drug release, distribution, drug-carrier interaction, permeability, and site specificity to advance gene targeting by nanocarriers [90, 91].

The metal-coated NPs are a multifunctional tool that can be employed for their optical, catalytic, and magnetic properties in conglomeration with computational techniques for gene therapy, imaging, and photothermal treatments [92, 93].

9.4.2 Computational Approach for Bone Drug Delivery

The delivery of drugs to the bone is very difficult to achieve by traditional methods due to various limitations such as drug distribution and non-target specificity. However, implants can be used to produce localized effects and to overcome the limitations, but PK studies are very difficult to achieve using 2D bone models, which are based on hit-and-miss analysis [94–96].

The computational simulation can be used to achieve the required data to design an optimized drug delivery. Hossein Mokhtarzadeh used Zetos 3D bioreactor to access the drug delivery to the trabecular bone and Abaqus, a computer-aided engineering software to obtain the mathematical simulation to acquire desirable parameters in the drug delivery system.

9.4.3 Computational Approach for Nasal Drug Delivery

Intranasal medications are extensively used nowadays but their efficiency depends on both the product as well as the patient. Various factors that hinder the efficacy of intranasal medications such as the patient's head position and breathing while administration of the medication. Optimization of the intranasal medication can be achieved by computational techniques, which include the development of a healthy nose model by the use of MRI [97, 98].

Computational fluid dynamics (CFD) is then employed, which mimics different disease conditions to develop models of the diseased nose. The computer simulation is then employed to employ the inspiratory air-flow field, which is used to obtain the movement of the drug in the simulated airflow field. The drug deposition, movement, and specificity are then determined by running simulations, and the data obtained can be used to achieve optimal intranasal drug delivery systems [97, 99].

9.4.4 Computational Approach for Polymeric Drug Delivery

Polymeric drug delivery systems have formidable potential in the field of chemotherapeutics. Nanoprecipitation can be used to entrap drugs in nanocarriers but the dominant complication is the entrapment efficiency of the polymers. MD and other computational simulations based on mathematical equations can be exploited to obtain superior drug entrapment efficiency [81, 100].

9.4.5 Computational Approach for Microsphere Drug Delivery

Microspheres have been utilized in recent years for their control release as well as site-specific properties. The microspheres are used to deliver site-specific drugs to the brain, liver, bone, etc., but the *in vitro* release profile is very difficult to study in 2D models. However, combining the mathematical models, which are available with the modernistic computational fluid dynamic models, can help to improve site specificity with improved controlled-release characteristics [101, 102].

The use of CFD can be used in combination with mathematical models to obtain a simulated brain, liver, or bone model, which can be used to perform enhanced *in vitro* studies. The model can help to determine the correct release profile of the drug as well as other factors such as local tissue binding, concentrated drug zones in the target organ, or the metabolism of the drug occurring at the site [101, 103].

9.4.6 Computational Approach for Liposomal Drug Delivery

The study of particles such as liposomes at the nanoscale is a very crucial task that requires a bulk of experimental data but then leaves certain blind spots that cannot be solved using the traditional approach [104]. The recent developments in MD have opened doors to study interactions that are orchestrating at nanoscales. The use of algorithms such as replica exchange and umbrella sampling can be used to study the free energy of a molecule to transit through a lipid layer [105]. Liposomal systems can be simulated using computational simulations such as coarse-grained MARTINI, dissipative particle dynamics model, and implicit solvent model [106]. Computational simulations have given us many unique drug deliveries like lipid membranes and NPs tethered together as performed by giving a complete modernistic drug delivery vehicle [107–109].

9.4.7 Computational Approach for Tumor Cord Drug Delivery

The delivery of the drug to the tumor is very unpredictable and intermittent owing to its vasculature and microenvironment [110]. PK resistance can be developed in the cells due to the inability of the drug delivery system to sustain the minimal amount of drug required for therapeutic response [111]. These can be caused by the failure of the drug to penetrate the tumor cells. Thus, computational modeling and simulation of tumor cords can be used to study various parameters, such as clinical PK, pharmacodynamics data, and dosing frequency, thus reducing cost, time, and money by accelerating the drug development process [112, 113].

9.4.8 Computational Approach for Gastroretentive Drug Delivery

Drug delivery gastroretentive drug delivery is a novel approach that acts as both a controlled-release and a site-specific system. The use of a gastroretentive approach such as microparticles can help the localization of drugs in the stomach, increasing their bioavailability and efficacy [114]. Computational simulations can be used to study parameters such as gastric emptying, gastroretentive microparticle motion, microparticle density, microparticle, and gastric juice viscosity, which influence the retention of drugs in the stomach [114–116].

A computational approach such as ultrasound imaging and scintigraphy technique can also be used to study the gastroprotective capabilities of the carrier as well as to check the parameters such as floating lag time, floating time, and drug release profile of the drug. Simulation of gastric emptying can be carried out using FLUENT CFD code based on the volume of fluid (VOF) technique and human stomach models can be simulated by combining techniques such as MRI with mathematical equations [115].

9.4.9 Computational Approach for Bioerodible Device-Based Drug Delivery

The practice of bioerodible devices has been reshaping surgical implants immeasurably as it requires only one surgical procedure, thus improving patient compliance. The utilization of computational approaches such as cellular automata and parallel iterations to develop models and simulations conclude the release of API from bioerodible matrix, surface erosion, release rate, drug loading, porosity, and drug distribution in a matrix along with drug release profile [117, 118].

The computational approach for dendrimer-based drug delivery dendrimers is immensely branched macromolecules wherein every branch is known as a generation. Computational simulations based on MARTINI parameters can be used to design stable dendrimers [119]. The computational techniques can also be used to explore their interactions with other molecules, solvents and counter ions, pH, the concentration of salt, deformation, and conformational changes in interaction with nanopores and membranes [90, 120].

The computational approach for CNT-based drug delivery CNTs has gained tremendous attention during the last decade due to their comprehensive physicochemical properties as well as their propensity to interact with biomembranes [121]. CNTs can be utilized in drug delivery for site-specific activity and to design a carrier based on CNTs computer simulations such as MD simulations and density functional theory–based simulations can be used to determine its interaction with the lipid bilayer, drug, and other biomolecules [122]. The computational approach in neuroscience and the advances in technologies such as CAT scan, MRI, and ultrasound have given us new ways to personalize drug delivery for individual patients by studying their unique responses to drugs at the molecular level [123]. The computational techniques can be used to simulate a model based on the genomic data and cellular response of drugs in patients to study parameters such as frequency of dosing, drug distribution, and drug release profile [124].

9.5 Future Prospects

Future developments in radiation, gene, and anticancer therapy may be facilitated by nanotechnology-based drug delivery systems. NPs are being employed in numerous ways researchers are attempting to extend the blood circulation time by coating the NPs' surfaces with red cell membranes rather than PEG, radioisotopes-based NPs that accumulate in tumors; changing the blood vessel permeability to enhance NP penetration; and preparing mAbs and tumor-specific vaccines. Multifunctional NPs are also in the trend that will identify malignant cells, distribute many medications at once, kill cancer cells with minimal side effects, and monitor and treat patients simultaneously. Recently, various methods have come over for designing the appropriate functionalized NPs, one of them is the computational approach that is catching the eye. The use of computational modeling and simulations can help us to understand the basics of the drug-carrier and drug biomolecule interactions. The computer-aided drug delivery and computational pharmaceutics hold the potential to revolutionize the paradigm of drug delivery systems. So, in the future, the main focus will be more on computational approaches that can be used to prepare the NPs for drug delivery.

9.6 Conclusion

Nanotechnology-based drug delivery systems are being fabricated for the enhancement of pharmacological actions and to achieve the maximum therapeutic properties of conventional drugs. The main objective of NP drug delivery is to enhance treatment efficacy, proper diagnostics applications, and monitoring while also lowering costs and improving performance, effectiveness, safety, and patient adherence. The current nano-based delivery systems focus on treating complicated diseases such as cancer, diabetes, heart disease, and ailments of the central nervous system.

Further research is also leading to the creation of multifunctional NPs that may treat multiple diseases simultaneously.

The use of computer-based simulation models serves as enormous support in designing a state-of-the-art drug delivery system. It can also be used to monitor, adjust parameters, and control the site specificity of the carrier molecule. The use of computational simulations can help us to better understand the interactions at nanoscales, which can enable us to design drugs and delivery systems that can help cure diseases that are not curable at present. Thus, computational models form a backbone in the formulation development process and designing novel dosage forms.

9.7 Acknowledgement

The authors are thankful to the Management of Medi-Caps University, Indore for the continuous motivation and constant support needed for the completion of this work.

9.8 Funding Information

No funding is required.

References

1. U.S. Department of Health & Human Services National Institutes of Health, Drug Delivery Systems, https://www.nibib.nih.gov/science-education/science-topics/drug-delivery-systems-getting-drugs-their-targets-controlled-manner#pid-1236, 2016.
2. Vargason, A.M., Anselmo, A.C., Mitragotri, S., The evolution of commercial drug delivery technologies. *Nat. BioMed. Eng.*, 5, 951–967, 2021.
3. Jain, K.K., An overview of drug delivery systems: Drug Delivery Systems. *Methods Mol. Biol.*, 2059, 1–54, 2019.
4. Witharana, C. and Wanigasekara, J., Drug Delivery Systems: A New Frontier in Nano-technology. *Int. J. Med. Res. Health Sci.*, 6, 9, 11–14, 2017.
5. Medindia: Health Information, Drug Delivery System, https://www.medindia.net/patients/patientinfo/drug-delivery-system.html, 2014.
6. Davis, S.S., Drug delivery systems. *Interdiscip. Sci. Rev.*, 25, 3, 175–183, 2000.
7. Zhang, Y., Chan, H.F., Leong, K.W., Advanced materials and processing for drug delivery: the past and the future. *Adv. Drug Deliv. Rev.*, 65, 1, 104–120, 2013.

8. Hardenia, A., Maheshwari, N. *et al.*, Scientific rationale for designing controlled drug delivery systems, in: *Basic Fundamentals of Drug Delivery*, pp. 1–28, Elsevier, Amsterdam, The Netherlands, 2019.
9. Kreuter, J., *Colloidal drug delivery systems*, CRC Press, Boca Raton, Florida, 2014.
10. Singh, S.K., Singh, S., Lillard Jr., J.W., Singh, R., Drug delivery approaches for breast cancer. *Int. J. Nanomed.*, 12, 6205, 2017.
11. Mitragotri, S., Burke, P.A., Langer, R., Overcoming the challenges in administering biopharmaceuticals: Formulation and delivery strategies. *Nat. Rev. Drug Discov.*, 13, 9, 65, 2014.
12. Adepu, S. and Ramakrishna, S., Controlled Drug Delivery Systems: Current Status and Future Directions. *Molecules*, 26, 5905, 2021.
13. Benoit, D.S.W., Overby, C.T., Sims Jr., K.R., Ackun-Farmmer, M.A., Drug delivery systems, in: *Biomaterials Science*, Fourth Edition, pp. 1237–1266, Academic Press, Cambridge, Massachusetts, 2020.
14. Lavik, E.B., Kuppermann, B.D., Humayun, M.S., Drug delivery, in: *Translational Basic Science*, pp. 734–745, 2013.
15. Bhowmik, D., Gopinath, H. *et al.*, Controlled release drug delivery systems. *Pharma Innov.*, 1, 10, 24–32, 2012.
16. Hussain, M.S., Mohit, Kaur, G., Pamma, P., Overview of controlled drug delivery system. *Adv. Biores.*, 12, 3, 248–255, 2021.
17. Hillery, A.M. and Hoffman, A.S., Historical introduction to the field of controlled drug delivery, in: *Drug Delivery: Fundamentals and Applications*, Second Edition, pp. 1–2, CRC Press, Boca Raton, Florida, 2016.
18. Stein, S.W. and Thiel, C.G., The history of therapeutic aerosols: A chronological review. *J. Aerosol Med. Pulm Drug Deliv.*, 30, 1, 20–41, 2017.
19. Urquhart, J., Controlled drug delivery: Therapeutic and pharmacological aspects. *J. Intern. Med.*, 248, 5, 357–376, 2000.
20. Yun, Y.H., Lee, B.K., Park, K., Controlled Drug Delivery: Historical perspective for the next generation. *J. Control. Release*, 219, 2–7, 2015.
21. Park, K., Control. drug delivery systems: Past forward and future back. *J. Control. Release*, 190, 3–8, 2014.
22. Gonella, A., Grizot, S, L. F., López Noriega, A., Richard, J., Long-acting injectable formulation technologies: Challenges and opportunities for the delivery of fragile molecules. *Expert Opin. Drug Deliv.*, 19, 8, 927–944, 2022.
23. Binauld, S. and Stenzel, M.H., Acid-degradable polymers for drug delivery: A decade of innovation. *Chem. Commun.*, 49, 21, 2082–2102, 2013.
24. Hoffman, A.S., The early days of PEG and PEGylation (1970s–1990s). *Acta Biomater.*, 40, 1–5, 2016.
25. Barenholz, Y., Doxil® — The first FDA-approved nano-drug: Lessons learned. *J. Control. Release*, 160, 2, 117–134, 2012.
26. Bobo, D., Robinson, K.J., Islam, J., Thurecht, K.J., Corrie, S.R., Nanoparticle-based medicines: A review of FDA-approved materials and clinical trials to date. *Pharm. Res.*, 33, 10, 2373–2387, 2016.

27. Mullard, A., FDA approves landmark RNAi drug. *Nat. Rev. Drug Discov.*, 17, 9, 613–613, 2018.
28. Abdelwahed, W., Degobert, G., Stainmesse, S., Fessi, H., Freeze-drying of nanoparticles: Formulation, process and storage considerations. *Adv. Drug Deliv. Rev.*, 58, 15, 1688–1713, 2006.
29. Chenthamara, D., Subramaniam, S., Ramakrishnan, S.G., Krishnaswamy, S., Essa, M.M., Lin, F.-H., Qoronfleh, M.W., Therapeutic efficacy of nanoparticles and routes of administration. *Biomater. Res.*, 23, 1, 20, 2019.
30. Pharmapproach, Parenteral Route of Drug Administration: Advantages and Disadvantages, https://www.pharmapproach.com/parenteralroute-of-drug-administration-advantages-and-disadvantages/, 2021.
31. Spitler, R., Zanganeh, S. *et al.*, Drug Delivery Systems: Possibilities and Challenges, in: *World Scientific Series: From Biomaterials Towards Medical Devices*, pp. 1–51, 2018.
32. Lansdowne, L. E., Drug Delivery., https://www.technologynetworks.com/drug-discovery/articles/drug-delivery-322035, 2020.
33. Illum, L., Nasal drug delivery — Recent developments and future prospects. *J. Control. Release*, 161, 2, 254–263, 2012.
34. Kublik, H. and Vidgren, M., Nasal delivery systems and their effect on deposition and absorption. *Adv. Drug Deliv. Rev.*, 29, 1–2, 157–177, 1998.
35. Le, J., Drug Administration, https://www.msdmanuals.com/en-in/home/drugs/administration-and-kinetics-of-drugs/drug-administration, 2022.
36. Gaudana, R., Ananthula, H.K., Parenky, A., Mitra, A.K., Ocular Drug Delivery. *AAPS J.*, 12, 3, 348–360, 2010.
37. Labiris, N.R. and Dolovich, M.B., Pulmonary drug delivery. Part I: Physiological factors affecting therapeutic effectiveness of aerosolized medications. *Br. J. Clin. Pharmacol.*, 56, 6, 588–599, 2003.
38. De Boer, A.G., Moolenaar, F., De Leede, L.G., Breimer, D.D., Rectal drug administration: clinical pharmacokinetic considerations. *Clin. Pharmacokinet.*, 7, 4, 285–311, 1982.
39. Alexander, N.J., Baker, E. *et al.*, Why consider vaginal drug administration? *Fertil. Steril.*, 82, 1, 1–12, 2004.
40. Giljohann., D.A., Seferos, D.S. *et al.*, Gold nanoparticles for biology and medicine. *Angew Chem. Int. Ed Engl.*, 49, 19, 3280–3294, 2010.
41. Wanigasekara., J. and Witharana, C., Applications of Nanotechnology in Drug Delivery and Design - An Insight. *Curr. Trends Biotechnol. Pharm.*, 10, 1, 78–91, 2016.
42. Hagens, W., II, Oomen, A.G., De Jong, W.H., Cassee, F.R., Sips, A.J.A.M., What do we (need to) know about the kinetic properties of nanoparticles in the body? *Regul. Toxicol. Pharm.*, 49, 3, 217–229, 2007.
43. Nikam, A.P., Ratnaparkhiand, M.P., Chaudhari, S.P., Nanoparticles – An Overview. *Int. J. Res. Dev. Pharm. Life Sci.*, 3, 5, 1121–1127, 2014.

44. Sovan, L.P., Utpal, J., Manna, P.K., Mohanta, G.P., Manavalan, R., Nanoparticle: An overview of preparation and characterization. *J. Appl. Pharm. Sci.*, 01, 06, 228–234, 2011.
45. Konwar, R. and Ahmed, A.B., Nanoparticle: An overview of preparation, characterization and application. *Int. Res. J. Pharm.*, 4, 4, 47–57, 2013.
46. Mohanraj, V.J. and Chen, Y., Nanoparticles – A Review. *Trop. J. Pharm. Res.*, 5, 1, 561–573, 2006.
47. Kumari, B., A review on nanoparticles: their preparation method and applications. *Indian Res. J. Pharm. Sci.*, 17, 1420–1426, 2018.
48. Ealia, S.A.M. and Saravanakumar, M.P., A review on the classification, characterisation, synthesis of nanoparticles and their application. *IOP Conf. Ser.: Mater. Sci. Eng.*, 263, 032019, 2017.
49. Varma, M.M., Kumar, K.T.S., Srivalli, I.D., A review on nanoparticles: synthesis, characterization and applications. *World J. Pharm. Med. Res.*, 7, 8, 169–179, 2021.
50. Rangasamy, M., Nano Technology: A Review. *J. Appl. Pharm. Sci.*, 01, 02, 08–16, 2011.
51. Khan, I., Saeed, K., Khan, I., Nanoparticles: Properties, applications and toxicities. *Arabian J. Chem.*, 12, 908–931, 2019.
52. Choudhury, A., Laskar, R.E. *et al.*, A Review on Nanoparticle: Types, Preparation and its Characterization. *Res. J. Pharm. Tech.*, 14, 3, 1815–1822, 2021.
53. Prabhakar, C.H. and Bala Krishna, K., A Review on Polymeric Nanoparticles. *Res. J. Pharm. Tech.*, 4, 4, 496–498, 2011.
54. Lohat, S.K., Kumar, S., Gaba, P., An Overview: Preparation Characterization and Applications of Nanoparticles. *J. Drug Deliv. Ther.*, 10, 6-s, 159–167, 2020.
55. Bhatia, S., Nanoparticles types, classification, characterization, fabrication methods and drug delivery applications, in: *Natural Polymer Drug Delivery Systems*, pp. 33–93, 2016.
56. Desai, N., Challenges in development of nanoparticle-based therapeutics. *AAPS J.*, 14, 2, 282–295, 2012.
57. Sandhiya, S., Dkhar, S.A., Surendiran, A., Emerging trends of nanomedicine–an overview. *Fundam. Clin. Pharmacol.*, 23, 3, 263–269, 2009.
58. O'Brien, M.E., Wigler, N. *et al.*, Reduced cardiotoxicity and comparable efficacy in a phase III trial of pegylated liposomal doxorubicin HCl (CAELYX/Doxil) versus conventional doxorubicin for first-line treatment of metastatic breast cancer. *Ann. Oncol.*, 15, 3, 440–449, 2004.
59. Li, L., Wartchow, C.A. *et al.*, A novel antiangiogenesis therapy using an integrin antagonist or anti-Flk-1 antibody coated 90Y-labeled nanoparticles. *Int. J. Radiat. Oncol. Biol. Phys.*, 58, 4, 1215–1227, 2004.
60. Farokhzad, O.C., Cheng, J. *et al.*, Targeted nanoparticle-aptamer bioconjugates for cancer chemotherapy. *in vivo. Proc. Natl. Acad. Sci. U S A*, 103, 16, 6315–6320, 2006.

61. Özcan, I., Bouchemal, K., Segura-Sánchez, F., Effects of sterilization techniques on the PEGylated poly (γ-benzyl-L-glutamate) (PBLG) nanoparticles. *Acta Pharm. Sci.*, 11, 7945, 2009.
62. Bozdag, S., Dillen, K., Vandervoort, J., Ludwig, A., The effect of freeze-drying with different cryoprotectants and gamma-irradiation sterilization on the characteristics of ciprofloxacin HCl-loaded poly(D,L-lactide-glycolide) nanoparticles. *J. Pharm. Pharmacol.*, 57, 699–707, 2010.
63. Kempner, E.S., Effects of High-Energy Electrons and Gamma Rays Directly on Protein Molecules. *J. Pharm. Sci.*, 90, 1637–1646, 2001.
64. Nel, A., Xia, T., Mädler, L., Li, N., Toxic Potential of Materials at the Nanolevel. *Science*, 311, 622–627, 2006.
65. Song, Y., Li, X., Du, X., Exposure to nanoparticles is related to pleural effusion, pulmonary fibrosis and granuloma. *Eur. Respir. J.*, 34, 559–567, 2009.
66. Mühlebach, S., Regulatory challenges of nanomedicines and their follow-on versions: A generic or similar approach? *Adv. Drug Deliv. Rev.*, 131, 122–131, 2018.
67. Paradise, J., Regulating nanomedicine at the food and drug administration. *AMA J. Ethics*, 21, 4, E347–E355, 2019.
68. Soares, S., Sousa, J., Pais, A., Vitorino, C., Nanomedicine: Principles, properties, and regulatory issues. *Front. Chem.*, 6, 360, 2018.
69. Resnik, D.B. and Tinkle, S.S., Ethical issues in clinical trials involving nanomedicine. *Contemp. Clin. Trials*, 28, 4, 433–441, 2007.
70. Bawa, R. and McGee, S., The Ethical Dimensions of Nanomedicine. *Med. Clin. N. Am.*, 91, 881–887, 2007.
71. Hussaarts, L., Mühlebach, S. *et al.*, Equivalence of complex drug products: advances in and challenges for current regulatory frameworks. *Ann. N. Y. Acad. Sci.*, 1407, 1, 39–49, 2017.
72. Astier, A., Barton Pai, A. *et al.*, How to select a nanosimilar. *Ann. N. Y. Acad. Sci.*, 1407, 1, 50–62, 2017.
73. Schellekens, H., Stegemann, S. *et al.*, How to regulate nonbiological complex drugs (NBCD) and their follow-on versions: points to consider. *AAPS J.*, 16, 1, 15–21, 2014.
74. Tinkle, S., McNeil, S.E. *et al.*, Nanomedicines: addressing the scientific and regulatory gap. *Ann. N. Y. Acad. Sci.*, 1313, 35–56, 2014.
75. Li, Y. and Hou, T., Computational simulation of drug delivery at molecular level. *Curr. Med. Chem.*, 17, 4482–4491, 2014.
76. Dua, P. and Pistikopoulos, E.N., Modelling and control of drug delivery systems. *Comput. Chem. Eng.*, 29, 2290–2296, 2005.
77. Leelananda, S.P. and Linder, S., Computational methods in drug discovery. *Beilstein J. Org. Chem.*, 12, 2694–2718, 2016.
78. Sharma, S., Sharma, E., Sharma, Y., A review: recent computational approaches in medicinal chemistry: Computer aided drug designing and delivery. *Pharma Innov. J.*, 6, 5, 05–10, 2017.

79. Dahlman, J., Barnes, C. *et al.*, *in vivo* endothelial siRNA delivery using polymeric nanoparticles with low molecular weight. *Nat. Nanotechnol.*, 9, 8, 648–655, 2014.
80. Van Lehn, R.C. and Alexander-Katz, A., Free energy change for insertion of charged, Monolayer Protected Nanoparticles into Lipid Bilayers. *Soft Matter*, 10, 648, 2014.
81. Haddish-Berhane, N., Rickus, J.L., Haghighi, K., The role of multiscale computational approaches for rational design of conventional and nanoparticle oral drug delivery systems. *Int. J. Nanomed.*, 2, 3, 315–331, 2007.
82. Liu, Y., Shah, S., Tan, J., Computational Modeling of Nanoparticle Targeted Drug Delivery. *Rev. Nanosci. Nanotechnol.*, 1, 66–83, 2012.
83. Raina, N., Goyal, A.K., Pillai, C.R., Rath, G., Development and characterization of artemether loaded solid lipid nanoparticles. *Indian J. Pharm. Educ. Res.*, 47, 2, 123–128, 2013.
84. Duncan, G.A. and Bevan, M.A., Computational design of nanoparticle drug delivery systems for selective targeting. *Nanoscale*, 7, 37, 15332–15340, 2015.
85. Heikkila, E., Martinez-seara, H. *et al.*, Cationic Au nanoparticle binding with plasma membrane-like lipid bilayers: potential mechanism for spontaneous permeation to cells revealed by atomistic simulations. *J. Phys. Chem.*, 118, 111–131, 2014.
86. Mishra, A., Vuddanda, P.R., Singh, S., Intestinal lymphatic delivery of praziquantel by solid lipid nanoparticles: Formulation design, *in vitro* and *in vivo* studies. *J. Nanotechnol.*, 2014, 351693, 1–12, 2014.
87. Metwally, A. and Hathout, R., Computer-Assisted Drug Formulation Design: Novel Approach in Drug Delivery. *Mol. Pharmaceutics*, 12, 2800–2810, 2015.
88. Lu, B., Smith, T., Schmidt, J.J., Nanoparticle–lipid bilayer interactions studied with lipid bilayer arrays. *Nanoscale*, 7, 7858–7866l, 2015.
89. Sun, Y., Yu, B. *et al.*, Enhanced antitumor efficacy of vitamin E TPGS emulsified PLGA nanoparticles for delivery of paclitaxel. *Colloids Surf., B*, 123, 716–723, 2014.
90. Ramezanpour, M., Leung, S.S.W. *et al.*, Computational and experimental approaches for investigating nanoparticle-based drug delivery systems. *Biochim. Biophys. Acta*, 1858, 7 Pt B, 1688–1709, 2016.
91. Jämbeck, J., Eriksson, E. *et al.*, Molecular dynamics studies of liposomes as carriers for photosensitizing drugs: development, validation, and simulations with a coarse-grained model. *J. Chem. Theory Comput.*, 10, 5–13, 2014.
92. Giulia, R. and Luca, M., Gold nanoparticles in model biological membranes: A computational perspective. *Biochim. Biophys. Acta*, 1858, 10, 2380–2389, 2016.
93. Janke, J., Bennett, W., Tieleman, D., Oleic acid phase behavior from molecular dynamics simulations. *Langmuir*, 30, 10661–10667, 2014.

94. Hossein, M., Moom, S. *et al.*, Computational and experimental model of nano engineered drug delivery system for trabecular bone. 11th World Congress on Computational Mechanics (WCCM XI). *5th European Conference on Computational Mechanics (ECCM V), 6th European Conference on Computational Fluid Dynamics (ECFD VI) (n.d.)*
95. Cardoso, L., Fritton, S.P. *et al.*, Advances in assessment of bone porosity, permeability and interstitial fluid flow. *J. Biomech.*, 46, 2, 253–265, 2013.
96. Spencer, T.J., Hidalgo-Bastida, L.A., Cartmell, S.H., Halliday, I., Care, C.M., In silico multi-scale model of transport and dynamic seeding in a bone tissue engineering perfusion bioreactor. *Biotechnol. Bioeng.*, 110, 4, 1221–1230, 2013.
97. Chen, X.B., Lee, H.P., Chong, V.F., Wang, D.Y., A computational fluid dynamics model for drug delivery in a nasal cavity with inferior turbinate hypertrophy. *J. Aerosol Med. Pulm. Drug Deliv.*, 23, 5, 329–338, 2010.
98. Basu, S., Frank-Ito, D.O., Kimbell, J.S., On computational fluid dynamics models for sinonasal drug transport: Relevance of nozzle subtraction and nasal vestibular dilation. *Int. J. Numer. Method Biomed. Eng.*, 34, 4, e2946, 2018.
99. Inthavong, K., Tian, Z.F., Tu, J.Y., Yangb, W., Xue, C., Optimising nasal spray parameters for efficient drug delivery using computational fluid dynamics. *Comput. Biol. Med.*, 38, 713–726, 2008.
100. Cai, K., He, X. *et al.*, Dimeric drug polymeric nanoparticles with exceptionally high drug loading and quantitative loading efficiency. *J. Am. Chem. Soc*, 137, 10, 458–3461, 2015.
101. Davis, Y., Lai, Y., Wang, C., Mathematical modeling and simulation of drug release from microspheres: Implications to drug delivery systems. *Adv. Drug Deliv. Rev.*, 58, 1274–1325, 2006.
102. Ostwald, R., Bartel, T., Menzel, A., An energy-barrier-based computational micro-sphere model for phase-transformations interacting with plasticity. *Comput. Methods Appl. Mech. Eng.*, 293, 232–265, 2015.
103. Kim, O.V. and Dunn, P.F., A microsphere-surface impact model for implementation in computational fluid dynamics. *J. Aerosol Sci.*, 38, 5, 532–549, 2007.
104. Gajjar, D., Patel, R., Patel, V., Patel, P.K.M., Novel hydroxyl terminated dendrimers as potential drug carriers: sustained release, hemolysis and cytotoxicity study. *Int. J. Appl. Pharm.*, 7, 1, 5–9, 2015.
105. Pathak, P.O., NagarsenkerM, M.S. *et al.*, Cholesterol anchored arabinogalactan for asialoglycoprotein receptor targeting: synthesis, characterization, and proof of concept of hepatospecific delivery. *Carbohydr. Res.*, 408, 33–43, 2015.
106. Shah, S.M., Goel, P.N. *et al.*, Liposomes for targeting hepatocellular carcinoma: use of conjugated arabinogalactan as targeting ligand. *Int. J. Pharm.*, 477, 128–139, 2014.

107. Hu, M., Stanzione, F., Sum, A.K., Faller, R., Deserno, M., Design Principles for Nanoparticles Enveloped by a Polymer-Tethered Lipid Membrane. *ACS Nano*, 9, 10, 9942–9954, 2015.
108. Bunker, A., Magarkar, A., Viitala, T., Rational design of liposomal drug delivery systems, a review: Combined experimental and computational studies of lipid membranes, liposomes and their PEGylation. *Biochim. Biophys. Acta*, 1858, 2334–2352, 2016.
109. Wytrwal, M., Bednar, J., Nowakowska, M., Wydro, P., Kepczynski, M., Interactions of serum with polyelectrolyte-stabilized liposomes: cryo-TEM studies. *Biointerfaces Colloids Surf., B*, 120, 152–159, 2014.
110. Fuso, N., Morosi, L., Zucchetti, M., Ballerini, A., Intratumor heterogeneity and its impact on drug distribution and sensitivity. *Clin. Pharmacol. Ther.*, 96, 224–238, 2014.
111. Altrock, P.M. and Michor, F., The mathematics of cancer: integrating. quantitative models. *Nat. Rev. Cancer*, 15, 730–745, 2015.
112. Hubbard, M.E., Jove, M., Loadman, P.M., Phillips, R.M., Twelves, C.J., Smye, S.W., Drug delivery in a tumour cord model: A computational simulation. *R. Soc. Open. Sci.*, 4, 5, 170014, 2017.
113. Barbolosi, D., Ciccolini, J., Lacarelle, B., Barlési, F., André, N., Computational oncology: mathematical modelling of drug regimens for precision medicine. *Nat. Rev. Clin. Oncol.*, 13, 242–254, 2016.
114. Hao, S., Wang, B., Wang., Y., Density-dependent gastroretentive microparticles motion in human gastric emptying studied using computer simulation. *Eur. J. Pharm. Sci.*, 70, 72–81, 2015.
115. Hao, S., Wang, Y., Wang, B., Zou, Q., Zeng, H., Chen, X., Liu, X., Liu, J., Yu, S., A novel gastroretentive porous microparticle for anti-Helicobacter pylori therapy: preparation, *in vitro* and *in vivo* evaluation. *Int. J. Pharm.*, 463, 1, 10–21, 2014.
116. Mandal, U., Chatterjee, B., Senjoti, F.G., Gastro-retentive drug delivery systems and their *in vivo* success: A recent update. *Asian J. Pharm. Sci.*, 11, 575–584, 2016.
117. Zygourakis, K. and Markenscoff, P.A., Computer aided design of bioerodible devices with optimal release characteristics: a cellular automata Approach. *Biomaterials*, 17, 2, 125–135, 1996.
118. Siepmann, J. and Gopferich, A., Mathematical modeling of bioerodible, polymeric drug delivery systems. *Adv. Drug Deliv. Rev.*, 48, 229–247, 2001.
119. Huynh, L., Neale, C., Pomès, R., Allen, C., Computational approaches to the rational design of nanoemulsions, polymeric micelles, and dendrimers for drug delivery. *Nanomedicine*, 8, 20–36, 2012.
120. Martinho, N., Silva, L.C. *et al.*, Practical computational toolkits for dendrimers and dendrons structure design. *J. Comput.-Aided Mol. Des.*, 31, 9, 817–827, 2017.

121. He, X.C., Lin, M. *et al.*, Advances in studies of nanoparticle–biomembrane interactions. *Nanomedicine*, 10, 1, 121–141, 2015.
122. Tabari, S.H., Jamali, Y., Poursalehi, R., Multi-Scale simulation of carbon nanotubes interactions with cell membrane: DFT calculations and molecular dynamic simulation. *Proc. Mater. Sci.*, 11, 423–427, 2015.
123. Weller, M., Pfister, S.M. *et al.*, Molecular neuro-oncology in clinical practice: a new horizon. *Lancet Oncol.*, 14, 9, e370–e379, 2013.
124. Herwig, R., Computational modeling of drug response with applications to neuroscience. *Dialogues Clin. Neurosci.*, 16, 4, 465–477, 2014.

10

Utilization of Computational Methods for Rational Development of Nanoemulsions, Polymeric Micelles, and Dendrimers Drug Delivery Systems

Swarupanjali Padhi* **and Rupa Mazumder**

Department of Pharmaceutics, Noida Institute of Engineering and Technology, Pharmacy Institute, Greater Noida, India

Abstract

With the advent of sophisticated instruments and new polymers, various novel drug delivery systems (in the nanoscale) can be designed accurately to overcome the physicochemical constraints of a therapeutic agent, which restricts their bioavailability. This chapter focuses on the use of theoretical aspects and computer software programs facilitating high throughput *in silico* screening of the suitable excipient among a wide range of polymers, surfactants, solvents, oils, and many other excipients for designing drug delivery systems. Different theoretical aspects, *viz*, solubility parameter, Flory-Huggins theory, structure-based partition coefficient estimation, and prediction of different types of interactions between the excipients and the therapeutic agent by molecular docking and dynamic simulations, are used for the progress of current therapeutic agent delivery systems. This theoretical research is considered an alteration to the tedious and expensive practical approach. When these theoretical aspects are corroborated with computer programs, a wide range of excipients can be analyzed, and the scientist can screen out the optimal one. Nano preparations, like nanoemulsions, block-polymer micelles, dendrimers, and lipid nanoparticles, are widely using these computational approaches to select the best carriers based on solubility and interaction with drug moiety, which dictates the payload of the drug in these nanoparticles. This chapter elaborates on the concept of solubility parameters and the calculation of solubility parameters by applying the group distribution approach given

**Corresponding author*: swppadhi@gmail.com; ORCID: 0000-0003-2200-5542

Rishabha Malviya, Sonali Sundram and Dhanalekshmi Unnikrishnan Meenakshi (eds.)
Drug Delivery Systems Using Quantum Computing, (283–324) © 2024 Scrivener Publishing LLC

by Smalls, Fedros, and Hoftyzer/Van Krevelen. These calculations are made easier by employing computational segments through various computer software packages. These are highly user-responsive and easy to operate, determining solubility parameters more conveniently. Some of the software commonly used are the Hansen Solubility Parameters in Practice (HSPiP) package and Maple software Molecular Modeling Pro. Both carrier-oriented and drug-oriented screening approach using MD simulation and docking has also been elaborated on in this chapter.

Keywords: Solubility parameter, Flory-Huggins (FH) concept, fragmental method, MD simulation, artificial intelligence, docking, machine learning

10.1 Introduction

The major research work in pharmaceutics revolves around the development of reformulated dosage forms to have a precise supply of therapeutic substances toward the biological position to have maximum efficacy of the API. In recent times, various nanoparticulates and nanocarriers have contributed to a larger portion of the prevailing drug delivery system. This nano-formulation is beneficial owing to its small dimension that regulates the ADME and pharmacodynamics of the therapeutic agent. Hence, most of the work related to the fabrication of nanoparticles revolves around the research and selection of raw materials, especially the lipids and polymers which are the main game changers in this type of preparation; another aspect that has to be considered is their stability. The effective lipid and polymeric nanocarrier preparation depends on the selection of the same because the API payload and entrapment efficiency depend on the interaction between the polymers/lipids and the therapeutic agent. The challenge associated with designing these dosage forms is the screening and deciding on the optimal excipients as a carrier as the success of the dosage form relies on the retention of cargo in the same. Success in the screening of excipients depends on the use of computational methods that have evolved as the best alternative to the experimental work in screening and selecting the appropriate ingredient in the optimum amount. Development in the *in silico* technique has paved a new way in enhancing the fabrication of nano-formulation with reduced laboratory work and expenses [1–3].

In this chapter, the focus is given on the overview of various reported computational procedures and models that were employed to foresee the drug payload, partition coefficient, and stability of various nanocarriers, *viz*, polymeric nanoemulsions, polymeric micelles, and dendrimers. The discussion also includes the constancy of these models verified by the experimental result of the same. Many models are used to describe the correlation between API and lipids/polymers for formulating a variety of drug-carrier systems. The chapter also explains vividly the application of the various stated methods. Some researchers emphasize the theoretical aspect, *viz*, various solubility models, and some focus on the proceeding use of various deep learning, machine learning (ML) tools, and MD simulation models to study the interaction between the API and the adjuncts. Solubility models are frequently used to predict the loading of drugs in the carriers. The application of molecular docking and molecular dynamic (MD) simulation nano preparation has improved the potential for designing various novel formulations. This chapter includes technology from the classical theoretical approach to the use of modernized computational technology [4–7].

Most of the drugs are weak electrolytes and have problems associated with solubility and permeability. The researchers are focusing on the development of such dosage forms, which would counter the abovementioned problem. Nanoparticles due to their size and polymeric/lipid excipients have emerged as a relevant alternative to conventional dosage forms, resolving the solubility and permeability problem, augmenting the beneficial therapeutic effect of the drugs, and targeting the same to the specific site. The advancement in sophisticated instrumentation, the availability of a large number of lipids and polymers, and a lot of research have led to rapid advancement in the nanotechnology arena. Many theoretical along computational approaches have to be considered for the rational designing and development of optimized nanoparticles based on the preformation prediction of the physicochemical parameters of active pharmaceutical ingredients (APIs) and excipients [8–10]. Some of the important physicochemical parameters that are necessary to be predicted for designing drug delivery systems are as follows:

a. Compatibility analysis among the therapeutic agent and the excipient as it plays a decisive role in selection of the same.
b. Intermolecular interaction between drug and excipient influences the incrementing drug loading, and retention of the drug, resulting in improved chemical stability of the formulation.
c. Particle size and surface properties impacting the nano preparation stability during storage.

10.2 Recognizing Optimal Drug-Excipient Pairs

In the process of rational designing of drug formulation, the physicochemical properties like lipophilicity, solubility, and compatibility can be predicted to recognize the optimum drug-excipient pairs along with their interaction strength.

10.3 Emerging of Nano-Formulation Using Theoretic Technique

Theoretical aspects are employed to accompany experimental data in the progression of the discovery of therapeutic agents to the development of various drug delivery systems. However, experimentation is perennially viable compared to the theoretical approaches, but then, as a large number of ingredients are available, more combination of the same is possible, involving a large number of experiments, which are expensive and time-intensive and cannot explain the drug formulation at the molecular level. Hence, theoretical methods can be used to study drug formulation at the molecular level and suggest various drug-excipient modifications along with their physicochemical property analysis.

10.3.1 Theoretical Approach

These models characterize probable associations in an appropriate mathematical form utilizing the group contribution method, which leads to the summation of fragments of molecules in structure-based molecular descriptors to predict pertinently physicochemical properties, *viz*, lipophilicity and solubility parameters (SPs). These can be estimated very fast for the huge quantity of composites employing automated computer programs. The various type of computational approaches is shown in Figure 10.1. In the subsequent segment, we will discourse on the methods applicable to the prediction of optimized drug material pairs. Focus is given to the study of the theoretical cohesive energy (Ecoh) of the interacting molecules, foreseeing drug loading and its retention in the dosage forms.

10.3.1.1 Solubility Parameter

The pharmacological action of the drug depends on its aqueous solubility and can be obtained by the computational method. The SP value indicates the cohesive forces acting in the substance. If there is a similarity between the SP value of the drug and that of the excipient like the polymers/lipids, then they will have mutual solubility. Hence, SP can be employed to place the ingredients based on their comparative capabilities to be miscible with the investigated drug [1]. The SP concept applies to the selection of

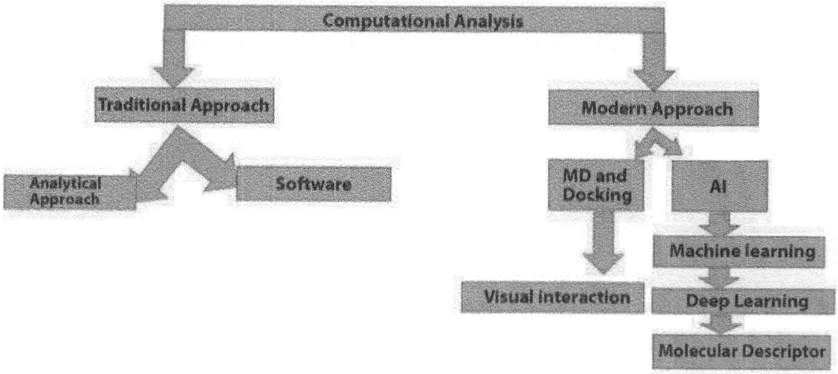

Figure 10.1 Different computational approaches.

solvents and non-solvents for polymers/lipids, required during the formulation process of nano preparation.

Ecoh of a constituent is expressed in terms of intensification in internal energy U per unit mole it. SP is expressed as the square root of Ecoh and can be calculated using Equation 10.1.

$$\delta = \left(\frac{E_{coh}}{V}\right)^{1/2} \tag{10.1}$$

Thermodynamic consideration was taken into account to express SP, and the key concepts of free energy, entropy, and enthalpy of mixing are used to explain the mutual solubility between two substances.

Hildebrand explained that, in the case of solvents, an association exists between the solubility and cohesive properties of the solvent. He projected SP as a term that explains the mutual interaction between the solvent and the polymer. According to the thermodynamic aspect, if two substances show mutual solubility, then ΔG_M, the value, should be negative. Hildebrand estimated the enthalpy of mixing by using Equations 10.2 to 10.6. ΔS_M should have a positive value and ΔH_M have limited positive value, and dissolution occurs beneath that point.

$$\Delta G_M = \Delta H_M - T\Delta S_M \tag{10.2}$$

where

ΔH_M = mixing enthalpy change on formation of the dual mixture, ΔS_M = entropy of mixing

$$\Delta G_M = \Delta H_M \text{-} T\Delta S_M = nRT\left[\frac{\varphi}{x} ln\varphi + (1-\varphi)\ln(1-\varphi) + \chi\varphi(1-\varphi)\right] \tag{10.3}$$

$$\Delta H_M = nRT\chi\varphi_1\varphi_2 = nRT\chi\varphi(1-\varphi) \tag{10.4}$$

$$\Delta H_{mix} = H_{12} - n_1 H_1 - n_2 H_2 \tag{10.5}$$

The solubility of the polymer with solvent or with any other substance depends on the structural similarity between both [7].

$$T\Delta S_M = nRT\left[\frac{\varphi}{x} ln\varphi + (1-\varphi)\ln(1-\varphi)\right] \tag{10.6}$$

n represents the number of lattice mole, φ_2 = volume fraction of substance 2, φ_1 = volume fraction of solvent (component 1), x = degree of polymerization, χ = Flory-Huggins interaction parameter, H_1= enthalpy of pure substance 1, H_2= enthalpy of pure substance 2, H_{12}= enthalpy of the binary blend of two pure substances 1 and 2.

It provided a high-speed testing means for the selection of lipid candidates. If $\Delta\delta_t$ (total SP of drug - total SP of lipids) is ≤ 5, then the drug would be miscible in the lipids to a greater significant level.

10.3.1.2 Assessment of Total SP

Hansen has explained the usages of total SP (δ_t) in the measurement of Ecoh of various ingredients (Hansen, 1967) [11]. The total SP comprises three factors.

$$\delta_t^2 = \delta_d^2 + \delta_p^2 + \delta_h^2 \tag{10.7}$$

where δ_t, δ_d, δ_p, and δ_h stand for total SP, partial SP related with dispersion forces, the polar force associated partial SP, and hydrogen bonding–associated partial SP, respectively. The chemical structure holds a significant part in establishing the mutual relationship between polymers and distinct solvents. Structural likeness favors solubility; hence, it is vital to have exhaustive knowledge about the chemical composition and the bonding of the substance. Researchers like Small (1953), Van Krevelen (1990), Fedors (1974), and others have recommended various approaches of group contributions to calculate various forces existing amidst the solvent and solute [12–14].

As an example, the Van Krevelen method of group contributions in the equations 10.8–10.10.

$$\delta_d = \frac{\Sigma F_d}{\Sigma V} \tag{10.8}$$

$$\delta_p = \frac{\sqrt{\Sigma F_p^2}}{\Sigma V} \tag{10.9}$$

$$\delta_h = \sqrt{\frac{\Sigma E_h}{\Sigma V}} \tag{10.10}$$

where F_d, F_p, and E_h stand for the group contribution for dispersion forces, polar forces, and hydrogen bond energy, respectively, and V represents molar volume.

Likewise, the mixing enthalpy (ΔH_M) per unit volume among composite 1 and composite 2 becomes

$$\Delta H_M = \varphi_1\varphi_2[(\delta_{d_1} - \delta_{d_2})^2 + (\delta_{p_1} - \delta_{p_2})^2 + (\delta_{h_1} - \delta_{h_2})^2] \tag{10.11}$$

where φ_1 and φ_2 represent volume fractions of element 1 and element 2, respectively.

The contribution of hydrogen bonding along with various polar relations is explained by the material's polarity (Xp).

$$X_p = 1 - \frac{\delta_d^2}{\delta_{total}^2} \tag{10.12}$$

The amount of energy essential for attaining solubility between the drug and lipids was estimated by calculating the mixing enthalpy (ΔH_M) between the drug and lipids. A single repeating monomer unit for a polymeric moiety was considered for calculation and lipids with a mixture of glycerides, the average of the main ratio was considered.

$\Delta\delta_t$, δ_t and X_p related to lipids and drugs were evaluated using Equations 10.7 to 10.12. The values of F_d, F_p, and E_h, and V of the definite functional moieties were referred from literature [11, 15–18].

SP of substance was estimated using Hildebrand and Hansen methods. According to the Hildebrand approach, the intermolecular attraction between the same molecules is the cohesion force. Any parameter used to calculate the Ecoh can be employed to determine SP, and the Hildebrand approach was best for the amorphous state. This approach was best applicable for nonpolar compounds as it considered the dispersion force only.

Another approach is Hansen SP which is based on the summation of various components of intermolecular force of attraction, *viz*, dispersion force as well as dipolar and hydrogen bonding component. This approach is applicable for predicting the miscibility of polar compounds, which can form hydrogen bonds or salt bridges. In Hansen SP, more data is required compared to that of Hildebrand SP. Two substances having an alike pattern of hydrogen bond and polar interaction are foreseen to be well-suited if the difference value in their Hildebrand SPs is <1.8 $cal^{1/2}cm^{-3/2}$. Compatibility among compounds can also be established depending on Hansen SP, if the variation of their Hansen SP ≤ Hansen, SP sphere radius [19].

Apart from the abovementioned method of direct calculation of SP, other indirect methods can be employed by which various other thermodynamic parameters can be calculated and then anticipate the compatibility between drug and polymers/lipids. Both Hansen and Hildebrand SPs can be utilized to calculate the theoretical value of enthalpies of mixing by considering the component's volume fractions in a mixture, and the same enthalpies of mixing can be applied to calculate the compatibility of a polymer/lipid or their hybrid nanoparticles [15]. The Hansen SP method was used by Li and co-researchers to find the enthalpies of verapamil-dextran sulfate conjugates with 15 lipids comprised of fatty acids, triglycerides, esters, and the mixture of the same. Referring to these calculated values, mono glycerol behenate and dodecanoic acid were recognized as the appropriate components for verapamil-dextran sulfate conjugate encapsulation. Furthermore, the theoretical value obtained was confirmed by corroborating with the experimental evaluated value, and based on the partition coefficient value, dodecanoic acid was selected [17].

10.3.1.3 Flory-Huggins (FH) Theory

The established Flory-Huggins (FH) theory expresses the thermodynamic performance of nonidealized polymer-solvent systems. It is a dimensionless parameter that suggests the association between the blend of polymer and a solvent expressed as enthalpy of mixing. If the value of X_{FH} is <0.5, then the compound is considered to be miscible. The X_{FH} value can be obtained experimentally by varying ΔH_{mix} related to the dual system [20].

$$\chi_{FH} = \frac{\Delta H_{mix}}{kTN_1\phi_2} \tag{10.13}$$

χ_{FH} = FH interaction parameter,
ΔH_{mix} = enthalpy change upon creation of a dual mixture,
k = Boltzmann constant,
T = absolute temperature,
N_1 = no of molecules of solvent, and
φ_2 = volume fraction of polymer.

$$\chi_{FH} = \frac{V_{ref}(\phi_1 CED_1 + \phi_2 CED_2 - CED_{12})}{RT} \tag{10.14}$$

$$\chi_{FH} = \frac{VA_{12}}{RT} + \beta \tag{10.15}$$

$$A_{12} = (\delta_1 - \delta_2)^2 \tag{10.16}$$

$$A_{12} = (\delta_{d1} - \delta_{d2})^2 + 0.25(\delta_{p1} - \delta_{p2})^2 + 0.25(\delta_{h1} - \delta_{h2})^2 \tag{10.17}$$

The value of B = 0 or 0.34 when Hansen and Hildebrand SPs are employed.

χ_{FH} = FH interaction parameter, V_{ref} = in the binary mixture the molar volume of the smaller molecule, φ_2 = volume fraction of substance 2, φ_1 = volume fraction of substance 1, and CED = Ecoh density. R = gas constant; T = absolute temperature; and δ_t, δ_d, δ_p, and δ_h stand for total SP, partial SP related with dispersion forces, the polar force associated partial SP, and hydrogen bonding–associated partial SP, respectively.

The X_{FH} value was calculated theoretically using MC [21] and MD [22] simulation, as given in Equation 10.13, or applying the group contribution method of Hildebrand and Hansen SPs, as given in Equations 10.14 to 10.17, though all the errors of concentration and all feasible conformation and hetero-interactions, independent SPs were proliferated to X_{FH}. It is important to consider these inaccuracies also because it was found from many studies that X_{FH} changed due to the significance of the proportion in which solute and solvent components are present in the mixtures. FH theory application has been newly expanded to foresee the miscibility among the drug and excipients. Dwan'Isa and his team estimated the value of X_{FH} for drug-polymer pairs using Hansen SPs. There was a methodical analysis of 19 APIs and 2 polymers consisting of hydrophilic PEG, and a copolymer of equimolar poly(ε-caprolactone) (hydrophobic) and trimethylene carbonate randomly dispersed. Micelles were formed using linear diblock copolymer MePEG-b-(PCL-co-TMC). The inner core was made of hydrophobic poly(ε-caprolactone) and trimethylene carbonate, and the outer hydrophilic shell was made up of MePEG. The X_{FH} value was calculated for the drug and hydrophobic core and again for drug and hydrophilic covering to foresee therapeutic agent payload in the above micelles. Created on root-sum-squared values of X_{FH}, rank order was determined. Experimentally therapeutic agent load was calculated for eight drugs (hydrophobic) with MePEG-b-(PCL-co-TMC) micelles. These drugs have aqueous solubilities ranging from 0.01 mg/mL to 1 mg/mL. Rank-order arrangement for three drugs, namely, indomethacin, cimetidine, and ketoprofen, exhibited good coordination among the theoretic and experimental outcomes. However, for the other five drugs, there were incorrect rank orders, which may be due to numerous reasons like the spontaneous collectively hypothetical score, restriction belonging to theoretical aspect associated with interfacial tension and rigidity of these drugs, and experimental conditions. From the outcome of the study, one point that is established is that values of X_{FH} are more important than only considering the hydrophobicity. This method is appropriate for designing libraries and screening and procurement of polymers [23].

Mahmud and coworkers did a similar type of study; they estimated the value of X_{FH} for cucurbitacin I (anticancer drug) and different aquaphobic principle-forming polymers, namely, PCL, poly(α-benzyl carboxylate-ε-caprolactone), and poly(α-cholesteryl carboxylate-ε-caprolactone) (PChCL) using Hansen SPs. Depending on the value of X_{FH}, poly(α-cholesteryl carboxylate-ε-caprolactone) was the utmost apposite for core-forming polymers, and PCL was less fitted for the same. This outcome showed good agreement with the investigational molar loading proportions of the therapeutic agent to central core-forming polymers. The drug loading was found to be 3, 8, and 15% in MePEO-b-PCL, MePEO-b-PBCL, and MePEO-b-PChCL micelles, respectively. Based on the negative value of X_{FH}, the researcher concluded that a more negative value indicates stronger inter-attraction and drug and excipient compatibility. And release rate is a kinetic property, hence cannot be predicted from X_{FH} as it is a thermodynamic parameter. The rate of uptake of the drug and release can be obtained from MD simulation [24].

Even though the FH model is a convenient tool, sometimes inconsistencies may be seen between the experimental data and theoretical foresee of interaction between drug and excipient. In the FH theory, it was postulation that the polymer segments are arbitrarily disseminated and there is a negligible attraction between the polymer and solvent. Hence, there will be no change in the volume of mixing when both polymer and solvent are mixed. The molecular weight (MW) of polymers cannot be determined using the FH theory. Therefore, to improve the prediction of drug-excipient interaction more accurately, the FH theory is required to be improved and extended further.

Mainly two approaches are considered:

i) Traditional Approach

a) Analytical Approach

In this method, both Hansen and Hildebrand SP approaches are employed in FH representations as a strong correlation exists among them (in the coming section below, we will discuss this in detail). For this calculation group contribution method, one of the commonly used analytical techniques to calculate SP was suggested by different scientists, *viz*, Fedros [14] and Hoftyzer/Van Krevelen. This method depends on the determination of cohesive force density, which considers various intermolecular forces existing between molecules, namely, polar, dispersion, and hydrogen bonds. Various chemical structures, functional groups, and chemical bonds were responsible for the presence of different forces [19, 25]. This can be better understood by using a few examples in which the researcher has used GCM-based SP calculation to theoretically predict numerous drug entrapments in nano-formulations fabricated by consuming various carriers. Ghitman and his coworkers determined the compatibility of the drug and polymeric matrix using the FH parameter. They determined the qualitative interaction between the therapeutic agent and polymer considering the thermodynamic parameter. The molar volume was determined employing the Fedors group contribution method [26].

$$R_g = \frac{R}{\sqrt{6}}, \text{ where } R \cong aN^2 \tag{10.18}$$

R_g = radius of twirling of rectilinear polymer, R = end-to-end distance, a = swelling exponent, and N = polymer's degree of polymerization.

$$\chi = \frac{(\delta_d - \delta_p)^2 v_d}{RT} \tag{10.19}$$

where δd and δp denote Hildebrand SPs of the drugs and polymeric matrix *v*d = the molar volume of the solubilized drug,

R = gas constant, and

T = temperature (Kelvin).

The Hansen SP was used to determine the SP of fatty acids from unsaturated and saturated oil structures, and the Barton compendium was used to determine the SP of fixed oil [27]. A gas chromatography technique was employed for the quantitative determination of fatty acids in vegetable oil. In the process of emulsification, oil also plays the role of HLB moderator. Sun and his co-researchers used methoxy polyethylene glycol-poly(*D*, *L*-lactide) (mPEG-PDLLA) as the polymeric carrier for drugs. They determined experimentally the drug loading efficiency of certain drugs in these polymeric micelles. However, these methods were found to be time-consuming. Therefore, they realized the need to adapt an alternative technology that could predict drug loading in much less time. The researchers used $mPEG_{5k}$-$PDLLA_{5k}$ as the polymer and estimated the loading of a series of drugs in it, using Hildebrand SPs, the *D* value considered from the 2D SP, and the FH interaction parameter (χ). They found that 2D SP obtained by the JB2013 group contribution technique was more suitable. When *D* value < 5.0 $(MJ/m^3)^{1/2}$, the solubility of the drug and $mPEG_{5k}$-$PDLLA_{5k}$ was satisfactory, and the drug loading competency was excellent. As the D value exceeded 8.0 $(MJ/m^3)^{1/2}$, there was no drug loading. Their research offered a qualitative foresee of the drug loading ability of $mPEG_{5k}$-$PDLLA_{5k}$ for water-insoluble drugs [28].

b) Software Approach

Presently, to ease the calculation of SP, various software packages are available that adapt the computational components to make the estimations faster. These are more user-responsive and cooperative, some examples of the software are the Hansen SP in Practice (HSPiP) package, Molecular

Modeling Pro, and Maple software. This software intends only to determine the calculated values and to predict the theoretical SP and does not play any role in improving or modifying its model and has other limitations while dealing with high polar and large molecules. Makoni and co-researchers considered formulating SLN using various solid lipids like Geleol, compritol ATO, stearic acid, perciol, cetyl palmitate, and five drugs, namely, minocycline HCl, didanosine, efavirenz, clarithromycin, and momelasone furoate. Preceding preparing optimized SLN, they calculated the SP difference between the investigated lipids and drug using Hiroshi Yamamoto's molecular breaking method in HSPiP software. They confirm the same even by performing relevant experiments. They suggested that some experimental data corroborated well with the SP expected by the software. A problem was encountered while dealing with large molecular structure compounds like clarithromycin. The researchers also concluded that only simple SP analysis will not give accurate results in selecting suitable lipids for a particular drug [29].

ii) Modern Approach

a) Computational Approach

The classical depiction of various solubility models required more advancement in the models to predict a better, clearer, and more reliable analysis; hence to achieve this, more integrating and accurate technology has to be developed. More complex modeling techniques were introduced to increase the ability of these solubility models. Models were developed using the concept of molecular docking and MD simulation calculations [30]. These MD models deal with both the drugs and the carriers dynamically with respect to time, and they involve the calculation of the satisfactory traveling of atoms along with the rotation, bending, and stretching of flexible bonds [31]. In MD simulations, the representative carrier molecule is not taken as a single entity, and the structures are repeatedly scanned considering all the manacles of the building blocks to get a 3D structure of interaction and energy-diminished carrier molecules. These carrier molecules are considered for nano preparation. This change in volume difference together with mobility concerns provides for equitable interactions between neighbors. Binding energy (G), among a drug and its carrier, can be determined by employing docking depending on the total volume and geometry of the carrier, considering all probable drug alignments in the interior of the transporter and showing precise material interactions. Patel and a coworker designed an initial relative experiment employing two methods depending on the abovementioned result to estimate the

FH interaction among two therapeutic agents and sequence of PEO-PCL copolymer (having various MWs); the method used were conventional methods that is GCM, and another method used was the latest developed MD simulation. For this study, various sources like Studio (MS) software were employed to study the MD simulation, applying to both the drugs and polymers in pure constituent form. The resulting route files were utilized to extract all molecular groupings CED and calculate the FH interaction parameter using Equation 10.19. The experimental validation was coherent with the MD simulation outcomes, as opposed to the analytical GCM, describing obtaining contradictory responses that examined the solubility of nimodipine and fenofibrate in liquid caprolactone.

The accuracy of contemporary techniques in predicting miscibility and FH interaction factors was supported by these data. Meunier *et al.* [32] employed the FH interaction factors to forecast the PLA core of PLA-PEG's loading capacity of anticancer medicines on nanoparticle formulation employing the same computational approach. Despite the author's dissatisfaction, a reasonably decent exponential association existed among the theoretical and investigational data acquired. The reason for this was more scattering in the association related to hydrophilic chemicals leading (less entrapment), particularly for the molecule (outlier), which had a larger XH value than expected. Alternative ways he proposed include more applicable algorithms capable of providing information regarding the potential energy of the comprehensive drug-carrier combination, as well as its pure components like mixing. This hypothesis was tested by Macháčková and his coworker [33]. They used modeling algorithms to assess the miscibility of cyclosporine-A with various polymers virtually designed, this was done employing the MS software's blends module, which works on a docking technology developed on the generalization principle of the FH theory. The mixing energy was obtained after estimating the energy and used in XH by substituting the numerator with Emix in Equations 10.2 to 10.6 and the XH value showed that both chitosan and cellulose were the best carriers for the model drug. The computational outcomes and the previous investigational results referred to by the researchers were in accord with one another. Always remember that those computational means are not a substitute for experiments, computational method facilitates or speeds up the process of selection of the set of optimized polymers which are finally further shortlisted using experimental methods [34]. Experimental methods are applied to authenticate the results obtained by the simulation results. Hence, these two approaches are complementary to each other.

10.4 Influence of Rigidity, Conformation, and Compatibility on Drug Loading

Along with solubility, other physical properties also influence the entrapment efficiency of the hydrophobic therapeutic agent and these parameters are conformation, rigidity, and the MW of the API and the excipient. Analysis of the abovementioned parameter can be done appropriately using the molecular simulation method. These methods can be used as an alternative technique to the analytical approach and help better understand the basic interfacial structure of drug-excipient or excipient-excipient pairs. It also discloses the vital relationship between the drug and excipients, which overrides their assemblage, and explains the physicochemical properties that can be altered accordingly to have a better effect on drug encapsulation and release.

Moreover, MD simulation can also be employed to estimate Hildebrand SPs but the same cannot be utilized in calculating Hansen SP. Molecular simulations can assess the conformations of both the individual components and their binary mixture. Researchers found that a simulation analysis revealed that the solubilities foresaw by MD simulations of an antitumor API, docetaxel, in different adjuncts, namely, vitamin E, β-caryophyllene, and triglycerides, showed that corroborate with the experimental results. For the estimation of χFH, the intermolecular interaction existing between the drug and excipient was used and the energy of mixing was found from simulation, as given in Equations 10.2 to 10.7. The range of deviation between the computational prediction and experimental result of solubilities of docetaxel was between 2 and 15% in triglycerides and vitamins. However, this simulation technique fails to provide the exact result of solubilities between the drug and β-caryophyllene and cannot be used for quantitative analysis. However, the rank order of solubility prediction was precise, which was enough for library upgradation. The imprecision of the foreseen solubility of docetaxel in β-caryophyllene can be attributed to the fact that the mixing ratio was assumed to be χFH = 0.5. It has been observed that with an increase in MW, the discrepancy between the SP value found by group contribution, and the simulation method increases, this signifies that solubility is dependent on the conformation and molecular arrangement [35–37].

10.5 MD Simulations and Docking

For accurate results, simulation and docking methods can be employed to develop perceptible prediction models. In the current scenario, these methods have emerged as an important tool used in modeling nanocarriers. These methods act as a suggestive tool that helps in the pre-formulation studies in screening the polymers/lipids for better and optimized formulation design.

10.5.1 Molecular Dynamics (MD)

MD simulation employed in many distinct scientific domains since its introduction more than 40 years ago [38] and in certain situations, it has become another tool with which to study a particular topic. The essential theories underpinning traditional MD simulations are briefly summarized in this chapter, after which we show the numerous methodologies and algorithms that even a non-skilled operator in a pharma research test center might employ to know various elements of miscibility. Although numerous outstanding assessments on both scientific and usage features of MD [39] are available, knowledge of what can and cannot be done with computational approaches for researching solubility (and why) is still lacking in the pharmaceutical community, in our perspective. To focus even more, we will only consider traditional, particle-based MD simulations with some of the conventional biomolecular forcefields.

MD is a computer imitation technique employed to analyze the movement of atoms and molecules with respect to a given fixed time to view the interaction between the drug and the excipient. Here, molecular mechanic force fields (FFs) or interatomic potentials were used to estimate the interaction between the particles (Figure 10.2). MD can be used in a variety of fields like biophysics, chemical physics, and material science. As it has to consider various molecular systems consisting of a large number of particles, the MD simulation employs various numerical methods, but these are complex systems and it is difficult to get error-free results. The error can be minimized by selecting accurate algorithms and parameters. If the system follows the ergodic hypothesis, the MD simulation can be employed to find the thermodynamic properties of the system. MD simulation provides an anticipated molecular motion on an atomic scale with time.

Molecular simulation is employed to inspect solubility along with the intra and intermolecular forces of a given system that requires considering the FF. Van der Waals and electrostatic forces were responsible for

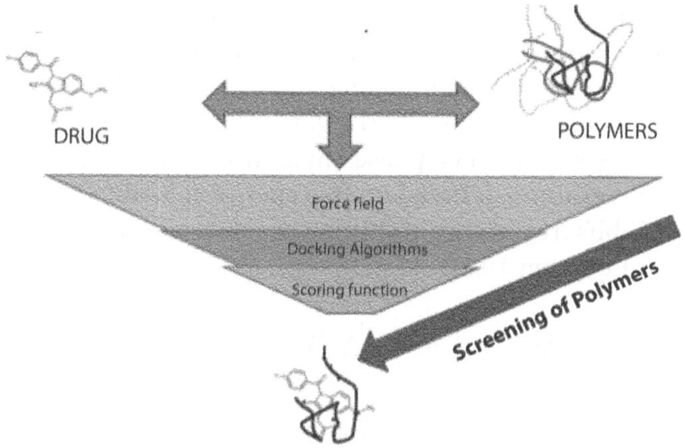

Figure 10.2 Use of molecular docking for selection of suitable polymer.

the interaction of the atoms and molecules inside an FF. Furthermore, the atoms held covalently are linked together by robust binding forces within each molecule. A representative biomolecular FF calculates each of these forces per atom and then adds them all up.

In a simulation or computational experiment, the FF specifies the physical interactions among atoms and molecules. The significant component is the utilization of these forces, which act between the atoms to move them, accomplished by solving Newton's motion equations with numerical methods. In practice, the forces are considered by calculating the gradient of each atom's interaction potentials. It is possible to get each atom's location in the system at a later time step using this approach; fresh forces were again estimated and employed to revise positions a second time, and it is continued likewise. By connecting each state to the preceding one, the time advancement of the entire approach can be traced. Numerous software packages are available to carry out the abovementioned types of simulations. Some of them include Amber, NAMD, HARvard Molecular Mechanics (CHARMM), ACEMD, Desmond, and GROningen MAchine for Chemical Simulations (Gromacs).

10.6 Solubility

The following focuses on the miscibility of a compound in a solvent with unlimited/solubility limit, where the solvent is present in such abundance that the properties of the solute are not significantly affected by changes in solvent quantity. Computationally, solubility at this boundary means that the absorption procedure may be replicated exclusively by taking into account any interactions among the solutes. Characteristically, the transmission of a solute through (i) a gas and liquid phase (solvation-free energy), (ii) a gas and water phase (hydration-free energy), or (iii) two liquid phases is examined. Based on free energy, solubility is defined as the variation in Gibb's free energy of shifting of solute among two phases α and β, as given in Equation 10.20.

$$\Delta G_i^{\alpha\beta}(T, p, X^\alpha, X^\beta) = -RT \ln K_i^{\alpha\beta} \tag{10.20}$$

$\Delta G_i^{\alpha\beta}$ = change Gibb's free energy of shifting of solute among two phases α and β,

X = composition at equilibrium,

R = gas constant,

p = pressure,

T = temperature,

K = partition coefficient is the ratio of the number of densities of solute i and phases.

The K values were discovered using the two techniques. For a variety of reasons, though, simulation computation of K is normally unreasonable. For instance, it might not be able to accurately collect the solute in the given set of two distinct phases on the way to obtain sufficient data to draw inferences confidently. As a result, many methods involving the transfer of free energy that can be employed to calculate solubility have been established.

For determining the difference in free energy by simulations, four potential approaches are frequently offered. These include (i) histogram-based techniques, (ii) nonequilibrium simulations, (iii) perturbation theory, and (iv) potential derivatives [38].

$$\Delta F_{ij} = -k_B T \, ln\left(\frac{Z_j}{Z_i}\right) \tag{10.21}$$

The Equation 10.21 shows the use of partitioning function to describe the free energy difference between two states, Zi and Zj are denoted as the statistical mechanic partition functions of the two states of interest, and kB represents the Boltzmann constant, and T temperature. Although in theory, these partition functions would give a comprehensive description of a system, they are frequently unknown, which is why it is challenging to determine the absolute free energies of real systems using computer simulations. These two states could, for instance, equate to an active pharmaceutical molecule being present or absent from a simulated box full of solvent molecules in terms of drug solubility [39, 40].

10.6.1 Free Energy Perturbation Calculations [41–44]

Free energy perturbation (FEP) is a technique built on the concepts of an arithmetic mechanic. FEP was utilized in computational science for the calculation of free energy from MD. FEP was first introduced by Robert W. Zwanzig in 1954. The FEP method was mainly implemented to calculate the free energy difference among two distinct systems, and the free energy can be correlated to physical properties like partition coefficient and solubility. These systems consist of an initial unperturbed system and a perturbed system. An example of a perturbed system includes the addition of extra particles or the creation of different energy functions. In these conditions, FEP can be employed to estimate the solvation-free energies. The free energy difference between the two systems can be estimated using Equation 10.22.

$$\Delta F_{ij} = -k_B T \, ln\left(e^{-\frac{E_j - E_i}{k_B T}}\right) \tag{10.22}$$

Ei and Ej are the energies related to two dissimilar states of the approach.

10.6.2 Widom Insertion Method [45]

The remaining chemical potential in the established group, where the total count of particles is kept fixed in the scheme, according to Widom is calculated using Equation 10.23.

$$\mu_{ex} = -k_B T \, ln\left(e^{-\varphi/kT}\right)_N \tag{10.23}$$

where kB = Boltzmann constant and
T = temperature of the system.

10.7 Screening

Many studies have confirmed that in pre-formulation carriers importance of MD simulations and docking are utilized to assess and select potential medications with high loading capacity for specific carriers. With reliable estimated outcomes that compare reasonably fine with investigational data, this technique gets up the entire formulation procedure incredibly time and expense-effective.

10.7.1 Carriers-Oriented Screening

Several carriers were considered as the investigational carriers for a particular drug and then the simulation and docking programs were running. According to the screening report, the best carrier was selected for the particular drug.

For instance, Yadav *et al.* [34] used MD simulation and docking to screen curcumin against five polymers. Chitosan surpassed other examined polymers with an exceptional ΔG model, according to observations. As a result, chitosan was chosen for the study's experimental design.

This was not unexpected, given that Dhanasekaran *et al.* [46] had previously reported similar findings. In chitosan c.f. chitin, docking studies revealed a lower ΔG. This was in line with the investigational conclusions, revealing the greatest ability of a carrier to carry medicine, i.e., around 100%. Likewise, Aparna *et al.* [47] for amphotericin B (AmB) delivery selected gelatin as a polymeric carrier based on an *in silico* result. Gelatin had the best docking score for AmB out of the eight amino functionalized polymers tested. The investigational EE and loading of AmB dosage in gelatin nano-formulation showed promising outcomes.

As aforementioned to the development of fluticasone nano-formulation (stable), Ahmed *et al.* [48] evaluated fluticasone against several polymers. The simulation and docking of MD models were completed. Compared to other simulated polymeric structures, PVP-stabilized PVP and HPMC-stabilized Eudragit complexes had the best ΔG findings (−35.22 and −25.17 kcal/mol, respectively). The EE percentage was >90% when these nanoparticles were prepared, supporting the *in silico* optimization.

In addition, Altintas *et al.* [49] used *in silico* screening to screen a specific E. coli endotoxin structure critical to 21 monomer building blocks and widely employed the creativity of molecularly embossed polymeric and nanoparticles (nanoMIPs). Itaconic acid, methacrylic acid, and acrylamide were chosen as the best candidates, with ΔG values of −52.24, −41.43, and

−39.87 kcal/mol, respectively. Surface plasmon resonance (SPR) signaling demonstrated that itaconic acid, methacrylic acid, and acrylamide have affinity patterns of 99.5 RU, 66.4 RU, and 35.6 RU, respectively. The results of the experimental investigation of nanoMIPs for endotoxin detection by SPR sensor were consistent with the computational modeling experiments and yielded similar patterns.

10.7.2 Drugs-Oriented Screening

In this case, the focus is given to studying the affinity of various drugs to a single carrier.

The established combined formulation of curcumin, paclitaxel, and vitamin D3 in PEG-tyrosine-derived polyacrylates-PEG carrier was tested using both MD simulations and docking calculations [50]. The maximal drug loading as calculated empirically was compared with the computationally generated ΔG findings for the three medicines utilized, and a significant linear connection (R2 = 0.93) was found.

Gayathri *et al.* [51] and Geetha *et al.* [52] studied the entrapment of antibacterial and anticancer drugs, respectively, in chitin nanocarriers in additional studies. The same APIs were docked, employing the identical chitin nanocarriers that were modeled. The articles, the values of DL, the three antibacterial medications (8.9, 5.6 and 3.5%), and the three anticancer drugs (3, 2, and 1%, respectively) surveyed an analogous tendency when compared to ΔG data (−8.1, −7.3, and −5.1 kcal/mol and −6.9, −6.5, and −5.4 kcal/mol). Antibacterial medications had an R2 of 0.85 and anticancer drugs had an R2 of 0.93, indicating outstanding associations between computational and experimental results. Nosrat M. Mahani [53] investigated the capacity of PLGA to accommodate nine powerful but unstable thiazoline derivatives with various alkyl and aryl substitutions in the same environment. All derivatives had negative binding energies, indicating good interaction with PLGA, according to the hybrid ONIOM calculation. In derivatives 3 and 7, we substitute R = CH3 and Ar = 1-methyl-1H-tetrazol-5-yl or 1-phenyl-1H-tetrazol-5-yl or 1-phenyl-1H-tetrazol-5-yl or 1-phenyl-1H-tetrazol-5-yl or 1-phenyl-1H-tetrazol-5-yl or 1-phenyl-1H-tetrazol-5. As a result, these moieties were chosen as the best candidates for creating stable, biologically active antimicrobial/anticancer preparations. When it comes to predicting the most steady stoichiometric insertion complex of naringenin with cyclodextrin, Tais *et al.* used MD simulations and docking investigations, with the determination to increase the solubility and dissolution of the water-insoluble therapeutic agent by enclosing it inside CD with minimal rate of dissociation.

Hathout *et al.* [54] used gelatin as the primary matrix in another example of multidrug modeling. The gelatin matrix simulated around 10 APIs was docked by MD, and the resulting ΔG data were compared to the loaded medicines' experimental masses.

Metwally and Hathout utilized two carriers, namely, tripalmitin and PLGA, to model the loading of 21 APIs to design nanocarrier systems employing MD simulation [55]. The PLGA system was used as a carrier for 10 APIs, while tripalmitin was used for 11 APIs. A clear relationship between the tripalmitin and PLGA loaded API docking ΔG data and for simulation CGenFF in GROMACs software and docking Argust Lab and AutoDock Vina was used for the preparation of SLN using tripalmitin and PN using PLGA. The researchers validated the consistency of the fabricated model using curcumin, and the drug loading was found to be 0.75 and 0.95 mg in 100-mg tripalmitin and PLGA, respectively, with exponential $R^2 = 0.87$ and 0.9 in SLN and PN, respectively, and % bias in loading was 12 and 2.03 in SLN and PN, respectively.

The effectiveness of modeling is also significantly impacted by various simulation techniques. Sizochenko and Leszczynski [56] used PLGA 50:50 as a carrier and AutoDock Vina for docking in a thorough analysis of 21 authorized medicines. Between the data on drug loading taken from the literature and the resulting G data, an R2 = 0.36 (R = 0.6) connection was found. This recorded result was significantly lower when linked to the correlation coefficient value in the Metwally and Hathout study (0.9) [54], which employed a different simulated FF but the same carrier type and docking procedure. An FF is what determines the default values for physical interactions, bond lengths, partial charges, and any force operating on or among atoms [40]. It specifies the underlying physics of the simulated system. Therefore, the accuracy and dependability of the modeled systems will depend on the level of the FF. Figure 10.3 provides a summary of these methods.

10.7.3 Interactions Imagining

In conjunction with the showing and carrier-choosing activities, simulations and docking are employed to provide insight regarding the method of binding processes. Hence, making it possible to view and evaluate the relationship between the API and the carrier at atomic and molecular level [57]. It was shown in the following instances how ΔG has been linked to various interactions in multiple types of research and how this relationship might help comprehend sudden loading conditions of certain chemicals in various carriers.

Brunacci *et al.* [58] used MD simulation using MD (AMBER FF in Ascalaph designer software), docking (AutoDock Vina), and interaction visualization (AutoDock visualizer tool) to illustrate probable relationships of dexamethasone in two polymers, PLGA 50:50 and oligobutyl morpholinediol (OBMD). They discovered that the amide group in OBMD serves as a hydrogen donor, allowing other hydrogen acceptor groups to function effectively. With one molecule of dexamethasone molecule, a single OBMD unit was able to generate three HBs.

A PLGA unit containing solely hydrogen acceptors and its ester groups, on the other hand, structures two HBs with dexamethasone. Following that, a molecular docking study revealed that dexamethasone had a better G with OBMD than with PLGA. The investigational validation outcomes, indicated by EE and DL, demonstrated that OBMD outperforms PLGA in the formation of dexamethasone-loaded polymeric nanoparticles.

Additional structural factors, such as largeness, tractability/stiffness, and even API isomerism, were able to all result in a substantial calculation of ΔG. The previous perfect ranking by Costache *et al.* [50] was attributable to the binding interactions as well to the drug largeness and flexibility. The three model medications' dimensions and elasticity were used to determine their orientation and, as a consequence, the drug tightly binds in the nanocarrier interior. Paclitaxel could in theory create 18 HBs and a lot of π-π contacts, but due to its huge size, the majority of them were intramolecular. This resulted in paclitaxel having a more rigid conformation, which allowed for increased surface binding and hence a lower DL. Hence, it is evident that physical interaction cannot be used alone to explain the binding affinity of nanosystems.

10.8 Artificial Intelligence and Machine Learning

The artificial intelligence (AI) process can simulate humanoid intellect in robots so that they can think and act like humans. These characteristics of AI provide it a unique capability to reason out leading to the execution of events, which are most likely to achieve a certain goal. AI deals with technologies that can do assignments like logic, reasoning, planning, learning, and perception [59]. It entails deploying computer-like technologies to imitate just the most basic human cognitive abilities using a variety of algorithms. According to Wilson and Km [60], AI may be used to do a variety of tasks, including understanding optimum explanations for

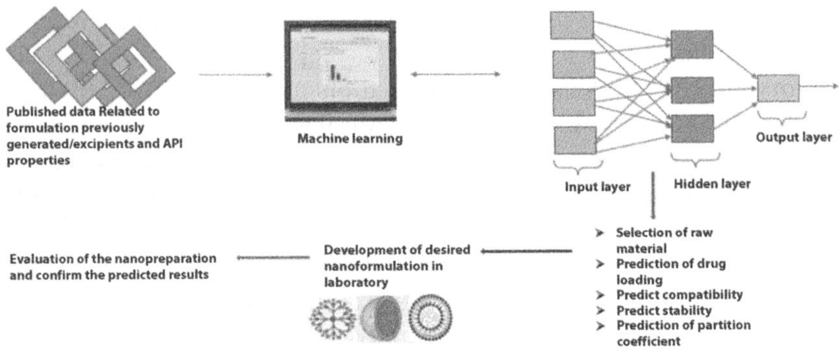

Figure 10.3 Use of machine learning in developing nano-formulation.

issues, detecting patterns, and even classifying data [61]. AI can assist in situation analysis by emulating the human mind in terms of understanding and analysis, and hence aid in problem-solving, a thought known as "machine learning" [62]. Artificial neural networks (ANNs) are theoretical frameworks specifically created for the execution of AI algorithms. They fully replicate the human brain as a network of interlinked neurons with weighted communication pathways between neurons (Figure 10.3) [62].

ML, both supervised and unsupervised, is a significant advancement in this subject. In terms of first, supervised ML is a branch of AI that can understand and model the correlation among a collection of independent variables and responses. As a result, a feature mapping equation can be used to forecast the results of additional data [63, 64]. Molecular descriptors and/or formulation characteristics could be supervised by ML model inputs (independent variables) in the realm of medicines and carriers.

Molecular descriptors are extensive numbers based on cheminformatics that aggregate hundreds of a compound's structural features [65]. It includes everything from simple bulk 1D data (polarizability, MW, partition coefficient, and so on) to 2D and 3D parameters (pharmacophoric points and molecular fingerprints). In general, ML techniques are less computationally costly than a combination of MD and docking. In this situation, ML systems can deliver outcomes that are either complimentary to or analogous to those obtained by MD and docking.

10.9 Partition Coefficient

The partition coefficient measures the lipophilicity of the therapeutic agents, it is a ratio of the concentration of a substance in the biphasic organic and aqueous phase at equilibrium. Computational methods are employed to estimate the partition coefficient in various systems, *viz*, octanol-water and hexadecane-water, which resemble the cell membrane of GIT and CNS respectively. The partition coefficient of the dosage form gives an understanding of the lipophilicity of the form as well as the permeability of the therapeutic agent through various biological membranes. The partition coefficient is expressed in terms of the logarithm (logP) [7]. Mannhold and Rekker established the group contribution theoretical approach for the first time and named it the fragmentation technique [66, 67]. The molecular structure, its topology can be used to estimate the various physiochemical parameters like logP using Equation 10.24.

Almost in all the micellar formulations, to have a better understanding of the integration of therapeutic agents in the micelle and external solution, the partition coefficient of the drug has to be calculated. Even the improvement in permeability of polymeric micelles across the membrane can be quantified by the partition coefficient concept. Various theoretical, as well as computational methods, have been projected to estimate the partition coefficient of drugs in polymeric micelles and for other nano preparation.

$$\log P = \sum_{i=0}^{n} a_i f_i + \sum_{i=0}^{M} k_j C_m \tag{10.24}$$

P = partition coefficient,
N = functional groups of the molecule,
f_i = hydrophobic fragmental constant,
a_i = incidence of functional group,
C_m = the correction factor ($C_m = 0.219$), and
k_j = the frequency of C_m.

LogP value can be assessed by employing the group contribution method. The value of logP is associated with the lipophilicity of the drug. Different techniques were established to calculate partition coefficients referring to the chemical structure of the therapeutic agent or the micelles. Quantitative structure–property relationships (QSPRs) is a popular method that is continuously progressing and is based on several algorithms and software programs related to QSPRs, *viz*, AlogPs and ADMET predictor, ACD/logD [68–70]. The ability of a prospective drug to accumulate inside the micelles can be obtained by logP, which is correlated with the thermodynamic aspect of the drug.

Alves and coworkers (2019) described the use of chemo-informatics in the rational designing of polymeric micelles [71]. A sequence of QSPR modeling approaches employing a genetic function approximation algorithm was implemented by Wu and his team to evaluate the loading capacity of polymeric micelles related to doxorubicin [72].

10.9.1 Prediction of Partition Coefficient Using Density Functional Theory (DFT)

DFT is a computational modeling technique employed to determine the physical properties of many-body systems at molecular and atomic levels using functionals of the electron cloud.

Out of various approaches, the use of quantum mechanic calculations and fragment approach is the most popular. The partition coefficient is estimated by calculating the transfer of free energy of molecules among two phases, whereas in the fragmentation method, the summation of quantities equivalent to individual constitutional molecular fragments (fragment constant) is considered. The main drawback of the method is that these are considered for some specific solvent pairs; hence this limits its application for other solvents. Saranjam and his coworkers used computational methods to predict logP, considering micelles systems by employing DFT in pure solvents. The DFT method is economical and time-saving. The researcher studied the partitioning behavior of 63 molecules and used single conformation for all the compounds; however, they used extended conformation for flexible molecules. DFT with B3LYP and M06-2X methods with 6-31++G ** basis set along with solvation method based on density was employed to optimize geometrically all compounds. SMD is a universal solvation model, which can be used for all types of solutes in all types of solvents.

The solvation-free energy of this model is comprised of the electrostatic part of the bulk and the cavity dispersion part. The advantage of this method is that it uses the DFT method and applies to any molecule, hence it is a universal method. The electronic structure program, Gaussian 16, was used for calculation, and free sign-stage molecule editor Avogadro was employed to get the initial structures of the compounds. For calculation, the researchers considered a temperature of 298.15 K and a pressure of 1 atm.

The partition coefficient of a drug is associated with the transfer of Gibbs free energy among the water and a particular phase (immiscible with water). Free energy was calculated by measuring absolute solvation energy in various investigated media by Equations 10.25 and 10.26.

$$\Delta G^0_{solv/wat} = \Delta G^0_{solvent} - \Delta G^0_{water} \tag{10.25}$$

$$logP = \frac{-\Delta G^0_{solv/wat}}{RT\ ln10} \tag{10.26}$$

The researchers compared the results of computationally projected logP and investigational partition coefficients in octanol/water using both functionals, *viz*, the B3LYP and M06-2X. These statistical analyses helped the researchers to develop the computation system for finding logP by correlating experimental values of particular micelles with various partition coefficients in particular solvent/water systems having similarity in physicochemical/polarity utilizing the DFT method [73–78].

10.9.2 Prediction of Partition Coefficient Using MD Simulations

Already, we have introduced MD simulation in predicting the solubility and drug loading capacity of micelles. Sandra Storm and his team investigated the partition coefficient of different solutes in self-assembled micelles of cationic (cetyltrimethylammonium bromide, CTAB) and anionic surfactants (sodium dodecyl sulfate, SDS) and water phase using MD simulations. They also investigated micellar size, density profiles, and shape of the micelles. A detailed discussion of this research work would enable the readers to investigate the logP parameter of various drugs in micelles of different types of surfactant taken in various concentrations in an aqueous medium. With the help of the COSMOmic software, the 3D structure obtained from MD simulations could be employed to predict the free energy summary and partition coefficient of micelles made of different solutes. The researcher also designed a comparative study between COSMO-RS and COSMOmic [1–3]. COSMOmic is an upgraded version of COSMO-RS. They also considered the pseudo-phase method considering the micellar phase and water phase. Both phases were assumed to be homogeneous; therefore, anisotropy was neglected. The purpose of this research was to establish the understanding of COSMOmic concerning the input of diverse micellar structures by MD simulation. All-atom MD simulations were employed to examine the self-aggregation in an orderly way of the surface active agent [77, 79–81].

The researchers selected the open-source GROMACS 4.5.3 MD package. The CHARMM36 FF has a great parametrization of amphiphilic molecules, especially lipids. The parameterization of SDS was taken from the previous data obtained by the FF CHARMM36 and was implemented without modification. CTAB parameters were obtained from associated lipid structures (DSPE and LPPC), and to obtain charge neutrality, Na^{+2}, and Br^{-2} were also taken into consideration [82–84]. The researcher placed a fixed concentration of surfactant monomer in the cubic box to get a preliminary configuration for the MD simulation employing the tool genbox of GROMACS. Followed by adding water. With the help steepest descent, the energies of the preliminary conformations can be diminished. The researcher first starts with short simulation in the NVT collaborative for 600 ps. Then, followed by the final simulation that was achieved in the NPT collaborative at 1-bar pressure with 298-K temperature, applying Parrinello-Rahman barostat and Nose-Hoover thermostat for 0.002 ps. An actual space cutoff of 1 nm was applied along with the particle mesh Ewald method, and the scientist studied and selected the long-ranged electrostatic interactions. The LINCS algorithm was utilized to compel the bonds attached to hydrogen atoms and used CHARMM36 FF.

After MD simulation, the selected structures are fed into the COSMOmic model, assuming that the micelle exists as a bilayer layered liquid having a fixed composition, hence the layers are expected to be homogeneous. Thus, the atomic distribution of micelle/water is necessary to be obtained from MD simulation or experimental method. In the COSMOmic model, molecular interaction was given in terms of polarization densities of the surface. The researcher then calculated the free energies and chemical potentials depending on their interaction at the surface segment using Equation 10.27.

$$\mu_M^X(r^X,d^X) = \mu_{M,CRS}^X(r^X,d^X) + \mu_{M,comb}^X(r^X,d^X) \tag{10.27}$$

r^x denotes the radial coordinate of X and d^x is a unit vector that denotes alignment of X about the radial vector of the micelle system. It assumed that, in every COSMOmic estimation, the interaction between solute molecules is considered to be zero on an infinite dilution. For polar part, $\mu_{M,CRS}^X$ is the dominating part of chemical potential.

Taking into account the spatial arrangements of solute, the more polar and the nonpolar parts are placed in the polar and nonpolar sections of the micellar system, respectively. The size and shape ratio of solute and solvent were taken into consideration. For the calculation of the partition of a given solute through all the n films and all n positionings within one layer, the total chemical potential was calculated. Equation 10.28 is used to calculate the total chemical potential.

$$Z_M^X = \sum_{i=1}^{n} Z_M^X(r_i) = \sum_{i=1}^{n} \sum_{j=1}^{n} exp\left\{\frac{-\mu_M^X(r_i, d_j)}{kT}\right\} \tag{10.28}$$

where k is the Boltzmann constant and T is the absolute temperature, and the possibility of obtaining solute X in layer I can be attained by using Equation 10.29. The free energy outline of the solute in the micelle structure can also be estimated from the probability using Equation 10.30.

$$p_M^X(r_i) = \frac{Z_M^X(r_i)}{Z_M^X} \tag{10.29}$$

$$G_M^X(r_i) = -kT\ ln\ p_M^X(r_i) \tag{10.30}$$

The researcher has considered the whole free energy profile to calculate the partition coefficient (K_M^X) of the drug. The partition coefficient related to size can be calculated according to Equation 10.31.

$$K_M^X = \frac{\sum_{i=1}^{n}\left(V(r_i)p_M^X(r_i) - V(r_n)p_M^X(r_n)\left(\frac{n_w(r_i)}{n_w(r_n)}\right)\right)}{V(r_n)\frac{n_g^w}{n_w(r_n)}p_M^X(r_n)} \tag{10.31}$$

$v(r_i)$ = layer i volume,
$n_w(r_i)$ = indicate the number of water molecules in layer I,
n_g^w = system containing the total quantity of water molecules, and
p_M^X = the relation among the probability of solute.

The partition coefficient K_M^X provides probabilities of a finding solute in the micelles.

The P^{MW} [L water/kg surfactant] value can be estimated using the equation, which can be completely associated with data procured by performing experiments.

$$P^{MW} = \frac{N_A V(r_n)}{n_{surfactant} M_{surfactant}} \frac{n_g^w}{n_w(r_n)} K_M^X \tag{10.32}$$

N_A = Avogadro constant, $n_{surfactant}$ refers to the number of surfactant molecules of the micelle, and $M_{surfactant}$ denotes the molecular mass of a surfactant.

For COSMOmic calculations, a representative surfactant conformer was selected by calculating all the SAS values of the surfactant molecules present in the micelle and estimating the average SAS value, followed by choosing the conformers having a value close to the average SAS value. Thus, these methods can be applied to predict the partition coefficient by computational method [79, 85]. Table 10.1 shows approaches used to model polymeric and lipid nanocarriers.

10.10 Machine Learning

AI is the study of how to make computers understand information instead of being specifically programmed. ML is a subpart of AI. The use of ML by pharmaceutical businesses has increased recently as a result of its success in fields like computer vision and machine translation. The use of ML can be used to examine the current dataset and identify the relationships between dependent and independent parameters, where the former may give formulation knowledge and experimental settings and the latter may highlight interesting formulation behaviors, as shown in Table 10.1. The prediction of formulations benefits from this method. Drug delivery systems including nanocrystals, solid dispersion (SD), cyclodextrin complex, and self-emulsifying drug delivery systems (SEDDSs) have all been successfully predicted using ML in prior investigations.

Using the physicochemical properties of the oils, surfactants, and cosurfactants as input, the guided ML model estimates the molar quantity of the oils, surfactants, and cosurfactants at which they can produce self-emulsifying preparation. This information was used to help select appropriate excipients for SEDDS formulation. Another computational method that can depict and look at the interactions between component molecules and the surroundings from a physiochemical aspect is the MD simulation. For therapeutic researchers toward gaining a systematic knowledge of dosage form characteristics, the MD technique has proven to be useful. The researchers establish various ML models and algorithms to predict

Table 10.1 Machine learning approaches used to model polymeric and lipid nanocarriers.

Descriptive parameters and inputs	Tools	Intended purpose	Ref
Molecular descriptors, Mwt, xLogP, TPSA, and fragment complexity	ANN model (jmp software) and (Bioclipse software)	Formulation of SLN and PN	[88]
Molecular descriptors, Mwt, xLogP, TPSA, and fragment complexity	Gaussian process model (jmp software) and (Bioclipse software)	Formulation of SLN	[53]
Molecular descriptors calculation, logP, SiRMS-lip, SiRMS-EO, and CCM	Molecular descriptors and M5P QSPR mode	Formulation of PN	[56]
Molecular descriptors calculation Selected descriptors (inputs) Mor29u, GATS5m, C-019, T(N...O) and MATS2m E_Strain, Reactive, SMR_VSA4, MINDO_HF, and SMR_VSA7 LgS10, WO6, DD4 and DRACAC	Drug molecular descriptors (DRAGON, MOE, and VolSurf+ programs) and MLR model (STATISTICA software)	Formulation of PN	[89]
Molecular descriptors hybrid of 185 1D-2D molecular descriptors, and eleven experimental conditions	Molecular descriptors (MOE software) and kNN or SVR model (ChemBench software)	Liposomes	[90]

(*Continued*)

Table 10.1 Machine learning approaches used to model polymeric and lipid nanocarriers. (*Continued*)

Descriptive parameters and inputs	**Tools**	**Intended purpose**	**Ref**
Molecular descriptors	DeepChem	Used python-based AI tool for performing various drug discovery assignment predictions	[91]
Molecular descriptors	PotentialNet	Employing graph convolutional neural network (CNN) to predict ligand-binding affinity	[92]
Molecular descriptors	ORGANIC	An effective molecular production means to generate molecules with required properties	[93]

the drug loading in different polymers and lipids and drug transportation across various membranes.

Wang and the co-researchers worked to build up an ML model, which was employed to develop LNP for mRNA vaccines. This opened a pathway for the development of LNP formulation by employing virtual screening models, and these predicting models would act as a guide for the selection of optimum lipids and polymers. The researchers developed an ML algorithm by collecting mRNA vaccine data samples to formulate for LNP by IgG titer. The animal experimental model also confirms the outcome of the predicted model [86, 87].

In this study, ML approaches were used to create a unique model to foresee SD formulation's stability (physical), which, otherwise, takes at least three or six months to determine stability by the traditional approach. Collected data points (646 points) explained the stability by using 20 molecular descriptors. The enhanced maximum dissimilarity technique

(MD-FIS) divided the data into the training group (60%), authentication set (20%), and testing set (20%). After comparing eight ML techniques, the random forest (RF) model produced the most accurate predictions (82.5%). Additionally, the RF models demonstrated how individually

Table 10.2 Using computational methods for rational development of nano-formulations.

Computational method	Development parameter	Reference
Machine learning	Developed a prediction model to gauge the solid dispersion formulation's physical stability. Twenty molecular descriptors were used to characterize the 646 stability data that were gathered. Eight ML approaches were considered. Among them, the random forest model gave the most accurate result. The physical stability predicted by ML was confirmed by experimental results.	[94]
The molecular descriptor of drug-polymer complexes/ chemical descriptor	Predicted the drug loading efficiency and loading capacity by building a QSPR model. Used a set of 61 molecules and 21 drugs and used 408 data points for Forty-one compounds for cheminformatic analysis.	[71]
Density functional theory	Predicted the partition coefficient of a substance in the sodium dodecyl sulfate micelles solution. To predict the free energy of a group of 63 molecules in 15 distinct solvents, various DFT calculation groups were used. The experimental partition coefficient result matches the predicted results.	[95]
Molecular Devices, USA, has different spectrum filters, and SoftMax	Predicted release, partition coefficient, and stability of doxorubicin in polymeric micelles	[96]

faded parameters contributed, which gave us the theoretic direction for SD formulations. Some works related to the use of computation method like ML is given in Table 10.2.

10.11 Conclusion

Analytical models are important for quick prescreening of delivery methods and drug compounds during development. Group contribution methods for calculating SPs and lipophilicity, as well as statistical methods such as FH pairwise interactions, are among the instruments available. However, when comparing substances with different chemical properties, there are inaccuracies connected with the theoretical computation of SP, logP, and FH. These theoretical models, however, are adequate for guiding early material design. The molecular simulations, on the other hand, give more dependable results and techniques of screening, hence more computationally enriched outcomes are available making the system appropriate for reconfirming and re-achieving ingredients, readily providing the greatest possible quality attributes in terms of drug loading, compatibility, and miscibility for given drug samples. Various types of software programs are already designed to screen the carriers as per the requirement according to given drugs. In this chapter, the focus has been provided on classical approaches and the upcoming molecular docking and simulation studies. The topic revolves around the selection of appropriate excipients for the drug to formulate various carrier-mediated drug delivery systems like nanoemulsions, polymeric micelles, and dendrimers drug delivery systems. For the rational designing of these formulations, the computational approach is the appropriate one to select the ingredients for the carrier drug delivery system in a cost-effective and time-saving manner. Furthermore, the results of computational approaches are also matched with the experimental outcomes. Nanoparticles possessing complex molecular arrangements can also be assessed using molecular simulations. Molecular simulations can also be used to assess performance-related aspects of drug delivery systems, such as nanoparticle size and morphology. However, sometimes it might provide inaccurate results, whereas more concrete, faster, and reliable results can be obtained by molecular simulations, but these are more affluent. Many researchers have made use of this computational system and simplified it by eliminating the water molecule or employing an increased simulation technique. The computational approach is still in its developmental stage and no systemic evaluations are available for such systems. Excessive data libraries of drugs and excipients are needed to be established. The relationships between the materials that constitute the core and shell frequently need to be optimized for a nano preparation composition to incorporate constant-size particles. Accomplished by experimentation by choosing complementary ingredients and influencing the proper mixing ratio, albeit this method takes a lot of effort. Current theoretical studies are focused on comprehending the atomistic level of component integration in nanoemulsions, thereby gathering data that will be beneficial for future rational design. The preferred materials for usage in diverse biomedical applications, such as the creation or coverings of biomaterials, drug delivery, and tissue engineering, have emerged as linear, amphiphilic diblock, and triblock copolymers. Drugs can be loaded and stored as cargo in the hydrophobic core of polymeric micelles. With or without medicines, molecular simulations have been used to examine the structure, dynamics, and self-aggregation characteristics of polymeric micelles. Depending on their surface characteristics and size, nanoparticles can be designed to experience improved cell uptake, bloodstream retention, and organ targeting. Importantly, theoretical techniques can forecast how these particles will behave in biological systems. For instance, the structural characteristics of nanoparticles and their impact on biological function have been studied using MD modeling. These theoretical analyses can offer direction for the logical design of medication delivery systems.

References

1. Torchilin, V.P., Multifunctional nanocarriers. *Adv. Drug Deliv. Rev.*, 64, 302–315, 2012.
2. Mishra, D.K., Shandilya, R., Mishra, P.K., Lipid-based nanocarriers: A translational perspective. *Nanomed. Nanotechnol. Biol. Med.*, 14, 2023–2050, 2018.
3. Kumari, A., Yadav, S.K., Yadav, S.C., Biodegradable polymeric nanoparticles based drug delivery systems. *Colloids Surf., B*, 75, 1–18, 2010.
4. Shrestha, H., Bala, R., Arora, S., Lipid-Based Drug Delivery Systems. *J. Pharm.*, 2014, 1–10, 2014.
5. Siepmann, J., Faham, A., Clas, S.-D., Boyd, B.J., Jannin, V., Bernkop-Schnürch, A., Zhao, H., Lecommandoux, S., Evans, J.C., Allen, C. *et al.*, Lipids and polymers in pharmaceutical technology: Lifelong companions. *Int. J. Pharm.*, 558, 128–142, 2019.
6. Li, Y. and Yang, L., Driving forces for drug loading in drug carriers. *J. Microencapsul.*, 32, 1–18, 2015.

7. Huynh, L., Neale, C., Pomès, R., Allen, C., Computational approaches to the rational design of nanoemulsions, polymeric micelles, and dendrimers for drug delivery. *Nanomed. Nanotechnol. Biol. Med.*, 8, 20–36, 2012.
8. Mulder, W.J.M., Strijkers, G.J., von Tilborg, G.A.F., Griffioen, A.W., Nicolay, K., Lipid-based nanoparticles for contrast-enhanced MRI and molecular imaging. *NMR BioMed.*, 19, 142–64, 2006.
9. Maas, J., Kamm, W., Hauck, G., An integrated early formulation strategy – from hit evaluation to preclinical candidate profiling. *Euro J. Pharm. Biopharm.*, 66, 1–10, 2007.
10. Gupta, U., Agashe, H.B., Asthana, A., Jain, N.K., Dendrimers: novel polymeric nanoarchitectures for solubility enhancement. *Biomacromolecules*, 7, 649–58, 2006.
11. Hansen, C.M., The three dimensional solubility parameter, in: *Danish Technical: Copenhagen*, vol. *14*, 1967.
12. Small, P.A., Some factors affecting the solubility of polymers. *J. Appl. Chem.*, 3, 2, 71–80, 1953.
13. Van Krevelen, D.W., Cohesive properties and solubility, in: *Properties of polymers*, 3rd edn, D.W. Van Krevelen (Ed.), pp. 189–225, Elsevier, Amsterdam, 1990.
14. Fedors, R.F., A method for estimating both the solubility parameters and molar volumes of liquids. *Polym. Eng. Sci.*, 14, 147–154, 1974.
15. Van Krevelen, D. and Nijenhuis, K.T., Cohesive Properties and Solubility, in: *Properties of Polymers*, pp. 189–227, Elsevier, Amsterdam, The Netherlands, 2009.
16. Rowe, R.C., Interaction of lubricants with microcrystalline cellulose and anhydrous lactose—a solubility parameter approach. *Int. J. Pharm.*, 41, 3, 223–226, 1988.
17. Li, Y., Taulier, N., Rauth, A.M., Wu, X.Y., Screening of lipid carriers and characterization of drug-polymer-lipid interactions for the rational design of polymer-lipid hybrid nanoparticles (PLN). *Pharm. Res.*, 23, 1877–87, 2006.
18. Latnikova, A., Grigoriev, D.O., Möhwald, H., Shchukin, D.G., Capsules made of cross-linked polymers and liquid core: possible morphologies and their estimation on the basis of Hansen solubility parameters. *J. Phys. Chem. C*, 116, 14, 8181–8187, 2012.
19. Venkatram, S., Kim, C., Chandrasekaran, A., Ramprasad, R., Critical Assessment of the Hildebrand and Hansen Solubility Parameters for Polymers. *J. Chem. Inf. Model.*, 59, 4188–4194, 2019.
20. Flory, P.J., Statistical thermodynamics of polymer solutions, in: *Principles of polymer chemistry*, 3rd ed, P.J. Flory (Ed.), pp. 495–540, Cornell University Press, Ithaca, NY, 1953.
21. Fan, C.F., Olafson, B.D., Blanco, M., Application of molecular simulation to derive phase diagrams of binary mixtures. *Macromolecules*, 25, 3667–76, 1992.

22. Case, F.H. and Honeycutt, J.D., Will my polymers mix? Applications of modeling to study miscibility, compatibility and formulation. *TRIP*, 2, 256, 1994.
23. Dwan'Isa, J.-P.L., Rouxhet, L., Préat, V., Brewster, M.E., Ariën, A., Prediction of drug solubility in amphiphilic di-block copolymer micelles: the role of polymer-drug compatibility. *Pharmazie*, 62, 499–504, 2007.
24. Mahmud, A., Patel, S., Molavi, O., Choi, P., Samuel, J., Lavasanifar, A., Self associating poly(ethylene oxide)-b-poly(α-cholesteryl carboxylate-ε caprolactone) block copolymer for the solubilization of STAT-3 inhibitor cucurbitacin I. *Biomacromolecules*, 10, 471–8, 2009.
25. Qian, F., Huang, J., Hussain, M.A., Drug–Polymer Solubility and Miscibility: Stability Consideration and Practical Challenges in Amorphous Solid Dispersion Development. *J. Pharm. Sci.*, 99, 2941–2947, 2010.
26. Ghitman, J., Stan, R., Vlasceanu, G., Vasile, E., Iovu, H., Predicting the drug loading efficiency into hybrid nanocarriers based on PLGA-vegetable oil using molecular dynamic simulation approach and Flory-Huggins theory. *J. Drug Deliv. Sci. Technol.*, 53, 101203, 2019.
27. Barton, A.F.M., *CRC Handbook of Solubility Parameters and Other Cohesion Parameters*, Second Edition (2nd ed.). Routledge, Taylor and Francis Group, Milton Park, Abingdon in the United Kingdom, 1991, https://doi.org/10.1201/9781315140575.
28. Sun, J., Wei, Q., Shen, N., Tang, Z., Chen, X., Predicting the Loading Capability of mPEG-PDLLA to Hydrophobic Drugs Using Solubility Parameters. *Chin. J. Chem.*, 38, 690–696, 2020.
29. Makoni, P.A., Ranchhod, J., Kasongo, K.W., Khamanga, S.M., Walker, R.B., The use of quantitative analysis and Hansen solubility parameter predictions for the selection of excipients for lipid nanocarriers to be loaded with water soluble and insoluble compounds. *Saudi Pharm. J.*, 28, 308–315, 2020.
30. Gupta, J., Nunes, C., Vyas, S., Jonnalagadda, S., Prediction of Solubility Parameters and Miscibility of Pharmaceutical Compounds by Molecular Dynamics Simulations. *J. Phys. Chem. B*, 115, 2014–2023, 2011.
31. Hollingsworth, S.A. and Dror, R.O., Molecular dynamics simulation for all. *Neuron*, 99, 1129–1134, 2018.
32. Meunier, M., Goupil, A., Lienard, P., Predicting drug loading in PLA-PEG nanoparticles. *Int. J. Pharm.*, 526, 157–166, 2017.
33. Macháčková, M., Tokarský, J., Capková, P., A simple molecular modeling method for the characterization of polymeric drug carriers. *Eur. J. Pharm. Sci.*, 48, 316–322, 2013.
34. Yadav, P., Bandyopadhyay, A., Chakraborty, A., Sarkar, K., Enhancement of anticancer activity and drug delivery of chitosancurcumin nanoparticle via molecular docking and simulation analysis. *Carbohydr. Polym.*, 182, 188–198, 2018.
35. Huynh, L., Leroux, J.C., Allen, C., Enhancement of docetaxel solubility via conjugation of formulation-compatible moieties. *Org. Biomol. Chem.*, 7, 3437–46, 2009.

36. Forrest, M.L., Zhao, A., Won, C.Y., Malick, A.W., Kwon, G.S., Lipophilic prodrugs of Hsp90 inhibitor geldanamycin for nanoencapsulation in poly(ethylene glycol)-b-poly(ε-caprolactone) micelles. *J. Control. Release*, 116, 139–49, 2006.
37. Huynh, L., Grant, J., Leroux, J.-C., Delmas, P., Allen, C., Predicting the solubility of the anti-cancer agent docetaxel in small molecule excipients using computational methods. *Pharm. Res.*, 25, 147–57, 2008.
38. Rahman, A., Correlations in the Motion of Atoms in Liquid Argon. *Phys. Rev.*, 136, A405–A411, 1964.
39. Allen, M.P. and Tildesley, D.J., *Computer simulation of liquids*, Oxford Academic, Walton Street, Oxford, 2017.
40. Hossain, S., Kabedev, A., Parrow, A., Bergströ̈m, C.A., Larsson, P., Molecular simulation as a computational pharmaceutics tool to predict drug solubility, solubilization processes and partitioning. *Eur. J. Pharm. Biopharm.*, 137, 46–55, 2019 Apr 1.
41. Bennett, C.H., Efficient estimation of free energy differences from Monte Carlo data. *J. Comput. Phys.*, 22, 245–268, 1976.
42. Chipot, C. and Pohorille, A., *Free energy calculations*, Springer, Salmon Tower Building New York City, 2007.
43. Christ, C.D., Mark, A.E., Van Gunsteren, W.F., Basic ingredients of free energy calculations: a review. *J. Comput. Chem.*, 31, 1569–1582, 2010.
44. Hansen, N. and Van Gunsteren, W.F., Practical aspects of free-energy calculations: a review. *J. Chem. Theory Comput.*, 10, 2632–2647, 2014.
45. Widom, B., Some topics in the theory of fluids. *J. Chem. Phys.*, 39, 2808–2812, 1963.
46. Dhanasekaran, S., Rameshthangam, P., Venkatesan, S., Singh, S.K., Vijayan, S.R., *In Vitro* and *In Silico* Studies of Chitin and Chitosan Based Nanocarriers for Curcumin and Insulin Delivery. *J. Polym. Environ.*, 26, 4095–4113, 2018.
47. Aparna, V., Melge, A.R., Rajan, V.K., Biswas, R., Jayakumar, R., Mohan, C.G., Carboxymethylated L-carrageenan conjugated amphotericin B loadedgelatin nanoparticles for treating intracellular Candida glabratainfections. *Int. J. Biol. Macromol.*, 110, 140–149, 2018.
48. Ahmed, S., Govender, T., Khan, I., Rehman, N.U., Ali, W., Shah, S.M.H., Khan, S., Hussain, Z., Ullah, R., Alsaid, M.S., Experimental and molecular modeling approach to optimize suitable polymers for fabrication of stable fluticasone nanoparticles with enhanced dissolution and antimicrobial activity. *Drug Des. Dev. Ther.*, 12, 255–269, 2018.
49. Altintas, Z., Abdin, M.J., Tothill, A.M., Karim, K., Tothill, I.E., Ultrasensitive detection of endotoxins using computationally designed nanoMIPs. *Anal. Chim. Acta*, 935, 239–248, 2016.
50. Costache, A.D., Sheihet, L., Zaveri, K., Knight, D.D., Kohn, J., Polymer–Drug Interactions in Tyrosine-Derived Triblock Copolymer Nanospheres: A Computational Modeling Approach. *Mol. Pharm.*, 6, 1620–1627, 2009.
51. Gayathri, N.K., Aparna, V., Maya, S., Biswas, R., Jayakumar, R., Mohan, C.G., Preparation, characterization, drug release and computational modelling

studies of antibiotics loaded amorphous chitin nanoparticles. *Carbohydr. Polym.*, 177, 67–76, 2017.
52. Geetha, P., Sivaram, A.J., Jayakumar, R., Mohan, C.G., Integration of *in silico* modeling, prediction by binding energy and experimental approach to study the amorphous chitin nanocarriers for cancer drug delivery. *Carbohydr. Polym.*, 142, 240–249, 2016.
53. Mahani, N.M., Quantum Chemical Study of Interaction of PLGA Polymeric Nanoparticles as Drug Delivery with Anti-Cancer Agents of Thiazoline. *Int. J. New Chem.*, 6, 34–42, 2019.
54. Hathout, R.M. and Metwally, A.A., Towards better modelling of drug-loading in solid lipid nanoparticles: Molecular dynamics, docking experiments and Gaussian Processes machine learning. *Eur. J. Pharm. Biopharm.*, 108, 262–268, 2016.
55. Hathout, R.M., Metwally, A., Woodman, T.J., Hardy, J.G., Prediction of Drug Loading in the Gelatin Matrix Using Computational Methods. *ACS Omega*, 5, 1549–1556, 2020.
56. Sizochenko, N. and Leszczynski, J., Drug-Nanoparticle Composites: A Predictive Model for Mass Loading. *J. Nanotoxicol. Nanomed.*, 2, 1–10, 2017.
57. Seeliger, D. and de Groot, B.L., Ligand docking and binding site analysis with PyMOL and Autodock/Vina. *J. Comput. Aided Mol. Des.*, 24, 417–422, 2010.
58. Brunacci, N., Neffe, A.T., Wischke, C., Naolou, T., Nöchel, U.; Lendlein, A. Oligodepsipeptide (nano)carriers: Computational design and analysis of enhanced drug loading. *J. Control. Release*, 301, 146–156, 2019.
59. Perez, J.A., Deligianni, F., Ravi, D., Yang, G.-Z., Artificial intelligence and robotics, in: *arXiv Prepr*, 2018, arXiv:1803.10813.
60. Wilson, B. and Km, G., Artificial intelligence and related technologies enabled nanomedicine for advanced cancer treatment. *Nanomedicine*, 15, 05, 433–5, 2020 Feb.
61. Adir, O., Poley, M., Chen, G., Froim, S., Krinsky, N., Shklover, J. *et al.*, Integrating artificial intelligence and nanotechnology for precision cancer medicine. *Adv. Mater.*, 32, 1901989, 2019.
62. Rong, G., Mendez, A., Assi, E.B., Zhao, B., Sawan, M., Artificial Intelligence in Healthcare: Review and Prediction Case Studies. *Engineering*, 6, 3, 291–301, 2020. https://doi. org/10.1016/j.eng.2019.08.015.
63. Osisanwo, F.Y., Akinsola, J.E.T., Awodele, O., Hinmikaiye, J.O., Akinjobi, J., Supervised Machine Learning Algorithms: Classification and Comparison. *Int. J. Comput. Trends Technol.*, 48, 128–138, 2017.
64. Jiang, T., Gradus, J.L., Rosellini, A.J., Supervised Machine Learning: A Brief Primer. *Behav. Ther.*, 51, 675–687, 2020.
65. Moriwaki, H., Tian, Y.-S., Kawashita, N., Takagi, T., Mordred: A molecular descriptor calculator. *J. Cheminf.*, 10, 1–14, 2018.
66. Mannhold, R. and Rekker, R.F., The hydrophobic fragmental constant approach for calculating log P in octanol/water and aliphatic hydrocarbon/water systems. *Perspect. Drug Discov. Des.*, 18, 1–18, 2000.

67. Rekker, R.E., Te Laak, A.M., Mannhold, R., On the reliability of calculated log P-values: Rekker, HanschLeo and Suzuki approach. *Quant. Struct. Act Relat.*, 12, 152–7, 1993.
68. Beck, B., Breindl, A., Clark, T., QM/NN QSPR models with error estimation: vapor pressure and logP. *J. Chem. Inf. Comput. Sci.*, 40, 1046–1051, 2000.
69. S̆oskič, M. and Plavsič, D., Modeling the Octanol–Water Partition Coefficients by an Optimized Molecular Connectivity Index. *J. Chem. Inf. Model.*, 45, 930–938, 2005.
70. Duchowicz, P., Castro, E., Toropov, A., Nesterov, I., Nabiev, O.M., QSPR modeling the aqueous solubility of alcohols by optimization of correlation weights of local graph invariants. *Mol. Diversity*, 8, 325–330, 2004.
71. Alves, V.M., Hwang, D., Muratov, E., Sokolsky-Papkov, M., Varlamova, E., Vinod, N., Lim, C., Andrade, C.H., Tropsha, A., Kabanov, A., Cheminformatics- driven discovery of polymeric micelle formulations for poorly soluble drugs. *Sci. Adv.*, 5, eaav9784, 2019.
72. Wu, Zhang, C., Lin, W., Chen, Q., Guo, X., Qian, Y., Zhang, L., Quantitative Structure-Property Relationship (QSPR) modeling of drug-loaded polymeric micelles via genetic function approximation. *PloS One*, 10, e0119575, 2015.
73. Tielker, N., Tomazic, D., Eberlein, L., Kast, S.M., Güssregen, S., The SAMPL6 challenge on predicting octanol–water partition coefficients from EC-RISM theory. *J. Comput. Aided Mol. Des.*, 34, 453–461, 2020.
74. Hodges, G., Eadsforth, C., Bossuyt, B., Bouvy, A., Enrici, M.H., Geurts, M., Kotthoff, M., Michie, E., Miller, D., Müller, J. *et al.*, A comparison of log K ow (n-octanol–water partition coefficient) values for non-ionic, anionic, cationic and amphoteric surfactants determined using predictions and experimental methods. *Environ. Sci. Eur.*, 31, 1–18, 2019.
75. Marenich, A.V., Cramer, C.J., Truhlar, D.G., Universal Solvation Model Based ocn Solute Electron Density and on a Continuum Model of the Solvent Defined by the Bulk Dielectric Constant and Atomic Surface Tensions. *J. Phys. Chem. B*, 113, 6378–6396, 2009.
76. Zhao, Y. and Truhlar, D., The M06 suite of density functionals for main group thermochemistry, thermochemical kinetics, noncovalent interactions, excited states, and transition elements: Two new functionals and systematic testing of four M06-class functionals and 12 other functionals. *Theor. Chem. Acc.*, 120, 215–241, 2008.
77. Nedyalkova, M., Madurga, S., Tobiszewski, M., Simeonov, V., Calculating the Partition Coefficients of Organic Solvents in Octanol/Water and Octanol/Air. *J. Chem. Inf. Model.*, 59, 2257–2263, 2019.
78. Turchi, M., Kognole, A.A., Kumar, A., Cai, Q., Lian, G., Mackerell, A.D., Predicting Partition Coefficients of Neutral and Charged Solutes in the

Mixed SLES-Fatty Acid Micellar System. *J. Phys. Chem. B*, 124, 1653–1664, 2020.

79. Klamt, A., Huniar, U., Spycher, S., Keldenich, J., COSMOmic: A Mechanistic Approach to the Calculation of Membrane–Water Partition Coefficients and Internal Distributions within Membranes and Micelles. *J. Phys. Chem. B*, 112, 12148–12157, 2008.
80. Pastor, R.W. and Mackerell, A.D., Development of the CHARMM Force Field for Lipids. *J. Phys. Chem. Lett.*, 2, 1526–1532, 2011.
81. Pronk, S., Pall, S., Schulz, R., Larsson, P., Bjelkmar, P., Apostolov, R., Shirts, M.R., Smith, J.C., Kasson, P.M., van der Spoel, D., Hess, B., Lindahl, E., GROMACS 4.5: a high-throughput and highly parallel open source molecular simulation toolkit. *Bioinformatics*, 29, 845–854, 2013.
82. MacKerell, A.D., Brooks, B., Brooks, C.L., III; Nilsson, L.; Roux, B.; Won, Y.; Karplus, M. CHARMM: The Energy Function and Its Parameterization, in: *Encyclopedia of Computational Chemistry*, John Wiley and Sons, Ltd, Hoboken, New Jersey, 2002.
83. Vermaas, J.V., Petridis, L., Ralph, J., Crowley, M.F., Beckham, G.T., Systematic parameterization of lignin for the CHARMM force field. *Green Chem.*, 21, 1, 109–22, 2019.
84. Horinek, D., Mamatkulov, S., II, Netz, R.R., Rational design of ion force fields based on thermodynamic solvation properties. *J. Chem. Phys.*, 130, 124507, 2009.
85. Henneré, G., Prognon, P., Brion, F., Nicolis, I., Molecular dynamics study of a phospholipid monolayer at a water/triglyceride interface: towards lipid emulsion modelling. *Chem. Phys. Lipids*, 157, 86–93, 2009.
86. Wei, W., Shuo, F., Zhuyifan, Ye., Hanlu, G., Jinzhong, L., Defang, Ouyang., Prediction of lipid nanoparticles for mRNA vaccines by the machine learning algorithm. *Acta Pharma. Sin.* B, 6, 12, 2950–2962, 2022.
87. Schneider, G. and Fechner, U., Computer-based de novo design of drug- like molecules. *Nat. Rev. Drug Discov.*, 4, 649663, 2005.
88. Metwally, A.A. and Hathout, R.M., Computer-Assisted Drug Formulation Design: Novel Approach in Drug Delivery. *Mol. Pharm.*, 12, 2800–2810, 2015.
89. Das, S., Roy, P., Islam, A., Saha, A., Mukherjee, A., Poly(DL-lactide-co-glycolic acid) Nanoparticle Design and Payload Prediction: A Molecular Descriptor Based Study. *Chem. Pharm. Bull.*, 61, 125–133, 2013.
90. Cern, A., Golbraikh, A., Sedykh, A., Tropsha, A., Barenholz, Y., Goldblum, A., Quantitative structure - property relationship modeling of remote liposome loading of drugs. *J. Control. Release*, 160, 147–157, 2012.
91. Ramsundar, B., Eastman, P., Walters, P., Pande, V., *Deep Learning for the Life Sciences: Applying Deep Learning to Genomics, Microscopy, Drug Discovery, and More*, O'Reilly Media, Inc, Shroff Publishers & Distributors Pvt. Ltd., Navi Mumbai, 2019.

92. Feinberg, E.N., Sur, D., Wu, Z., Husic, B.E., Mai, H., Li, Y. *et al.*, PotentialNet for molecular property prediction. *ACS Cent. Sci.*, 4, 1520–1530, 2018.
93. Sanchez-Lengeling, B., Outeiral, C., Guimaraes, G.L., Aspuru-Guzik, A., Optimizing distributions over molecular space. An objective-reinforced generative adversarial network for inverse-design chemistry (ORGANIC), ChemRxiv; University of Cambridge, 2017.
94. Han, R., Xiong, H., Ye, Z., Yang, Y., Huang, T., Jing, Q., Lu, J., Pan, H., Ren, F., Ouyang, D., Predicting physical stability of solid dispersions by machine learning techniques. *J. Control. Release*, 311, 16–25, 2019.
95. Saranjam, L., Fuguet, E., Nedyalkova, M., Simeonov, V., Mas, F., Madurga, S., Prediction of Partition Coefficients in SDS Micelles by DFT Calculations. *Symmetry*, 13, 9, 1750, 2021 Sep 19.
96. Ponta, A., Fugit, K.D., Anderson, B.D., Bae, Y., Release, partitioning, and conjugation stability of doxorubicin in polymer micelles determined by mechanistic modeling. *Pharm. Res.*, 32, 5, 1752–63, 2015 May.

11

Molecular Simulation: A Promising Tool for *In Silico* Design of Drug Delivery Formulations

Prem Shankar Mishra[1*], Rakhi Mishra[2] and Deepika Sharma[1]

[1]*Department of Pharmacy, School of Medical and Allied Sciences, Galgotias University, Greater Noida, Uttar Pradesh, India*
[2]*Noida Institute of Engineering and Technology (Pharmacy Institute), Greater Noida, Uttar Pradesh, India*

Abstract
To maximize therapeutic effectiveness, medication design must regulate drug release and deliver it to the right site(s). Molecular modeling is a useful bioinformatic tool for researching interspecies interactions at the atomic level and answering issues unanswered by experimental data. It is utilized for CAD drug research and design. A molecular simulation approach is a viable tool for computer-aided drug discovery and development that analyzes interactions between multiple species in varied systems down to the atom level. Atoms and molecules in the system interact over a specified duration, showing its dynamic "evolution" MD simulations are used to demonstrate the characteristics and intermolecular interactions of drug delivery systems (DDSs) such as medicines, DNA, membranes, polymers, proteins, peptides, nanoparticles, and other small molecules. Although simulation predictions cannot be directly compared to real observations, they give useful correlations between expected values and experimental observables. Simulations reveal molecular data beyond the reach of even the most advanced experimental equipment. This technique faces two challenges: determining the molecular mechanism underlying medication activity in the body and developing computers and computer programs fast enough to do simulations. Furthermore, using these techniques for problems existing over considerably large lengths and timescales is now achievable because of gradual improvements in processing performance and effective modeling methodologies. It will need a coordinated effort on the part of

**Corresponding author*: misra.prem@rediffmail.com

Rishabha Malviya, Sonali Sundram and Dhanalekshmi Unnikrishnan Meenakshi (eds.)
Drug Delivery Systems Using Quantum Computing, (325–352) © 2024 Scrivener Publishing LLC

computational scientists, experimentalists, and industry individuals working in drug delivery to identify specific areas where these simulations might be particularly helpful in order to progress the field of drug delivery in the future. The future of simulation software approaches looks very promising and will play a significant role in the field of drug development because of continuous improvement in computing power and algorithm design.

Keywords: Simulation, drug discovery, computer-aided, atom level

11.1 Introduction

It is generally agreed upon that in the pharmacy, active pharmaceutical molecules should be converted into appropriate formulations, drug delivery systems, or dosage forms, depending on the problem. NME R&D effectiveness, measured in terms of the number of NMEs brought to market by the pharmaceutical industry as a result of their R&D investment, has been steadily decreasing over the past several years [1].

Over 60 years since the 1950s, modern pharmaceutics, which is intimately related to pharmaceutical formulations and drug delivery technologies, has seen a major shift [2]. Artificial intelligence (AI) and multi-scale modeling approaches have been integrated into drug delivery systems by a new field known as "computational pharmaceutics" in the previous decade, presenting the enormous potential to transform the existing formulation development paradigm [3].

This trend may be traced to the exponential growth in computing power, which has allowed for the widespread use of similar approaches in the pharmaceutical industry. As we shall see, the bulk of this toolkit consists of techniques for fitting experimental data to a mathematical model that yields a numerical response, such as the optimal structure of a therapeutic molecule or the optimal formulation of a drug delivery system [4]. Mechanistic knowledge is a subset of computational techniques that provides much more than a response, i.e., an ideal value or collection of values, as opposed to simply an answer, i.e., an explanation of what is occurring in the system to get the results. A simplified depiction of the system as a three-dimensional cartoon depicting the current condition. This output, known as a simulation, has significantly more power than a simple conclusion of what is best for the application at hand; it can lead to a well-informed design process that is more efficient, permits more intuitive leaps from its previous state perspective, and provides insight that transcends the application studied [5, 6]. These computational techniques can reduce the time required for laboratory studies and provide preliminary and accurate assessments of a drug's efficacy in a particular carrier [7]. Pharmaceutical drug delivery systems use molecular dynamic simulations (MDS) to simulate the behavior of atomic molecules in order to understand molecular architectures. Drug/excipient complexes' physical characteristics can be measured using MDS without the need for pricey research [8, 9]. Recently, isotropic lipid–based nanocarriers have drawn a lot of interest as potential transporters for substances that cannot be dissolved in water. Different isotropic lipids are combined to form dynamic lipid systems, such as solid lipid nanoparticles (SLNs), self-emulsifying systems, and lipidic micelles. They are able to adapt the therapeutic effectiveness of problematic drugs to meet the needs of patients. The composition of the nanostructure, drug solubilization, and system stability is challenging to forecast, assess, and contextualize with any degree of accuracy [10, 11].

As a result, MDS may be a beneficial tool for figuring out the internal microstructure of lipid-based delivery systems and drug localization.

11.1.1 Simulation of Molecular Dynamic

Since its introduction more than four decades ago, molecular dynamics (MD) has been employed in a range of research fields [12], and in certain cases, it has evolved into a new instrument for researching a specific issue. This section describes the several tools and methods that a novice user in a pharmaceutical laboratory may employ to comprehend the various aspects of solubility. We first quickly explain the fundamental concepts underlying traditional MD simulations. However, there have previously been other outstanding studies of MD's technical and application aspects, and this is the first to combine both perspectives [13, 14]. It is still our experience that the pharmaceutical industry lacks a general understanding of what can and cannot be done to study solubility using computational methods, as well as the reasons why. In order to hone the focus even further [15, 16], we confine ourselves to predominantly classical, particle-based MD simulations applying any of the commonly utilized biomolecular forcefields. In addition, we briefly contrast MD with many other methodologies, such as Monte Carlo approaches as COSMO.

Various acclaims from technical and practical domains in a classical system consisting of discrete molecules, atoms move as a result of forces exerted by the rest of the system. To use molecular simulation to study solubility, it is necessary to have a force field that can accurately simulate the intermolecular and intramolecular interactions in the system of interest [17]. van der Waals (vdW) and electrostatic forces are the primary mechanisms by which atoms and molecules interact with one another within a given force field [18–22], however in our experience,

The complexity of validating the force fields and molecules employed in the process of identifying acceptable interaction variables (through parameterization) for a new molecule varies widely. The force field of a simulation or computer experiment is the set of physical connections between molecules [23–25].

The method of rearranging atomic structure by using interatomic forces is shown in Figure 11.1. Newton's kinematics are quantitatively solved to make that possible. The forces are determined by determining the gradient of the interaction potentials between individual atoms. The precise location of each atom in the system may be calculated using this approach. These revised coordinates may be used as inputs for a new force calculation that can be used to refine the coordinates once more [26, 27].

The following methods listed below can be used to calculate energy.

(i) Histogram-based techniques [28]. (ii) Nonequilibrium modeling [29]. (iii) Theory of perturbations [30]. (iv) Potential derivative [31].

11.1.2 A Molecular Simulations Using a General System

Table 11.1 shows the methodology for carrying out the MDS operation for the examination of a pharmaceutical lipid-based system. The initial geometrics of the target ligand (drug) and additional components (such as the lipid carrier, membrane, solvent, and conditions) are created or acquired from a database (Materials Studio, Gaussian suite), the information was obtained from a data source (for example, Protein Data Bank), and so on. The network's individual bound and unbonded atomic pairs are then listed using these geometrics, and the system's force-field parameters are

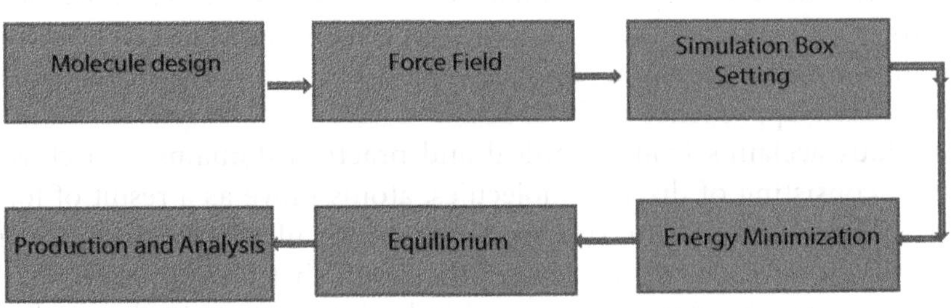

Figure 11.1 Automatic simulation.

Table 11.1 General steps for performing MDS (molecular dynamics simulations).

1. Initial geometrics	• Generation of trajectories, PDF files • Generation of topology files
2. Simulation setup	• Preparation of atoms, molecules, and boxes. • Molecule fitting
3. Energy minimization	• Setting up energy molecule configuration and minimization for thermodynamic interaction.
4. MD simulation	• Simulation production • Record coordinates
5. Analysis	• Molecule and system characteristics determination • Intermolecular interactions

established [32–34]. The next stage is to create a simulation box and put molecules of various species within it in decreasing concentrations. High interatomic repulsion caused by the MDS of atoms/molecules in the simulation box leads to numeric instability. Therefore, while doing an energy minimization or short MDS, it is recommended that a much shorter phase duration (time phase) be employed than that used in the actual simulation. Following energy minimization, the thermodynamic equilibrium state is used to determine the MDS time frame. Normal equilibration times range from 10 to 100 ns [35, 36]. Equilibrium may be quantified by observing the convergence of the potential energy, atomic density, and molecular association selectivity. The general simulation methodologies are shown for drug design in Figure 11.2.

The parameters of a stable formulation are then monitored by analysis, which is principally prescribed for MDS. The period and stability of the entire investigation are essentially a result of the computer systems and operating setups. The majority of Linux-based operating systems are compatible with MDS [37].

The molecular docking method is often used to model a lipid-based system for MDS and compute the free energy of binding to investigate how medicine interacts with a specific lipidic carrier system. Thus, the drug loading in the lipidic carrier system is linked to the binding energy [38].

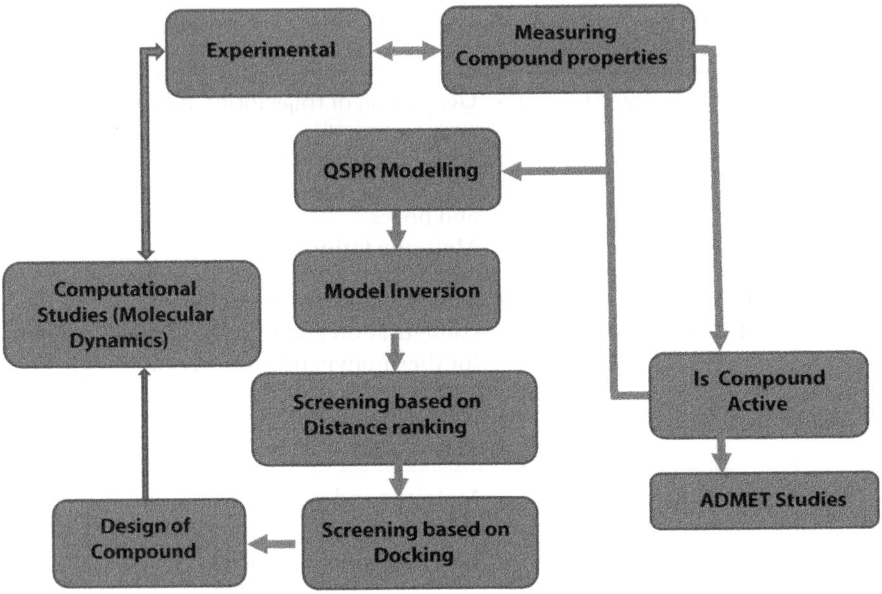

Figure 11.2 General drug discovery methodology.

The molecular structure of lipid droplets may be characterized using MDS, which reveals details on the relative spatial accumulation of different types of lipid molecules inside the droplets as well as the impact of surfactant molecules on the internal structure of the droplets. These models might also predict how lipid chain length would affect droplet patterns and how the mass ratio of solid to liquid lipids will affect lipid distribution inside the lipid matrix [39–41].

11.2 Software Available for Performing MDs

The development of appropriate computer programs for molecular simulations is a labor-intensive process that requires the cooperation of specialists from several domains. Fortunately, a concerted effort by scientists worldwide has produced a wide range of open-source and commercial programs that are extremely competent and successful (Table 11.2) [42]. Materials Studio, Marvin, PRODRUG, Swiss Param, ATB, MKTOP, Avogadro, Gaussian suite, and other programs and web platforms are used individually to build molecules, calculate force field parameters, analyze data, and imagine consequences (e.g., visual MDs, PYMOL,

Table 11.2 Commonly used molecular dynamics simulation software.

Software	Computing requirements	Description	Visualization	Package
GROMACS	Working with Linux. High-end configuration required	MD simulator with web mechanics. force field setup with MARTINI	Molecular graphics viewer	GROMACS 4.02
AMBER	Working with Linux and graphics processing unit technology	MD simulator with molecular mechanics, force field setup	Visual molecular dynamics	AMBER TOOLS 14
NAMD	Working with Linux. High-end configuration required	MD simulator with molecular mechanics, built-in Gaussian suite	J Mol	NAMD 2.0
DESMOND	Working with Linux and Windows compatible with graphics processing unit technology	MD simulator, realistic and quantitative simulations	Visual molecular dynamics	DESMOND 4.0
ACEMD	Working with Linux and graphics processing unit technology	MD simulator, HTMD, Python system, quantitative simulations	J Mol	ACEMD 3.0

and CHIMERA). These programs produce dynamic results for simulated systems and give crucial information on crucial quality characteristics of delivery systems. However, these programs have some limitations that restrict their use [43, 44].

The current force fields need to be improved upon in order to improve the accuracy of free-energy estimates. Other electronic interactions, such as polarization and charge transfer, are either completely or partially disregarded. Second, it is impossible to perform routine simulations for more than a microsecond due to the enormous processing demands. Since actual activities like protein folding only last a few seconds, current computer systems are unable to do such computations [45, 46].

As a result, the best software package for simulations is determined by the input variables, simulation time, and processing requirements of the delivery systems in question.

Therapeutic efficacy at the site of action may be increased, and the toxicity to nontarget tissues can be reduced if the appropriate molecules and/or delivery techniques can be identified and used in the delivery of drug research and medicine development [47].

The ability of a drug to propagate cellular membranes and the intensity to which selection of candidates or chemical alteration, method of administration, dosage regimen, or delivery method structure can preferentially modify membrane permeation and drug distribution are important in the role of delivery of drugs in achieving this goal [48–50].

The following are features that should be present in an ideal carrier for targeting drugs [51].

(1) Drugs are only delivered to the intended organ, tissue, cell, and compartments.
(2) Plasma dispersion at the interface level.
(3) Long-term management of drug depots.
(4) Delivery of medications primarily to diseased or cancerous tissue.
(5) Sustained regulated medication delivery.
(6) Drug delivery without compromising on vital biological processes.
(7) Drugs at therapeutic concentrations at the site of action.
(8) Reducing the amount of unbound medication that enters the circulation.
(9) Suppression of plasma enzyme inactivation.
(10) Elimination and eventual biodegradation with low toxicity.
(11) Diverse chemical compositions (diagnostics and therapeutics).
(12) Prepared with minimal effort.

To reduce the expense of drug design and development, computational approaches to forecasting drug delivery before trials are becoming more and more appealing. To research medication distribution, significant advancements in MD modeling approaches have been developed. Such techniques are very useful for designing and improving medication delivery systems. Problems that are difficult to explore in drug delivery laboratory experiments may often be resolved with the use of simulations. We now have a theoretical knowledge of the drug delivery process because of the recent advances made possible by computational models of targeted drug delivery, such as our most recent MD analysis of dendrimer drug delivery [52–54].

According to the research of Moradnia *et al.*, many theoretical studies have been conducted lately to assess the efficacy of SWCNTs as potential host nano-vehicles for cancer diagnosis and therapy. Using computational modeling or simulation, vdW, electrostatic forces, and binding free energy for drug interactions were calculated and analyzed [55, 56].

In this context, MD simulation has been extensively used to study the functioning of complicated nano-sized systems in fields as varied as cellular biology and materials research [57–59]. In addition to its usefulness in assessing the mechanical and thermal properties of nanostructures [60], MD may also be used to learn more about the atomic-level interactions in nano-sized complex systems that include bio-molecules [61].

In the modern era, the encapsulation procedures were simulated using MD using the LAMMPS program (March 2018 version [62–64]). Previous research showed that the popular CHARMM27 force field, used often in drug delivery modeling, was used to ascertain the interaction characteristics of oxygen and hydrogen atoms in molecules of water (TIP3P) [65, 66].

11.3 Applications of Molecular Simulation in Drug Delivery System

Standard MD prediction, in which atoms are considered the smallest unit, is called an all-atom simulation, and it is founded on the concepts of Newtonian kinematic mechanics. The empirical force field is used to characterize atomic interactions. The computer accomplishes the simulation of macroscopic qualities by adding random preliminary velocities to all molecules in the system by the Boltzmann distribution law, solving the differential equation numerically to acquire the velocity and coordinate information at every point [67–69].

There has been some recent research that used MD simulation to look at the steps involved in making and dissolving solid dispersion formulations.

The features of the drug inclusion complex may be investigated with the use of molecular modeling [70], which provides a theoretical foundation for stable structures, absorption coefficient, and chiral differentiation [71, 72].

11.3.1 Lipid Bilayer Drug Diffusion and Penetration

Following the Meyer-Overton principle, the ability to penetrate the phospholipids membrane has a major impact on a drug's bioavailability [73]. Hydrophobic medications are lipophilic and so easily pass through the lipid membrane. Hydrophobicity and lipophilicity may be determined by the quantity of hydrophobic and lipophilic sites. You may obtain a molecule's hydrogen-bond donors and acceptors by looking at its chemical makeup. The orientation of the molecules is one of a multitude of information found in molecular simulations. the energetic interactions of the drug (defined in terms of the bilayer normal) with the bilayer components and the drug molecule [74–76]. These studies have been effective across a variety of drug and membrane chemistries. Additionally, transporter molecules may be shown in action in molecular simulations, which can also reveal other sources of Meyer-Overton rule exceptions. In conclusion, molecular simulations may provide precise structural information and the behavior of lipid membranes and the channel formation process [77].

11.3.2 Solubility of Drugs

Soluble drugs are more likely to be absorbed because they are present in higher quantities in the digestive tract's fluids. However, due to the aqueous nature of stomach and intestinal fluids, hydrophilic medications have a low membrane permeability due to their high solubility in these fluids [78]. Solubility and permeability of therapeutic molecules are therefore key properties to characterize. The four groups that make up the biopharmaceutics classification (BCS) are as follows: type I (high permeability, good solubility), type II (high permeability, low solubility), type III (low permeability, high solubility), and type IV (low permeability and solubility) [79–81]. One of Lipinski's five criteria for determining whether or not a chemical has drug-like properties is the number of hydrogen-bond donors and acceptors. Based on the "like dissolves like" principle, an alternative theoretical explanation of solubility states that two molecules are deemed soluble if their solubility values are just marginally different. Even while the permeability characteristics may be determined from the existing experimental data or retrieved first from the chemical structure using a group-contribution technique, molecular simulation can provide a more precise estimation, particularly for innovative drug candidates [82–84]. Moreover, only molecular simulations can offer a comprehensive assessment of variation in bioavailability caused by changes in pH and saline levels in the gastrointestinal tract.

11.3.3 Carrier-Drug Miscibility

Many different types of carrier proteins (excipients) are employed to increase the solubility of lipophilic drug medications or to keep drug supersaturation constant. The choice of such excipients can be aided by methods for assessing drug-excipient interactions and molecular simulations. Excipients are, in general, substances that combine physically with drug molecules [85–87].

Furthermore, the interaction between the drug and excipient should not be too strong, or else the drug molecules will not be able to escape. Assessment of drug-excipient interactions, especially for polymeric excipients, may be done with the help of Flory-Huggins interaction parameters, which can be generated using molecular simulations [88].

Molecular simulations have the potential to demonstrate the phase formation of excipient-drug complexes, exhibiting composition-dependent variations in miscibility and complex formation. The suitability of the drug phenytoin with cellulose polymeric materials [hydroxypropyl methylcellulose, HPMC, and hydroxypropyl methylcellulose acetate succinate (HPMCAS)] was investigated by Jha and Larson (2014) through analytical Molecular simulation [89]. Drug-polymer complexation was found to be correlated with drug accumulation inhibitory activity using radial distribution; the drug diffusion coefficient through the use of carriers has been used to quantify the drug release rate [90].

Since the final solvent-free state is equivalent to the dosage form, they were also able to reverse the dissolving process by extracting water molecules in sequence and then re-equilibrating. An increasing number of researchers are studying pH-responsive excipients, which bond with medicinal compounds in the stomach and then deliver them at such a stronger pH inside the bowel. Recent results for both the doxorubicin-graphene and doxorubicin-nanodiamond systems suggest that molecular simulation may be a useful tool for developing such excipients [91–93].

11.3.4 Drug Crystallization

The drug molecules in several oral dosage forms can crystallize, and poorly soluble medications can also do so inside the body. Excipients prevent crystallization because crystalline versions of medication have poorer absorption than amorphous versions of the same substance. Molecular models of nucleation and growth are theoretically and economically prohibitive because crystals are structures with long-range spatial correlations, necessitating simulations of bigger systems, and because nucleation is an uncommon occurrence (requires very long simulations or the use of smart algorithms) [94].

While models of crystal development and dissolution have come a long way in recent years, they are not yet capable of nucleating crystals of big medicinal molecules [95]. Computational research might be utilized to compare substances' crystallization inhibition potential, locate crystal faces with quicker growth rates, and foretell the crystallization behaviors of new pharmaceuticals. Important to note in crystal modeling is the systematic, but aggressive growth induced by atomistic simulations with the correct parameterization, which constitutes the efficient coarse-graining processes [96, 97].

11.3.5 Drug Loading and Release

When discussing the carrier-mediated release of drugs, the terms "drug solubility" and "drug delivery" are often used to describe the incorporation of the drug into the carrier and the subsequent release of the drug into the body, respectively. To swing the flow of medicines in a different direction, we must change the environment, considering that the osmotic pressures inside and outside the carrier differ, which causes the drug loading and release processes [98]. Since the release of drugs in the body arises in an aqueous condition, drug loading is often performed using a nonaqueous solvent. Studies of drug release profiles from vehicle matrices have found great success using continuous methodologies based on the Higuchi equation, although these studies often ignore the structural properties of the carrier. If the drug loading/release process includes the creation of drug excipient nanostructures, then molecular simulations may provide a detailed description of the loading/release mechanism [99–101].

Due to the large difference between both the predicted and actual length scale, molecular simulation results do not directly compare with practically acquired loading/release data of pharmacological formulations, in contrast to continuous techniques. The drug diffusivity can be indirectly calculated using the carrier and then used as a linear model in the simulation or collaboration with experimental observations. Molecular simulations can provide an appropriate description of the drug's loading or releasing mechanism, especially if it involves the production of drug and excipient nanostructures [102, 103]. Molecular simulation for medical distribution is shown in represented in Table 11.3.

Table 11.3 Typical research on medication distribution using molecular simulations [104].

System	Simulation software and method	Simulation time	Properties
Felodipine with HPMC	Amber, MD	30–100 ns	Miscibility, phases behaviors, the dissolution factor, the Flory-Huggins interactions parameter, and hydrophobic interactions
Phenytoin with polymer excipients	GROMACS, MD	90 ns	Drug accumulation and polymer-drug interactions
Subjection of paclitaxel-containing DPPC bilayers to shockwave and nanobubbles	GROMACS, MD	600 ns	Drug-bilayer engagement and membrane penetration
Phenytoin with HPMCAS	GROMACS, atomistic MD, CG BD	10 μs	The effects of polymer-drug interactions, drug aggregation, and phase behavior
Danazol in simulated GI environment	GROMACS, MD	200 ns	Interactions between phase behavior, drug agglomeration, and solubility

(*Continued*)

Table 11.3 Typical research on medication distribution using molecular simulations [104]. (*Continued*)

System	Simulation software and method	Simulation time	Properties
Phenytoin crystal	GROMACS, atomistic and CG MD, well-tempered metadynamics	100 ns	Free energy of adsorption, drug crystal polymorphs, and drug crystal development
Nanocarriers are made of pristine graphene (PG) and graphene oxide (GO) with doxorubicin for cancer treatment	Materials Studio	1–3 ns	The potential energy of drug-graphene interactions, drug loading, and release

11.4 Molecular Dynamics Simulations for Nanomedical Applications

The use of various levels of coarse-graining in MD simulations has provided mechanistic insight into the creation of innovative nanoparticle-based medication delivery systems. Researchers have examined the MD of a wide variety of nanoparticles, along with polymeric nanoparticles, gel nanostructures, polymersomes, some nanocrystals, solid-lipid nanoparticles, other micelles, nanomaterials, carbon dots, nanotubes, nanographene, genomic nanotubes, nanodiamonds, peptide nanoparticles, gold nanoparticles, and silver nanoparticles [104, 105].

Molecular modeling of vesicle-based drug delivery systems, such as niosomes, ufasomes, polymeric vesicles (polymerases), and glyceryl monostearate vesicles, has shed light on their potential. A great deal of computational research using MD modeling has also been done on the functionalization approach to nanoparticles. Either attach a protein to a polymer or dissolve it in water to observe how the protein behaves by looking at the potentials that have already been established [106, 107].

An introduction to molecular modeling for the study of protein-polymer conjugation was proposed by Lin and Colina (2019). The simulation of drug distribution by nanomedicine has been studied with a wide range of nanoparticle types and many different drugs [108–110]. Drugs and their application evaluated by MD simulation are detailed in Table 11.4.

Table 11.4 In the context of the delivery of the drug, the following pharmaceuticals have been evaluated using MD simulations.

Drugs	Applications
5-Fluorouracil	Anticancer drug
Camptothecin	Chemotherapy agent
Carmustine	Chemotherapy agent
Chlortetracycline	Antibiotic
Fulvestrant	Breast cancer drug
Exemestane	Breast cancer drug
Gemcitabine	Chemotherapy agent
GF-17	Antimicrobial peptide
Ibuprofen	Medications, both anti-inflammatory and painkillers
Itraconazole	Antifungal drugs
Metronidazole	Antibiotic
Sunitinib	Renal carcinoma medication
Picoplatin	Colorectal cancer drug
Streptozotocin	Neuroendocrine tumors drug
Interferon-alpha	Compound with antitumor and antiviral properties

11.5 Molecular Dynamics Insight Examples

11.5.1 Bloodstream Behavior and Coronal Protective Polymer

As a nanoparticle enters the bloodstream, hydrodynamic forces cause a corona of bloodstream proteins to develop on its surface. To get rid of macrophages, a portion of such peptides triggers a fundamental structural response that is very selective [111]. The best approach for modeling nanoparticle transport in the dysfunctional tumor vasculature by Gao *et al.* (2020) is a combination of theoretical calculations and computational fluid dynamics (CFD), a discretized continuum model that was described and used to model nanoparticle behavior in the bloodstream and the impact of size and shape. The exceedingly complex and mostly unstudied process of protein corona formation [112]. It is clear that the nanoparticle's surface characteristics have an effect on this, and that the protective polymer corona's modulation of its contact with bloodstream proteins is the method by which it lengthens the bloodstream's life span [113].

The associations of two therapeutic polymers, PEG and polysarcosine, with such a group of proteins found in the circulation were investigated using experimental analysis and MD. Research revealed that the interactions are not amino acid–specific and instead have a definite trend focused on the charge and polarization of the amino acid and also the properties of the interaction between the polymer and water. This evidence was in addition to its direct interaction between both the polymer and the protein. Their approach may pave the way for a logical design strategy for the creation of novel polymer materials that work more successfully as protective polymer corona when combined with experimental data [114, 115].

Lee *et al.* directly investigated the interaction of liposomes with bloodstream proteins using the coarse-grained MARTINI model. They further utilized simulation with atomic precision to investigate the impact of nanoparticle electrostatics in the development of protein corona [116]. PEGylation has been widely researched in the context of nanoparticles other than liposomes using MD simulations. Ramezanpour *et al.* claim that a larger proportion of PEG-cholane compounds to the provided VIP-palm peptides is necessary for supramolecular [117]. Casalini *et al.* claim that a larger proportion of PEG-cholane compounds to the provided VIP-palm peptides is necessary for large molecules [118].

11.5.2 Controlled Release and Drug Loading

Several researchers have investigated, for a plethora of nanoparticle forms, whether nanoparticles may store medications and release them when externally triggered. Typically, hydrophobic and located in a nonpolar zone, the investigated substances are nonpolar [119].

Different drug loading problems for liposomes and polymersomes transporting their drug carrier have mostly been simulated. These problems include hydrophobic medicines in the liposome membrane, carbon nanotubes, nanographene, peptide carriers, PAMAM dendrimers, polymeric nanoparticles, and micelles, whereas 5-fluorouracil [120, 121], cucurbitacin, carmustine, chaconne, picoplatin, porphyrins, ibuprofen, paclitaxel, and albendazole have all been investigated in addition to doxorubicin, which is the most often recommended therapy in these model systems. It is a common practice to build these nanoparticles such that they deliver a potential therapeutic load in proportion to a pH shift [122].

The effects of pH variation may be modeled using molecular simulation. The partial charges on the atoms in an MD simulation are used to simulate pH, allowing the system to be brought to equilibrium with partial charges that correspond to a neutral pH. The partial charges can then be altered to simulate pH variations and the system's reactions to them, such as the release of medications. It is an unusual aspect of the work done in this sector employing MD modeling that all three degrees of coarse-graining are used: the DPD, the MARTINI model, and atomic MD. When MD modeling and experimentation were used to study itraconazole in a liposome, it was able to offer specific information that was not possible via simulation [123, 124].

The poor solubility of the antifungal drug itraconazole decreases its absorption. Incorporating drugs inside liposomes, as has been done in a few different trials, is a potential solution to the problem of limited solubility.

Because cholesterol is known to maintain the stability of lipid bilayers, it is widely utilized to improve the characteristics of liposomes [125]. According to the research of Bulbake *et al.* in 2017, in 9 out of the 15 liposome-based formulations that have achieved clinical approval, cholesterol is present. Cholesterol was added to the liposome-itraconazole mixture as a result. Because itraconazole and cholesterol would not interact very well and separate into tiny domains, MD simulations showed that this is not the optimal strategy [126].

11.6 Conclusion

The quick progress in drug design using computational methods has made it possible to envision several uses for the MD approach in drug delivery systems. To evaluate drug delivery systems, researchers have investigated a lot of functions of molecular simulation approaches. In this approach, these computations might be used to gather crucial information. According to our knowledge, the bulk of free energy estimations discussed in the literature were simulated with a single drug without taking the drug-loaded medium into account.

Moderately, data may be obtained on a certain drug's drug release, solubility, loading capacity, etc., into a particular DDS.

It is feasible to obtain several DDS development characteristics by molecular simulation. This method might be used for new DDS systems. They present a great opportunity for the development of drug delivery devices that might benefit from MD simulations. This computational technique may lead to new uses for existing medications as well as novel therapies like gene delivery.

References

1. Wang, W., Ye, Z., Gao, H., Ouyan, D., Computational pharmaceutics - A new paradigm of drug delivery. *J. Control. Release*, 338, 119–136, 2021.
2. Shariatinia, Z., Molecular Dynamics Simulations on Drug Delivery Systems, in: *Modeling and Control of Drug Delivery Systems*, pp. 153–182, Academic Press, Elsevier, 2021, ISBN 9780128211854.
3. Abdel-Halim, H., Traini, D., Hibbs, D., Gaisford, S., Young, P., Modelling of molecular phase transitions in pharmaceutical inhalation compounds: An *in-silico* approach. *Eur. J. Pharm. Biopharm.*, 78, 1, 83–89, 2011.
4. Tommaso, C., Not only *in silico* drug discovery: Molecular modeling towards *in silico* drug delivery formulations. *J. Controlled Release*, 332, 2021, 390–417, 2021.
5. Van Gunsteren, W.F. and Berendsen, H.J.C., Computer Simulation of Molecular Dynamics: Methodology, Applications, and Perspectives in Chemistry. *Angew. Chem. Int. Ed. Engl.*, 29, 992–1023, 1990.
6. Monticelli, L. and Tieleman, D.P., Force fields for classical molecular dynamics. *Methods Mol. Biol.*, 924, 197–213, 2013.
7. Hubbard, R.E., Molecular graphics and modeling: Tools of the trade, in: *Guidebook on Molecular Modeling in Drug Design*, vol. pp, N.C. Cohen, (Ed.), pp. 19–54, Academic Press, New York, NY, USA, 1996.

8. Hess, B., Kutzner, C., van der Spoel, D., Lindahl, E., GROMACS 4: Algorithms for Highly Efficient, Load-Balanced, and Scalable Molecular Simulation. *J. Chem. Theory Comput.*, 4, 435–447, 2008.
9. Brooks, B.R., Brooks III, C.L., Mackerell, A.D., Jr., Nilsson, L., Petrella, R.J., Roux, B., Won, Y., Archontis, G., Bartels, C., Boresch, S. *et al.*, CHARMM: The biomolecular simulation program. *J. Comput. Chem.*, 30, 1545–1614, 2009.
10. De Vivo, M., Masetti, M., Bottegoni, G., Cavalli, A., Role of molecular dynamics and related methods in drug discovery. *J. Med. Chem.*, 59, 9, 4035–4061, 2016.
11. Dror, R.O., Dirks, R.M., Grossman, J.P., Xu, H., Shaw, D.E., Biomolecular simulation: A computational microscope for molecular biology. *Annu. Rev. Biophys.*, 41, 1, 429–452, 2012 Jun 9.
12. Ganesan, A., Coote, M.L., Barakat, K., Molecular dynamics-driven drug discovery: Leaping forward with confidence. *Drug Discov. Today*, 22, 249–269, 2017.
13. Gao, Y. and Olsen, K.W., Molecular dynamics of drug crystal dissolution: Simulation of acetaminophen form I in water. *Mol. Pharmaceutics*, 10, 3, 905–917, 2013.
14. Cheng, Y., Pei, Q.X., Gao, H., Molecular-dynamics studies of competitive replacement in peptide–nanotube assembly for control of drug release. *Nanotechnology*, 20, 145101, 2009.
15. Gubbins, K.E. and Moore, J.D., Molecular modeling of mater: Impact and prospects in engineering. *Ind. Eng. Chem. Res.*, 49, 7, 3026–3046, 2010.
16. Gupta, J., Nunes, C., Vyas, S., Jonnalagadda, S. Prediction of solubility parameters and miscibility of pharmaceutical compounds by molecular dynamics simulations. *J. Phys. Chem. B*, 115, 9, 2014–2023, 2011.
17. Lipinski, C.A., Lombardo, F., Dominy, B.W., Feeney, P.J., Experimental and computational approaches to estimate solubility and permeability in drug discovery and development settings. *Adv. Drug Deliv. Rev.*, 64, 4–17, 2012.
18. Huang, W., Lee, S.L., Yu, L.X., Mechanistic approaches to predicting oral drug absorption. *AAPS J.*, 11, 2, 217–224, 2009.
19. Salo-Ahen, O.M., Alanko, I., Bhadane, R., Bonvin, A.M., Honorato, R.V., Hossain, S., Juffer, A.H., Kabedev, A., Lahtela-Kakkonen, M., Larsen, A.S., Lescrinier, E., Molecular dynamics simulations in drug discovery and pharmaceutical development. *Processes*, 30, 9, 1, 71, 2020 Dec.
20. Martínez, L., Villalobos, R., Sánchez, M., Cruz, J., Ganem, A., Melgoza, L.M., Monte Carlo simulations for the study of drug release from cylindrical matrix systems with an inert nucleus. *Int. J. Pharm.*, 369, 1–2, 38–46, 2009.
21. Hossain, S., Kabedev, A., Parrow, A., Bergström, C.A., Larsson, P., Molecular simulation as a computational pharmaceutics tool to predict drug solubility, solubilization processes and partitioning. *Eur. J. Pharm. Biopharm.*, 137, 46–55, 2019 Apr 1.

22. Sakkiah, S., Chandrasekaran, M., Lee, Y., Kim, S., Lee, K.W., Molecular modeling study for conformational changes of Sirtuin 2 due to substrate and inhibitor binding. *J. Biomol. Struct. Dyn.*, 30, 235–254, 2012.
23. Brooijmans, N. and Kuntz, I.D., Molecular recognition and docking algorithms. *Annu. Rev. Biophys. Biomol. Struct.*, 32, 335–373, 2003.
24. Pantsar, T. and Poso, A., Binding Affinity via Docking: Fact and Fiction. *Molecules*, 23, 1899, 2018.
25. Riniker, S., Allison, J.R., van Gunsteren, W.F., On developing coarse-grained models for biomolecular simulation: A review. *PCCP*, 14, 36, 12423–30, 2012.
26. Husic, B.E. and Pande, V.S., Markov State Models: From an Art to a Science. *J. Am. Chem. Soc.*, 140, 2386–2396, 2018.
27. Teobaldi, G. and Zerbetto, F., Molecular dynamics of a dendrimer-dye guest-host system. *J. Am. Chem. Soc.*, 125, 7388–7393, 2003.
28. Gurtovenko, A.A., Lyulin, S.V., Karttunen, M., Vattulainen, I., Molecular dynamics study of charged dendrimers in salt-free solution: Effect of counterions. *J. Chem. Phys.*, 124, 094904, 2006.
29. Suek, N.W. and Lamm, M.H., Effect of terminal group modification on the solution properties of dendrimers: A molecular dynamics simulation study. *Macromolecules*, 39, 4247–4255, 2006.
30. Li, Y. and Hou, T., Computational Simulation of Drug Delivery at Molecular Level. *Curr. Med. Chem.*, 17, 36, 4482–4491, 2010.
31. Pan, A.C., Borhani, D.W., Dror, R.O., Shaw, D.E., Molecular determinants of drug–receptor binding kinetics. *Drug Discov. Today*, 18, 13–14, 667–673, 2013.
32. Alder, B.J. and Wainwright, T.E., Studies in Molecular Dynamics. I. General Method. *J. Chem. Phys.*, 31, 459–466, 1959.
33. Karplus, M. and McCammon, J.A., Molecular dynamics simulations of biomolecules. *Nat. Struc. Biol.*, 9, 646–652, 2002.
34. Berthelsen, R., Klitgaard, M., Rades, T., Müllertz, A., *In vitro* digestion models to evaluate lipid-based drug delivery systems; present status and current trends. *Adv. Drug Deliv. Rev.*, 142, 35–49, 2019.
35. Liu, X., Shi, D., Zhou, S., Liu, H., Liu, H., Yao, X., Molecular dynamics simulations and novel drug discovery. *Expert. Opin. Drug Discov.*, 13, 23–37, 2018.
36. Lbadaoui-Darvas, M., Garberoglio, G., Karadima, K.S., Cordeiro, M.N., Nenes, A., Takahama, S., Molecular simulations of interfacial systems: Challenges, applications and future perspectives. *Mol. Simul.*, 49, 12, 1229–66, 2023 Aug 13.
37. Arcangeli, C., Lico, C., Baschieri, S., Mancuso, M., Characterization of blood–brain barrier crossing and tumor homing peptides by molecular dynamics simulations. *Int. J. Nanomedicine*, 14, 10123, 2019.

38. Jones, D.E., Lund, A.M., Ghandehari, H., Facelli, J.C., Molecular dynamics simulations in drug delivery research: calcium chelation of G3.5 PAMAM dendrimers. *Cogent Chem.*, 2, 1229830, 2016.
39. Missner, A. and Pohl, P., Meyer-Overton rule: Predicting membrane permeability of gases and other small compounds. *ChemPhysChem*, 10, 9–10, 1405–1414, 2009.
40. Yousefpour, A., Modarress, H., Goharpey, F., Amjad-Iranagh, S., Interaction of PEGylated anti-hypertensive drugs, amlodipine, atenolol and lisinopril with lipid bilayer membrane: A molecular dynamics simulation study. *Biochim. Biophys. Acta, Biomembr.*, 1848, 8, 1687–1698, 2015.
41. Barton, A.F.M., *CRC handbook of solubility parameters and other cohesion parameters*, 2nd ed., CRC Press, Boca Raton, FL, 1991.
42. Fedors, R.F., A Method for Estimating Both The Solubility Parameters and Molar Volumes of Liquids. *Polym. Eng. Sci.*, 14, 2, 147–154, 1974.
43. Marsac, P.J., Shamblin, S.L., Taylor, L.S., Theoretical and practical approaches for prediction of drug–polymer miscibility and solubility. *Pharm. Res.*, 23, 10, 2417–2426, 2006.
44. Han, R., Ye, Z., Zhang, Y., Cheng, Y., Zheng, Y., Ouyang, D., Predicting liposome formulations by the integrated machine learning and molecular modeling approaches. *Asian J. Pharm. Sci.*, 18, 3, 100811, 2023.
45. Adnan, A., Lam, R., Chen, H., Lee, J., Schaffer, D.J., Barnard, A.S., Liu, W.K., Atomistic simulation and measurement of pH dependent cancer therapeutic interactions with nanodiamond carrier. *Mol. Pharm.*, 8, 368–374, 2011.
46. Becerra, D., Jois, P.R., Hall, L.M., Coarse-grained modeling of polymers with end-on and side-on liquid crystal moieties: Effect of architecture. *J. Chem. Phys.*, 158, 22, 224901, 14 June 2023.
47. Greiner, M., Elts, E., Briesen, H., Insights into pharmaceutical nanocrystal dissolution: A molecular dynamics simulation study on aspirin. *Mol. Pharm.*, 11, 9, 3009–3016, 2014.
48. Mandal, T., Marson, R.L., Larson, R.G., Coarse-grained modeling of crystal growth and polymorphism of a model pharmaceutical molecule. *Soft Matter*, 12, 39, 8246–8255, 2016.
49. Jha, P.K. and Larson, R.G., Assessing the efficiency of polymeric excipients by atomistic molecular dynamics simulations. *Mol. Pharm.*, 11, 5, 1676–1686, 2014.
50. Siepmann, J. and Peppas, N.A., Higuchi equation: Derivation, applications, use and misuse. *Int. J. Pharm.*, 418, 1, 6–12, 2011.
51. Xiang, T.X. and Anderson, B.D., Molecular dynamics simulation of amorphous Hydroxypropylmethylcellulose and its mixtures with felodipine and water. *J. Pharm. Sci. Res.*, 106, 3, 803–816, 2017.
52. Moghadam, S. and Larson, R.G., Assessing the efficacy of poly(Nisopropylacrylamide) for drug delivery applications using molecular dynamics simulations. *Mol. Pharm.*, 14, 2, 478–491, 2017.

53. Lu, X., Yuan, B., Zhang, X., Yang, K., Ma, Y., Molecular modelling of transmembrane delivery of paclitaxel by shock waves with nanobubbles. *Appl. Phys. Lett.*, 110, 2, 023701, 2017.
54. Huang, W., Mandal, T., Larson, R.G., Computational modelling of hydroxypropylmethylcellulose acetate succinate (HPMCAS) and phenytoin interactions: A systematic coarse-graining approach. *Mol. Pharm.*, 14, 3, 733–745, 2017.
55. Birru, W.A., Warren, D.B., Han, S., Benameur, H., Porter, C.J.H., Pouton, C.W., Chalmers, D.K., Computational models of the gastrointestinal environment. Phase behaviour and drug solubilization capacity of a type I lipid-based drug formulation after digestion. *Mol. Pharm.*, 14, 3, 580–592, 2017.
56. Mahdavi, M., Rahmani, F., Nouranian, S., Molecular simulation of pH-dependent diffusion, loading, and release of doxorubicin in graphene and graphene oxide drug delivery systems. *J. Mater. Chem. B*, 4, 46, 7441–7451, 2016.
57. Hedayati, M.H., Norouzian, D., Aminian, M., Teimourian, S., Ahangari Cohan, R., Sardari, S., Molecular design, expression and evaluation of PASylated human recombinant erythropoietin with enhanced functional properties. *Protein J.*, 36, 36–48, 2017.
58. Cohan, R.A., Madadkar-Sobhani, A., Khanahmad, H., Roohvand, F., Aghasadeghi, M.R., Hedayati, M.H., Design, modelling, expression, and chemo selective PEGylation of a new nano size cysteine analogue of erythropoietin. *Int. J. Nanomed.*, 6, 1217–1227, 2011.
59. Xu, D., Smolin, N., Shaw, R.K., Battey, S.R., Tao, A., Huang, Y., Molecular insights into the improved clinical performance of PEGylated interferon therapeutics: a molecular dynamics perspective. *RSC. Adv.*, 8, 2315–2322, 2018.
60. Yang, C., Lu, D., Liu, Z., How PEGylation enhances the stability and potency of insulin: a molecular dynamics simulation study. *Biochemistry*, 50, 2585–2593, 2011.
61. Mu, Q., Hu, T., Yu, J., Molecular insight into the steric shielding effect of PEG on the conjugated staphylokinase: biochemical characterization and molecular dynamics simulation. *PLoS One*, 8, e68559, 2013.
62. Kacar, G., Molecular understanding of interactions, structure, and drug encapsulation efficiency of pluronic micelles from dissipative particle dynamics simulations. *Colloid Polym. Sci.*, 297, 1037–1051, 2019.
63. Dehaghani, M.Z., Yousefi, F., Seidi, F., Bagheri, B., Mashhadzadeh, A.H., Naderi, G., Esmaeili, A., Abida, O., Habibzadeh, S., Saeb, M.R., Rybachuk, M., Encapsulation of an anticancer drug Isatin inside a host nano-vehicle SWCNT: a molecular dynamics simulation. *Sci. Rep.*, 11, 1, 18753, 2021 Sep 21.
64. Wolski, P., Narkiewicz-Michalek, J., Panczyk, M., Pastorin, G., Panczyk, T., Molecular dynamics modeling of the encapsulation and de-encapsulation of the carmustine anticancer drug in the inner volume of a carbon nanotube. *J. Phys. Chem. C*, 121, 18922–18934, 2017.

65. Katiyar, R.S. and Jha, P.K., Molecular simulations in drug delivery: Opportunities and challenges. *Comput. Mol. Sci.*, 8, 1358, 2018.
66. Ark, S. and Schulten, K., Calculating potentials of mean force from steered molecular dynamics simulations. *J. Chem. Phys.*, 120, 5946–5961, 2004.
67. Ghadri, Z., Raissi, H., Shahabi, M., Farzad, F., Molecular dynamics simulation study of glycine tip-functionalisation of single-walled carbon nanotubes as emerging nanovectors for the delivery of anticancer drugs. *Mol. Simulat.*, 46, 111–120, 2020.
68. Farzad, F. and Hashemzadeh, H., Probing the effect of polyethylene glycol on the adsorption mechanisms of Gem on the hexagonal boron nitride as a highly efficient polymerbased drug delivery system: DFT, classical MD and well-tempered metadynamics simulations. *J. Mol. Graph. Model.*, 98, 107613, 2020.
69. Asadzadeh, H., Moosavi, A., Arghavani, J.H., The effect of chitosan and PEG polymers on stabilization of GF-17 structure: a molecular dynamics study. *Carbohydr. Polym.*, 2020237, 116124, 2020.
70. Kacar, G., Thermodynamic stability of ibuprofen loaded poloxamer micelles. *Chem. Phys.*, 533, 110713, 2020.
71. Zannoni, C., Arcioni, A., Cavatorta, P., Fluorescence depolarization in liquid crystals and membrane bilayers. *Chem. Phys. Lipids*, 32, 3–4, 179–250, 1983.
72. Kumar, G., Kant, A., Kumar, M., Masram, D.T., Synthesis, characterizations and kinetic study of metal organic framework nanocomposite excipient used as extended-release delivery vehicle for an antibiotic drug. *Inorgan. Chim. Acta*, 496, 119036, 2019.
73. Farmanzadeh, D. and Ghaderi, M., A computational study of PAMAM dendrimer interaction with trans isomer of picoplatin anticancer drug. *J. Mol. Graph. Model.*, 80, 1–6, 2018.
74. Dehneshin, N., Raissi, H., Hasanzade, Z., Farzad, F., Using molecular dynamics simulation to explore the binding of the three potent anticancer drugs sorafenib streptozotocin and sunitinib to functionalized carbon nanotubes. *J. Mol. Model.*, 25, 159, 1–15, 2019.
75. Gupta, S. and Biswas, P., Effect of pH on size and internal structure of poly (propylene imine) dendrimers: a molecular dynamics simulation study. *J. Phys. Chem. B*, 122, 9250–9263, 2018.
76. Gao, Y., Shi, Y., Fu, M., Feng, Y., Lin, G., Kong, D. *et al.*, Simulation study of the effects of interstitial fluid pressure and blood flow velocity on trans vascular transport of nanoparticles in tumor microenvironment. *Comput. Meth. Prog. Biol.*, 193, 105493, 2020.
77. Settanni, G., Schäfer, T., Muhl, C., Barz, M., Schmid, F., Polysarcosine and poly(ethylene-glycol) interactions with proteins investigated using molecular dynamics simulations. *Comput. Struct. Biotechnol.*, 16, 543–550, 2018.
78. Magarkar, A. and Karakas, E., Molecular Dynamics Simulation of PEGylated Bilayer Interacting with Salt Ions: A Model of the Liposome Surface in the Bloodstream. *J. Phys. Chem. B*, 116, 14, 4212–9, 2012

79. Ambrosio, E., Podmore, A., Gomes Dos Santos, A.L., Magarkar, A., Bunker, A., Caliceti, P., Control of peptide aggregation and fibrillation by physical PEGylation. *Biomacromolecules*, 19, 3958–3969, 2018.
80. Cern, A., Barenholz, Y., Tropsha, A., Goldblum, A., Computer-aided design of liposomal drugs: *in silico* prediction and experimental validation of drug candidates for liposomal remote loading. *J. Control. Release*, 173, 125–131, 2014.
81. Grillo, D.A., Albano, J.M.R., Mocskos, E.E., Facelli, J.C., Pickholz, M., Ferraro, M.B., Mechanical properties of drug loaded diblock copolymer bilayers: a molecular dynamics study. *J. Chem. Phys.*, 148, 214901, 2018.
82. Barraza, L.F., Jiménez, V.A., Alderete, J.B., Effect of PEGylation on the structure and drug loading capacity of PAMAM-G4 dendrimers: a molecular modelling approach on the complexation of 5-flourouracil with native and PEGylated PAMAM-G4. *Macromol. Chem. Phys.*, 216, 1689–1701, 2015.
83. Róg, T. and Vattulainen, I., Cholesterol, sphingolipids, and glycolipids: what do we know about their role in raft-like membranes? *Chem. Phys. Lipids*, 184, 82–104, 2014.
84. Bulbake, U., Doppalapudi, S., Kommineni, N., Khan, W., Liposomal Formulations in Clinical Use: An Updated Review. *Pharmaceutics*, 9, 2, 12, 2017.
85. Poojari, C., Zak, A., Dzieciuch-Rojek, M., Bunker, A., Kepczynski, M., Róg, T., Cholesterol reduces partitioning of antifungal drug itraconazole into lipid bilayers. *J. Phys. Chem. B*, 124, 2139 2148, 2020.
86. Poojari, C., Wilkosz, N., Lira, R.B., Dimova, R., Jurkiewicz, P., Petka, R. *et al.*, Behavior of the DPH fluorescence probe in membranes perturbed by drugs. *Chem. Phys. Lipids*, 223, 1047–84, 2019.
87. Pantsar, T., KRAS(G12C)–AMG 510 interaction dynamics revealed by all-atom molecular dynamics simulations. *Sci. Rep.*, 10, 119–92, 2020.
88. Nussinov, R., Tsai, C., Jang, H., Oncogenic Ras Isoforms Signaling Specificity at the Membrane. *Cancer Res.*, *78*, 593–602, 2018.
89. Walden, D.M., Bundey, Y., Jagarapu, A., Antontsev, V., Chakravarty, K., Varshney, J., Molecular Simulation and Statistical Learning Methods toward Predicting Drug–Polymer Amorphous Solid Dispersion Miscibility, Stability, and Formulation Design. *Molecules*, 26, 182, 1–23, 2021.
90. Meanwell, N.A., Improving drug candidates by design: A focus on physicochemical properties as a means of improving compound disposition and safety. *Chem. Res. Toxicol.*, 24, 1420–1456, 2011.
91. Palucki, M., Higgins, J.D., Kwong, E., Templeton, A.C., Strategies at the interface of drug discovery and development: Early optimization of the solid-state phase and preclinical toxicology formulation for potential drug candidates. *J. Med. Chem.*, 53, 5897–5905, 2010.
92. Rahim, M.A., Jan, N., Khan, S., Shah, H., Madni, A., Khan, A., Jabar, A., Khan, S., Elhissi, A., Hussain, Z., Aziz, H.C., Sohail, M., Khan, M., Thu, H.E.,

Recent Advancements in Stimuli Responsive Drug Delivery Platforms for Active and Passive Cancer Targeting. *Cancers (Basel)*, 13, 4, 1–52, 2021.
93. Thorat, N.D. and Bauer, J., Functional smart hybrid nanostructures based nanotheranostic approach for advanced cancer treatment. *Appl. Surf. Sci.*, 146809, 2020.
94. Wang, F., Xiao, J., Chen, S., Sun, H., Yang, B., Jiang, J., Zhou, X., Du, J., Polymer Vesicles: Modular Platforms for Cancer Theranostics. *Adv. Mater.*, 30, e1705674, 2018.
95. Kabedev, A., Hossain, S., Hubert, M., Larsson, P., Christel, A.S., Bergstrom, Molecular Dynamics Simulations Reveal Membrane Interactions for Poorly Water-Soluble Drugs: Impact of Bile Solubilization and Drug Aggregation. *J. Pharm. Sci.*, 110, 176–185, 2021.
96. Piana, S. and Gale, J.D., Understanding the Barriers to Crystal Growth: Dynamical Simulation of the Dissolution and Growth of Urea from Aqueous Solution. *J. Am. Chem. Soc.*, 127, 1975–1982, 2005.
97. Parks, C., Koswara, A., Tung, H.H., Nere, N.K., Bordawekar, S., Nagy, Z.K., Ramkrishna, D., Nanocrystal Dissolution Kinetics and Solubility Increase Prediction from Molecular Dynamics: The Case of α-, β-, and γ-Glycine. *Mol. Pharm.*, 14, 1023–1032, 2017.
98. Mandal, T., Huang, W., Mecca, J.M., Getchell, A., Porter, W.W., Larson, R.G., A framework for multi-scale simulation of crystal growth in the presence of polymers. *Soft Matter*, 13, 1904–1913, 2017.
99. Sosso, G.C., Chen, J., Cox, S.J., Fitzner, M., Pedevilla, P., Zen, A., Michaelides, A., Crystal Nucleation in Liquids: Open Questions and Future Challenges in Molecular Dynamics Simulations. *Chem. Rev.*, 116, 7078–7116, 2016.
100. Montalbán, M.G., Chakraborty, S., Peña-García, J., Verli, H., Villora, G., Horacio Pérez-Sánchez, F., Díaz-Baños, G., Molecular insight into silk fibroin-based delivery vehicle for amphiphilic drugs: Synthesis, characterization and molecular dynamics studies. *J. Mol. Liq.*, 299, 112–156, 2020.
101. Kocaaga, B., Seniha Guner, F., Kurkcuoglu, O., Molecular dynamics simulations can predict the optimum drug loading amount in pectin hydrogels for controlled release. *Materials*, 31, 103268, 2022.
102. Tian, Y., Shi, C., Sun, Y., Zhu, C., Sun, C.C., Mao, S., Designing Micellar Nanocarriers with Improved Drug Loading and Stability Based on Solubility Parameter. *Mol. Pharm.*, 12, 3, 816–825, 2015.
103. Wang, F., Xiao, J., Chen, S., Sun, H., Yang, B., Jiang, J., Zhou, X., Du, J., Polymer Vesicles: Modular Platforms for Cancer Theranostics. *Adv. Mater.*, 30, e1705674, 2018.
104. Dutta, S., Corni, S., Brancolini, G., Molecular Dynamics Simulations of a Catalytic Multivalent Peptide–Nanoparticle Complex. *Int. J. Mol. Sci.*, 22, 7, 3624, 2021.
105. Bunker, A., Poly(Ethylene Glycol) in Drug Delivery, Why Does it Work, and Can We do Better? All Atom Molecular Dynamics Simulation Provides Some Answers. *Phys. Proc.*, 34, 24–33, 2012.

106. Lin, P. and Colina, C.M., Molecular simulation of protein–polymer conjugates. *Curr. Opin. Chem. Eng.*, 23, 44–50, 2019.
107. Ko, J.H. and Maynard, H.D., A guide to maximizing the therapeutic potential of protein–polymer conjugates by rational design. *Chem. Soc. Rev.*, 47, 8998–9014, 2018.
108. Shu, J.Y., Panganiban, B., Xu, T., Peptide-polymer conjugates: from fundamental science to application. *Annu. Rev. Phys. Chem.*, 64, 631– 657, 2013.
109. Liu, M., Tirino, P., Radivojevic, M., Phillips, D.J., Gibson, M., II, Leroux, J.-C., Gauthier, M.A., Molecular Sieving on the Surface of a Protein Provides Protection Without Loss of Activity. *Adv. Funct. Mater.*, 23, 2007–2015, 2013.
110. Gumz, H., Boye, S., Iyisan, B., Krönert, V., Formanek, P., Voit, B., Lederer, A., Appelhans, D.A.-O., Toward Functional Synthetic Cells: In-Depth Study of Nanoparticle and Enzyme Diffusion through a Cross- Linked Polymersome Membrane. *Adv. Sci.*, 6, 1801299, 2019.
111. Taylor, P.A. and Jayaraman, A., Molecular Modeling and Simulations of Peptide-Polymer Conjugates. *Annu. Rev. Chem. Biomol. Eng.*, 11, 1, 257–276, 2020.
112. Amaro, R.E., Baron, R., McCammon, J.A., An improved relaxed complex scheme for receptor flexibility in computer-aided drug design. *J. Comput. Aided Mol. Des.*, 22, 693–705, 2008.
113. de Oliveira, C.A.F., Hamelberg, D., McCammon, J.A., On the application of accelerated molecular dynamics to liquid water simulations. *J. Phys. Chem. B*, 110, 22695–22701, 2006.
114. Bunker, A., Magarkar, A., Viitala, T., Rational design of liposomal drug delivery systems, a review: Combined experimental and computational studies of lipid membranes, liposomes and their PEGylation. *Biochim. Biophys. Acta, Biomembr.*, 1858, 10, 2334–2352, 2016.
115. Lehtinen, J., Magarkar, A., Stepniewski, M., Hakola, S., Bergman, M., Róg, T., Yliperttula, M., Urtti, A., Bunker, A., Analysis of cause of failure of new targeting peptide in PEGylated liposome: molecular modeling as rational design tool for nanomedicine. *Eur. J. Pharm. Sci.*, 46, 3, 121–130, 2012.
116. Lee, H., Effects of nanoparticle electrostatics and protein-protein interactions on corona formation: conformation and hydrodynamics. *Small*, 16, 1906598, 2020.
117. Ramezanpour, Leung, S.S.W., Delgado-Magnero, K.H., Bashe, B.Y.M., Thewalt, J., Tieleman, D.P., Computational and experimental approaches for investigating nanoparticle-based drug delivery systems. *Biochim. Biophys. Acta, Biomembr.*, 1858, 7, 2016.
118. Casalini, T., Not only *in silico* drug discovery: Molecular modeling towards *in silico* drug delivery formulations. *J. Control. Release*, 332, 390–417, 2021.
119. Siepmann, J. and Siepmann, F., Mathematical modeling of drug delivery. *Int. J. Pharm.*, 364, 328–343, 2008.

120. Koetting, M.C., Peters, J.T., Steichen, S.D., Peppas, N.A., Stimulus-responsive hydrogels: theory, modern advances, and applications. *Mater. Sci. Eng., R*, 93, 1–49, 2015.
121. Nel, A.E., Madler, L., Velegol, D., Xia, T., Hoek, E.M.V., Somasundaran, P., Klaessig, F., Castranova, V., Thompson, M., Understanding biophysicochemical interactions at the nano-bio interface. *Nat. Mater.*, 8, 543–557, 2009.
122. Yu, L.X., Amidon, G., Khan, M.A., Hoag, S.W., Polli, J., Raju, G.K., Woodcock, J., Understanding pharmaceutical quality by design. *AAPS J.*, 16, 771–783, 2014.
123. Riniker, S., Fixed-charge atomistic force fields for molecular dynamics simulations in the condensed phase: an overview. *J. Chem. Inf. Model.*, 58, 565–578, 2018.
124. Ramananarivo, S., Ducrot, E., Palacci, J., Activity-controlled annealing of colloidal monolayers. *Nat. Commun.*, 10, 1, 3380, 2019 Jul 29.
125. Marrink, S.J., Risselada, H.J., Yefimov, S., Tieleman, D.P., de Vries, A.H., The MARTINI force field: coarse grained model for biomolecular simulations. *J. Phys. Chem. B*, 111, 7812–7824, 2007.
126. Thorat, N.D. and Bauer, J., Functional smart hybrid nanostructures based nanotheranostic approach for advanced cancer treatment. *Appl. Surf. Sci.*, 146809, 2020.

12

Controlling the Drug Release Rate and Targeted Drug Delivery to the Desired Site by Molecular Simulation

Deepika Bairagee[1]*, Sunita Panchawat[2], Neelam Jain[3] and Sirisha Pingali[4]

[1]Department of Pharmacy, Acropolis Institute of Pharmaceutical Education and Research, Indore, Madhya Pradesh, India
[2]Department of Pharmaceutical Sciences, Mohan Lal Sukhadia University, Udaipur, Rajasthan, India
[3]Oriental College of Pharmacy and Research, Oriental University, Indore, Madhya Pradesh, India
[4]Consultant QC Scientist, IPT, Lonza Biologics, Slough, United Kingdom

Abstract

Molecular simulations may be used to design *in silico* drug delivery formulations, reducing the requirement for *in vitro* and *in vivo* testing because they anticipate formulation qualities prior to synthesis. These simulations provide important molecular information that is often unavailable through modern experimental methods. Although these methods were originally limited to predicting single-molecule behavior (e.g., drug alignment in a cell membrane), progress in computer promptness and cost-effective computation offloading have enabled them to be applied to manifestations occurring at much significantly bigger length scales (e.g., carrier–drug chemical bonding) with a relatively small supercomputing budget and effort. We present a non-mathematical overview of biochemical modeling techniques and their application in medication administration, with a focus on atomistic and coarse-grained Monte Carlo (MC) and molecular dynamics (MD) approaches, but eliminating medication docking investigations. Performance measures and challenges using these capabilities in the early stages of drug delivery, along with instances of molecular simulations in biomedical applications, are discussed.

**Corresponding author*: bairagee.deepika@gmail.com
Deepika Baraigee: ORCID: https://orchid.org/0000-0001-7877-9315
Neelam Jain: ORCID: https://orchid.org/0000-0002-4885-7930

Rishabha Malviya, Sonali Sundram and Dhanalekshmi Unnikrishnan Meenakshi (eds.)
Drug Delivery Systems Using Quantum Computing, (353–388) © 2024 Scrivener Publishing LLC

Although molecular simulations are expected to fulfill a significant improvement in the progress of drug delivery, computational scientists, experimentalists, and corporate workers active in the sector must work together to find particular areas in which these simulations could be especially helpful.

***Keywords*:** Quantam computing, drug delivery, molecular simulation, drug release, *in-vivo*, molecular dynamics, drug alignment, cell membrane

Abbreviations

3D	3 Dimensional
ADMET	Adsorption, Distribution, Metabolism, Excretion and Toxicity
aMD	Accelerated Molecular Dynamics
BCS	Biopharmaceutics Classification Method
CFD	Computational Fluid Dynamics
CG	Coarse-Grained
CNT	Carbon nanotubes
CPU	Control Processing Unit
DFT	Density Functional Theory
DNA	Deoxyribonucleic acid
GIT	Gastrointestinal Tract
GPUs	Graphics Processing Units
HF	Hatree-Fock
HPMC	Hydroxypropyl-methylcellulose
HPMCAS	Hydroxypropyl-methylcellulose acetate succinate
IPDU	Isophorone Diurethane
kMC	kinetic Monte Carlo
MC	Monte Carlo
MD	Molecular Dynamics
NMR	Nuclear Magnetic Resonance
PBLG	Poly gamma benzyl-l-Glutamate
PEG	Poly ethylene glycol
PLGA	Poly Lactic-co-Glycolic Acid
QM	Quantum-Mechanical
SWCNTs	Single-walled carbon nanotubes

12.1 Introduction

"The building of a mainframe simulation is necessary for the sake of investigation," according to the Oxford Dictionary (2017). Undeniably, processor imitations have proven to be quite effective at simulating reality, and have proven to be useful tools for understanding complicated systems and testing theories, particularly in situations when tests are impractical, costly, or dangerous. The wonderfully animated videos of the simulation output, on the other hand, add to the illusion, as they frequently reveal far more information than can be validated by testing. That is also particularly true for nanotechnology simulations, that even the most sophisticated imaging techniques can only acquire a portion of the structural and dynamical information provided by a molecular simulation. Nevertheless, the fact that fresh investigations are confirming a growing number of prior computer simulation conclusions has increased our faith in computer simulations. There has been a paradigm shift with the use of computational models in recent decades, moving away from understanding complex systems and toward evolutionary computation (*in silico*) design techniques, which include machine learning screening of components for an application purpose by attempting to predict the mechanical and physical properties of biomaterials aforementioned in synthesis [1].

"Molecular simulation" or "molecular modeling" refers to any theoretical and mathematical approaches (Table 12.1) used to analyze the behavior of a particle or a structure of molecules [2]. The movement of a chemical is fundamentally linked to the occurrence of its atoms, which is linked to the movement of its atomic components. For an accurate description of molecular activity, quantum-mechanical (QM) approaches that efficiently answer the Schrödinger equation are necessary. QM approaches, on the other hand, turn out to be computationally costly and often unneeded in establishing the properties of interest in sophisticated atoms and molecules. The frequency of a substance's atoms is essentially tied to the movement of its atomic components, which is linked to the frequency of the substance's atoms. QM techniques that successfully explain the Schrödinger equation are required for an accurate representation of biochemical activity. On the other hand, QM techniques prove to be expensive and frequently needless for finding the belongings in complicated chemical systems. Atoms in molecules move because of forces from nearby atoms (sometimes referred to as the "force field") that are transmitted by atoms in identical or nearby molecules or by fragments in the system as a whole. The atomistic Monte Carlo (MC) methodology differs from the MD method in terms of methodology,

Table 12.1 List of common molecular simulation approaches used in medication administration.

Simulation method	Variants	Applications
QM methods	Density functional theory (DFT), Hartree–Fock (HF) theory, semiempirical and QM/MM methods [5]	Docking, force-field parameterization, geometry optimization, and potential energy
MC	Atomistic Monte Carlo [6]	Docking, adsorption/binding energy, and free energy
	Coarse-grained Monte Carlo [7]	Membrane translocation, gel carrier swelling, and self-assembly
	Lattice Monte Carlo, kMC [8]	Crystallization and drug release from excipient matrices
MD	Atomistic molecular dynamics [9]	Drug aggregation, crystallization, glass transition, drug miscibility, hydrogen bonding, diffusivity, membrane permeability, carrier–drug miscibility, and carrier–drug interaction.
	Coarse-grained molecular dynamics [10]	Drug release from excipient matrices, membrane translocation, and self-assembly

albeit using the above assumptions and the idea of the gravitational field [3]. Boltzmann weights are used to sample a network of components' permutations. Intelligent trial moves (steps) are utilized to build sequential topologies made up of randomized atom deformations that are approved with a mathematical probability on the test method's global electricity shift. In contrast to MD, which statistically integrates equations of motion using short multidivisional, MC computations omit a perception of time, hence the formations measured in an MC simulation do not understand the actual course of molecules. However, when both the MC and MD techniques are employed regularly and properly, they provide identical results in terms of anticipating specific settlement (for example, a couple of statistical parameters) [1].

Just for clarity, we will speak of both MC and MD simulations as MC/MD computations. Nevertheless, viewers should be aware that the foundation techniques are radically different. Although processing capacity scales analogously due to both the abundance of clock cycles in MD simulations and the sequence of activities in MC simulations for a known set dimension, the phase margin in an medical electronic device and the timestep in a simulation investigation have different meanings. The study's most basic time scales (measured in nanoseconds) are related to dynamic simulation. Conversely, in MD simulations, the projected time and the number of steps are related to the number of time-steps and stages, respectively.

In a separate section, the benefits and drawbacks of MC and MD simulations will be examined. It is worth noting that in comparison to MC simulations, MD simulations have been employed more frequently for medication delivery concerns. As a result, we focus primarily on MD simulations and skip the discussions on the topical advances in MC simulations that have yet to be applied to drug delivery challenges [1, 4].

12.1.1 MC and MD Processes

Midway through the twentieth century, the emergence of the first computers coincided with the development of MC and MD procedures. On the other hand, potential uses of these approaches were incomplete to toy models like as the "hard sphere" model for modeling fully flexible collisions or spheres interrelating with a Lennard–Jones potential for simulating liquid Argon dynamics [11]. It has been only gradually feasible to implement more complicated atomic models of composite fragments (like proteins) due to exponential growth in processor speed and the development of effective computational tools. System sizes of 15 nm and simulation periods of 100 ns (for MD) are normally doable with today's modern computing power. To comprehend the computational complexity of these simulations, consider a situation of N atoms with a pairing connection strength among all couples of atoms. Calculating N2, pair-interaction forces or energy would be one aspect of the simulation. Such computations must always be done after every simulation stage and the simulation must always be run again until required features converge to an average value since pair interactions change as inter-atomic ranges grow ("thermodynamic equilibrium"). Recognizing that not every pair interaction needs to be computed for narrow engagements might result in significant speedups since interaction energies are substantially smaller than infrared radiation (kBT) over long distances (e.g., van der Waals interactions).

To completely analyze all N-particle systems modes (for MC simulations) or minimize errors in the arithmetical integration of the equation of motion in the MD/MC process, the research period must always be strictly constrained (for MD simulations). MD simulations of atomistic systems generally include 1 million clock cycles during 1 fs to study the set of parameters for 1 ns. Simulations must be performed for a period of time that is considerably longer than the time scales of critical events to identify certain thermodynamic properties [1, 12].

12.1.2 Correlation of Continual Methods with Molecular Simulation

In drug delivery challenges, discontinuous methodologies (Table 12.2), including those concerning the methodical and arithmetical treatment of convective heat calculations or computational fluid dynamics (CFD) simulations, have been used with some effectiveness [11]. To better understand the ADMET (adsorption, distribution, metabolism, excretion, and toxicity) appearances of psychoactive mixtures, partition simulations of drug release via compounds, in particular, have been very successful. As a result, they are frequently used in the formulation of cosmetics and the determination of dosage levels. There is no imposed restriction to the period and length scales that may be reproduced as extended as even the modeling criteria are satisfied, making similar approaches useful tools if reliable parameters of the model can be obtained from these *in vitro/in vivo* experiments [1, 13].

Such methods, on the other hand, contain two significant problems. For starters, since these continuous methodologies do not incorporate individualist specifics, data relating to formulated interaction is individually gathered at a high stage, prohibiting exact molecular insights. As little more than a consequence, while structural simulations can be employed before synthesis in the designing phase, these approaches are primarily used during the formation. Furthermore, the necessary investigations take a lot of money and time. Many *in vitro* and *in vivo* studies, particularly those into the deaths of food animals, are required since several processes show poor *in vitro* to *in vivo* linkage. Improved *in vitro* and *in vivo* correlations could be achieved in atomistic simulations, which decrease the number of trials needed [14] if the fundamental biochemistry is recognized.

Table 12.2 Comparison of molecular simulation and continuous simulation methodologies [1].

Attribute	Continuous simulations	QM calculations	Atomistic simulations	Coarse-grained simulations
Simulation output	Profiles of velocity, temperature, and concentration of species	Charge distribution, electronic structure, and potential energy	Coordinates and momenta of atoms, thermodynamic properties	Coordinates and momenta of coarse-grained particles, thermodynamic properties
Assumptions	Continuum approximation	Mean field approximation, neglect of electron corrections	The motion of atoms approximated by the motion of nuclei obeying Newtonian mechanics	Effective degrees of freedom
Governing equations	Transport equations (fluid flow, heat transfer, mass transfer)	Schrödinger equation	Newton's laws of motion, statistical mechanics	Newton's law of motion, Langevin equation, statistical mechanics
Model parameters	Viscosity, heat transfer coefficient, and diffusivity	None	Force-filed parameters (bonded and nonbonded)	Effective force-field parameters (bonded and nonbonded)
Time scale	Typically < 1 ns (within current computational capabilities)	None	Typically < 100 ns (within current computational capabilities)	Depends on coarse-graining resolution (between atomistic and continuum)
Length scale	Typically > 1 μm (continuum approximation should hold)	Typically ~ 1 nm (within current computational capabilities)	Typically < 30 nm (within current computational capabilities)	Depends on coarse-graining resolution (between atomistic and continuum)

12.1.3 Simulation Scheme

A typical atomistic MC/MD simulation's rounds are shown in Figure 12.1. Free energy computations and grand-canonical cluster simulations, in particular, necessitate a slightly more complicated methodology, which will be described individually in the following sections. A structural construction program (such as Materials Studio, Marvin, and GaussView) or a resource (such as the Protein Data Bank and ZINC [15]) is being used to generate

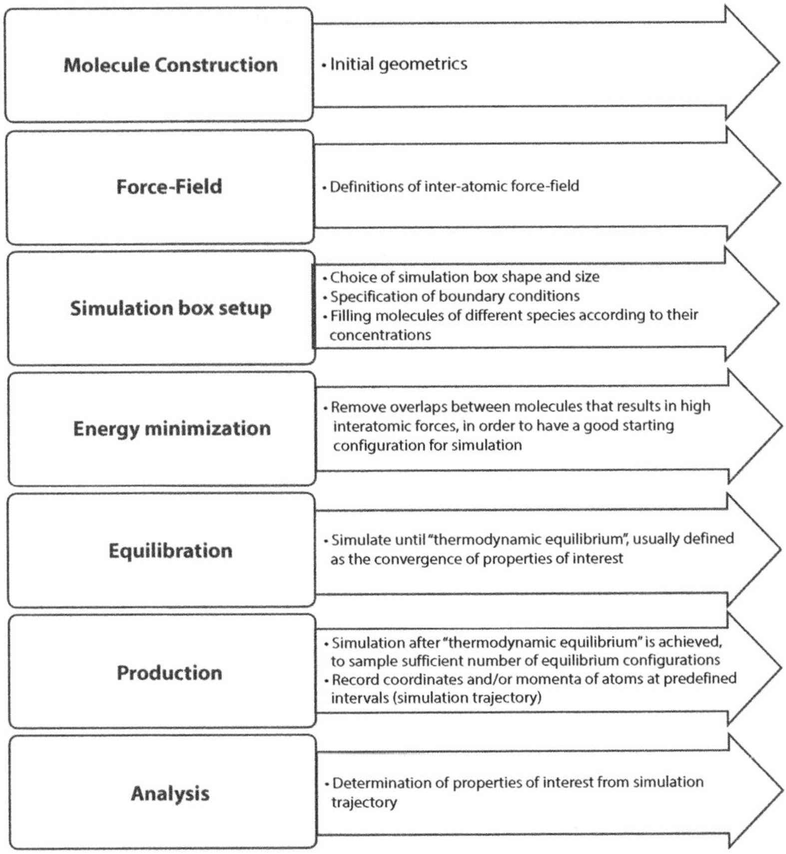

Figure 12.1 The fundamental steps in atomistic simulations MC/MD.

the foundation architecture of the drug and several additional substances under inquiry (such as transporters, transmembrane components, and solvents).

The equilibration and production stages of MC/MD simulations begin with the arrangement acquired and subsequently dynamism lowering. Equilibration simulations are used to reach "thermodynamic equilibrium," as the term implies. Meanwhile, the whole amount of water molecules in the system is far lower than in a remote or macro-environment; the term "thermodynamic equilibria" is used generically here. In general, variations in thermodynamic variables that diminish with the opposite linear root of a number of nodes are exaggerated. A thermodynamic property of interest is also typically assessed and recorded, and when it seems to converge on or deviate from an average score, it is said to have "equilibrated." This method captures physiological relaxation modes at period scales smaller than the simulated period, but not slower relaxation modes, for instance, small molecules (e.g., of macromolecules). Because our capacity to obtain data is limited by computer processing power, the maximum simulation time is almost always much shorter than the experimental period. The entire relaxed characteristic of biomolecules is very far from perfect. Comparable limits exist in the study such as crystal formation. However, such processes can already be mimicked for minuscule particles [16].

Starting with an equilibrated setup, the production simulation commences. They used this setup to collect data from numerous optimum setups to calculate the desirable belongings. This setups both MC and MD simulations capture the variations in atom locations over time, allowing structural metrics like molecule size and pair distribution function to be calculated. On either side, MD simulations keep track of the rotational motion of atomic orbits, which can be utilized to calculate dynamic parameters like the diffusion rate. Predictions of the means and variations of thermodynamics need to be appropriate in order for the sampled configuration to be mutually independent, which means that the time interval between two specimens must be considerably greater than the correlation time. The connection is additionally assumed to just be ergodic, which means that 'night before going to bed' features derived from simulations for a sampling element match the spatial-averaged values for the entire system. The ergodic assumption is acknowledged by the efficacy of MC/MD approaches in forecasting equilibrium features for a wide variety of systems [1, 17]. However, it is disputed whether such an assumption holds for a simulated system.

12.1.4 Comparison Between MC and MD Processes

As previously stated, in many circumstances, MC and MD approaches may be utilized interchangeably to forecast specific thermodynamic features that are around over a great number of scheme configurations throughout a manufacturing run. However, their approaches to exploring phase space, that is, how they get multiple configuration settings, are fundamentally different. The trajectory of the system's molecules, which is generated by resolving differentiated equations of motion, is shown in a set that is generated at a subsequent time in an MD simulation. However, on the other hand, an MC motion is a random movement of a system's atom or group of atoms that is formally acknowledged with a likelihood, with excellent potential allocated to movements that lead to lower energy or smaller energy changes ΔE = Enew - Eold. This is a global procedure unlike force computations in MD simulation, which compute the energy for the overall network. Most MC simulations utilize the Metropolis criteria, which approve movement that leads to a negative ΔE with a possibility equivalent to exp(ΔE/kBT), as predicted by the Boltzmann distribution [1]. Movements that result in good E are allowed with a likelihood similar to exp(ΔE/kBT).

12.1.5 Size and Boundary Conditions of the System

The simulated box, as stated previously, must be big enough just to replicate the real system, while yet being small enough to be simulated using current computing resources. In reality, modern quantum computers can manage simulation box sizes of 10–30 nm for billions of years [18, 19] of 100 ns simulations (for MD). Any use of different boundary conditions in nanoscale simulations makes it easier to investigate bulk system behavior. Whenever we presume that the simulation's setup is accurate, periodic boundary conditions (PBCs), as shown in Figure 12.2, are commonly

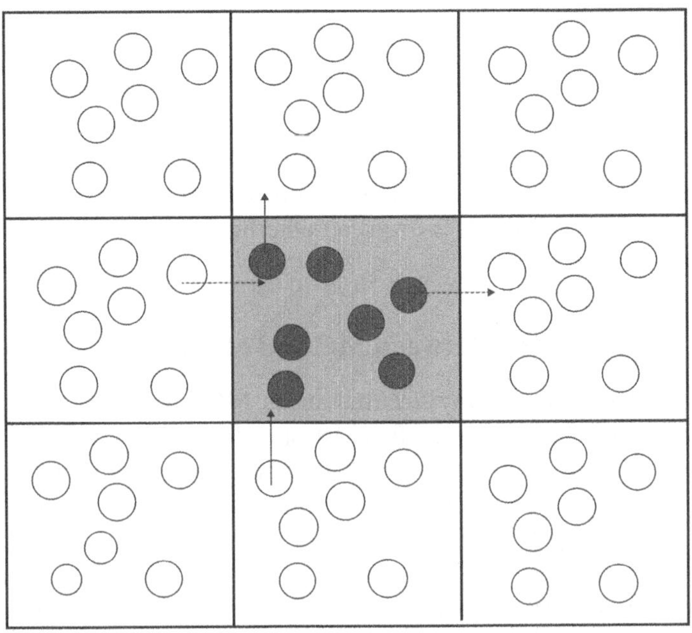

Figure 12.2 A two-dimensional (2D) illustration of a regular border. The central cell is a representation of the simulation box (filled with yellow). In the simulation box, occupied nodes represent atoms, while empty circles show their periodic image in other cells. Bold and dotted lines at the boundary represent the motion between two particles, and when one component leaves the simulated box, its duplicate joins the box from the extreme side.

used. Occasionally, the container is replicated in an endless lattice. This involves the usage of a three-dimensional (3D) lattice since the simulation box is necessary to be both space-filling and proportionate. A simulation box can be cubic but not spherical since it does not take up the area. Due to the appearance of a "crystalline" organization at distance scales far beyond the extent of the network, PBC results in such finite-size aberrations. To reduce these artifacts, the simulation must be done for high enough system scales. This can be made sure of in practice by performing simulations with different scheme dimensions larger until the necessary characteristics coincide. Nevertheless, similar convergence rarely happens for those other qualities (such as diffusion), but instead, a clear feature climbing with scheme dimension is established. We must be skeptical of simulation conclusions since that may not be possible to simulate schemes above a confident scale to prove converging with the number of nodes in a number of other situations (Box 12.1). Because biomolecules, including polymers and proteins, are too big to model and can only be explored partially, their models are more complicated. In such circumstances, the simulation box should be far greater than the largest potential extent of the component in the extended state [20].

12.1.6 Coarse-Graining

The efficiency of computing is much enhanced when molecules are represented in a coarse-grained (CG) fashion instead of an atomistic manner (Box 12.2). Unified atom components (such as GROMOS and OPLS-UA) give a smaller coarse-graining by substituting a united atom chosen to represent the communication center of CHn for carbon and n hydrocarbon atoms, which avoids explicitly integrating H atoms connected to aliphatic and aromatic carbon atoms. The solvent molecules are handled just like in the all-atom protective shield, meanwhile, the H atoms connected to oxygen are all still specifically addressed. As a result, the computational effectiveness of connected atoms has been most clearly demonstrated in solid-state simulations of large molecular-weight hydrocarbons. Additionally, including large amounts of data in a liquid state, the MARTINI water model offers significant processing advantages even though it is composed of four water molecules. By eliminating the solution particles and taking into consideration the solvent impact in the interparticle force parameterization between molecules, the dry MARTINI force field [21] advances it even further. The bead-spring models, which have shown to be efficient tools for qualitative thoughtful of polymer motion, have stimulated increasingly severe coarse-graining procedures. In these models, the coarse-graining level, the quantity of "beads," and the interaction between beads are all loosely described in order to capture the basic scalability rules applied to enormous molecular weights. A common process for creating such parameterization goes through the following stages. Firstly, whenever a CG model is built, the interparticle force form is chosen. Secondly, simulations for a small-scale system are run using both atomistic and CG methods. Thirdly, the "target" values of a property are established using the outcomes of both simulations. After that, in order to achieve the "target" values, CG model limits or the geometry of inter-particle services are adjusted. This procedure is done until computer graphics calculations can accurately replicate the "target" standards. The produced CG model can be utilized in place of an essentialist simulation when it is not possible.

12.1.7 Free Energy Calculation

We are interested in free energies [22] rather than equilibrium characteristics derived using traditional MC/MD simulations for various applications in drug administration. Consider the formation of drug–receptor complexes as an example [23], in which the equilibrium characteristic of the adsorption system is done by the complex's binding affinities. The thermodynamics schemers, which include integrating the energy alterations along a predetermined path of a connection parameter, are widely used to calculate these free energies. For instance, we dramatically improve the drug–receptor connection from 0 (unbound state) to 1 (bound state). Sluggish solvent equilibration limits the creation of such complexes, making these computations time-consuming when predicting free energy deviations in a resolution stage. As demonstrated in Figure 12.3, the concept of a thermodynamic cycle can be utilized to quickly determine binding energy for drug–receptor binding. The free electricity changes of uncontrolled drug and receptor desolvation, as well as the free energy changes of bound drug–receptor complex solvation, are added to the free energy changes of thermodynamic integration in a vacuum. The estimation of drug and carrier solvation-free energy is another, more basic use of this idea. Two other techniques for determining free energy fluctuations used include an unfairness potential and the metadynamics strategy. According to Filion, Hermes, Ni, and Dijkstra [24], these methods are enormously beneficial for examining unexpected occurrences like crystal nucleation and growth.

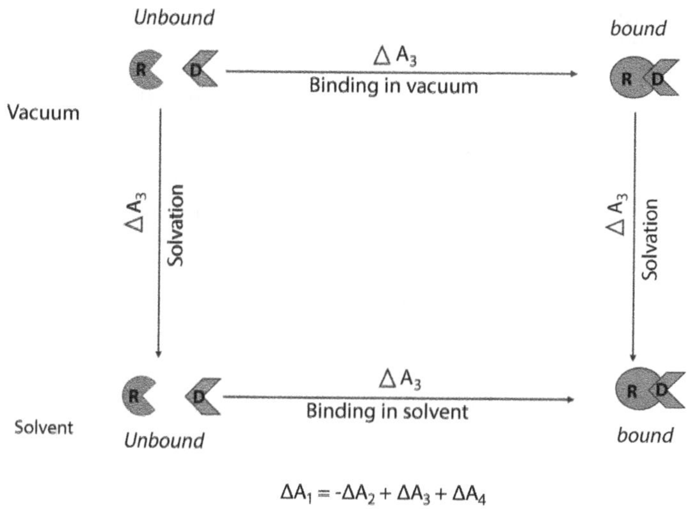

Figure 12.3 Thermodynamic cycle can be utilized to quickly determine binding energy for drug–receptor binding.

12.1.8 Force Fields

In molecular simulations, a wide range of force fields is used, some of which are given in Table 12.3. Although the practical form and model limitations of distinct communications and strictures vary significantly among force fields, they generally include the subsequent communications:

1. In a molecule, the bonds stretch between linked pairs of atoms.
2. A molecule's three successive atoms bend angles at different angles.
3. A "correct" dihedral in a molecule is characterized as a bond spin (torsion) among four subsequent atoms.
4. Out-of-plane twisting sustains certain molecules' planar forms, even in the case of benzene, which would be represented by an "inappropriate" dihedral consisting of a number of its atoms.
5. Due to partial covalent bonding, Coulombic contact occurs among couples of particles from an identical or dissimilar fragment(s).
6. When two atoms from the same or different molecule come into contact, this is known as a van der Waals interaction.

Table 12.3 List of frequently utilized force fields.

Force filed family	Variants	Original developer
MARTINI	MARTINI	Siewert J. Marrink, University of Groningen and Peter Tieleman, University of Calgary
MMFF	MMF and MMFF94	Merck Research Laboratories
GROMOS	GROMOS43a1, 43a2, 43b1, 45a3, 53a5, and 53a6	Wilfred van Gunsteren, University of Groningen and ETH Zurich
OPLS	OPLS-AA and OPLS-UA	William Jorgensen, Purdue University and Yale University
AMBER	GAFF, GLYCAM, ff94, ff96, ff98, ff99, ff02, ff02EP, and so on	Peter Koltman, University of California, San Francisco
Consistent force fields	COMPASS, COMPASS II, CFF, CFF91, and PCFF	Molecular Simulations Inc. (now Biovia)
UFF	UFF	William Andrew Goddard III, California Institute of Technology
CHARMM	CGenFF, CHARMM22, CHARMM27, and CHARMM36	Martin Karplus, Harvard University

Particular electromagnetic forces may also comprise cross words containing more than one of the above-mentioned interaction kinds. Although they are all components of a molecular dynamics protective shield, angles, dihedrals, and coverage probability are not essential to the QM model. Rather, they arise as a consequence of the translation between the QM description and the atomistic representation. Fractional charge is a measurement of the negative charge distribution around an atom that takes on negative (positive) values for electron-deficient (electropositive) atoms. As a result, QM calculations can be used to determine the force-field parameters fully. However, in actuality, these assessments take a lot of time, and the force-accuracy field decreases as a consequence of the limits of the empirical data utilized for force-field parameterization and the compromises made when the force-field was created (e.g., handling of van der Waals connections, usage of charged objects, disregard of polarization effects). In order to replicate specific experimentally discovered structural features, coercing creation frequently includes a mix of QM calculations and actual fitting of coercing characteristics. As a

Table 12.4 List of most widely used molecular simulation software.

Software	Distribution	Maintained by	Features
MacroModel	Commercial	Schrodinger LLC	MM, MD, MC
NAMD	Open-Source	Klaus Schulten, the University of Illinois at Urbana-Champaign	MD
GROMACS	Open-Source	University of Groningen, Netherlands	MD
LAMMPS	Open-Source	Sandia National Laboratories	MD
CHARMM	Commercial	Martin Karplus and others (Academic version), Biovia (Commercial version)	MD
AMBER	Commercial	David Case, Rutgers University, and others	MD
Materials Studio	Commercial	Biovia	MD, MC, QM, MM
DESMOND	Commercial	D.E. Shaw Research	MD
DL_POLY	Commercial	W. Smith and T. R. Forester, Darebury Laboratory	MD
ACEMD	Commercial	Acellera Solutions	MD

result, the words in a force field that correspond to the above-mentioned interactions should not be taken as precise measures of such interactions. Instead, think of the final functional form as a useful mathematical statement with optimal perfect restrictions [1].

12.1.9 Simulation Software

It takes a lot of work and involvement of academics from several fields to create top-notch computer programs for molecular simulations. Thankfully, a global scientific endeavor has produced a variety of extremely affordable and powerful development tools and software (Table 12.4). Additionally, a wide variety of web portals and implementation details are used separately for designing and constructing compounds (such as Substances Studio and Marvin), attempting to determine compression variables (such as SwissParam, ATB, PRODRG, MKTOP, and OBGMX), and visualizing outcomes (such as SwissParam, ATB, PRODRG, MKTOP, OBGMX, VMD, PYMOL, RASMOL, and CHIMERA). The user-friendliness of these programs has encouraged widespread usage of molecular simulations, raising concerns that findings are being created and published without a thorough grasp of algorithmic subtleties, leaving them vulnerable to error [1, 25, 26].

12.2 MD Simulations: Existing Limits

Notwithstanding, molecular dynamics simulations are all still constrained by two significant problems [27, 28]: the force fields employed now have to be improved, and high computational stresses prevent routine simulations from long lasting than a millisecond, often chiefly due to completely inadequate sampling of conformational notions. Think about how long it would take 24 CPUs to simulate a relatively modest system in 1 ms (roughly 1000 atoms).

The utilization of phenomena that are just approximate representations of the quantum-mechanical certainty that controls the atomic domain presents additional difficulties in addition to the high processing demands of the simulation. The binding of transitional copper atoms, for example, is a system where subatomic particles are present, and simulators can anticipate several significant chemical motions with accuracy.

To address this issue, some scientists have combined quantum mechanics arithmetic operations with conventional molecular-dynamics phenomena.

Quantum mechanical laws are used to simulate the movements and interactions of active sites of enzyme or other comparatively small places of focus, while the kinetics of small molecules are being used to accelerate the motions of the larger system. This hybrid approach has been effectively used in a variety of systems, although it is far from the "ideal" employment of quantum mechanics to explain the complete design. For example, a "QM [quantum mechanical] region" was recently reported in a simulation of *Desulfovibrio desulfuricans* and *Clostridium pasteurianum* [Fe–Fe] hydrogenases, incorporating a good area of the nutrient content thought to be catalytically crucial, while the rest of the protein was simulated using conventional computer vision modeling. The simulations demonstrated a significant amount of electron transport in the QM area, as well as a bond-breaking and bond-formation event that was not observed in a typical force field. The proposed reaction mechanism is then supported by experimental data [29].

Another quantum-mechanical phenomenon that is generally disregarded is electronic polarization, which is caused by electrons moving from one atomic nucleus to another within chemically linked groupings of atoms. During simulations of classical molecular mechanics, every atom is given a predetermined partial energy prior to the experiment. In reality, atoms' partial charges are more dynamic and sensitive because the electron clouds that surround them are constantly shifting in response to their surroundings. Regardless of the fact that there is a universal agreement regarding the importance of taking into account electronic polarization, it has taken 30 years of research to establish a broadly recognized polarizable force field, and MD simulations using the present beams are rare [30]. But a variety of polarizable dynamics are being developed right now [31], and if they become available in the future, precision rate may go up.

Instead of ignoring computational challenges, molecular dynamics research is constrained by the often simulated shorter periods. The simulations must investigate every one of the molecule's various related figures in order to accurately mimic thermodynamic characteristics and/or specify all legally enforceable geometries pertinent to pharmaceutical research. Regrettably, many biological phenomena, such as drug-binding receptor conformational changes, occur over far longer time scales than could be reproduced. Simulations are typically restricted to billionths of a second or fewer, with a few significant outliers [32].

Several approaches to this problem have previously received limited adoption. Large energy barriers are artificially decreased in accelerated molecular dynamics (aMD). Although a few distortions are unavoidable, this approach enables proteins to switch between conformations that would otherwise be unavailable due to traditional molecular dynamics' time scales. Molecular dynamics or other approaches can then be used to examine these novel conformations further.

The simple allowance of conventional molecular mechanics simulations has also been overcome using cutting-edge technology. The majority of the calculations needed for these simulations are handled by audio-visual games and interactive charts programs. As a consequence, graphics processing units (GPUs), which were first created to allow faster video games, may be utilized to expect higher molecular dynamics simulations [28].

However, many engineers have developed new chips, especially for these simulations rather than rewriting molecular-dynamics code to track on specialized graphics mainframes. Another power that is perceived by this approach is the research team led by DE Shaw. They developed Anton, a microprocessor capable of doing microsecond simulations around the clock. In Anton simulations lasting longer than 1 ms, the winding and unwinding of proteins, as well as drug-binding activities, have been effectively recorded [33]. Irrespective of whether there are truly problems, these and many other future solutions will make great headway in surpassing the current conformational collection restrictions.

12.3 MD Simulation and Drug Discovery

MD simulations and the perceptions they provide into protein motion are frequently used in drug discovery, despite the inadequacies of conventional force fields and conformational sampling. The dynamics of proteins cannot be inferred from a particular protein shape, any more than from a single photograph of a runner's stride. NMR, X-ray crystallography, and homology modeling, signal transduction, and medication binding are indeed to be very vital processes for researching complicated molecular structure. A continually shifting macromolecule instead of a singular, fixed form is seen once a tiny molecule behaves similar to how such a medicine (for instance, a ligand) approaches its destination (for instance, a receptor) in liquid.

When this happens, the receptor fits together into a pretty stagnant complex formation, much similar to the lock and key model, and protein mobility is constrained. The community of all conceivable various crystal structures moves at those that may best accommodate binding since the ligand may only bind and fix a tiny portion of the numerous conformations measured by its dynamic receptor. It is possible for the ligand to bind and cause conformational changes that are not visible whenever the receptor is present. In any case, the majority of small-molecule medicines are bound by receptor movements. The information that molecular dynamics simulations may give on these activities has indeed been put to use in a number of ways [28].

12.4 Applications in Drug Discovery

Molecular simulations have been used in thousands of research in recent years for drug delivery applications. The drug–receptor binding or docking investigations account for a substantial portion of this research, which is not the subject of this study. Scheme scope and simulation duration have risen dramatically during the years, allowing researchers to study systems with vast numbers of molecules. Furthermore, while previous research concentrated on the local features of solitary molecules (e.g., molecular surface area, drug orientation in a bilayer), contemporary research has shifted its attention to the worldwide behavior of an assembly of particles (e.g., phase behavior, crystallization). The following sections cover five major areas where molecular simulations can be used in medication delivery [34, 35].

12.4.1 Drug Diffusion and Lipid Bilayer Permeation

The ability of a medicine to cross a lipid bilayer membrane has a significant influence on the drug's bioavailability. According to the Meyer–Overton rule, hydrophobic medications are lipophilic and may pass through the semipermeable bilayer with ease. A molecule's chemical composition can be used to determine how many H-bond providers and acceptors are present. Two data, as well as information that may be discovered through molecular simulations, include the location of the drug molecule (definite in terms of the bilayer normal) and the energetic exchanges of the medication with multilayer components. A range of pharmacological and membrane chemistries have benefited from this type of research. Molecular simulations can also demonstrate the role of transporter molecules in departures from the Meyer–Overton rule and find new sources of deviations. Furthermore, molecular simulations can disclose the channel generation process, as well as provide detailed information on multilayer structure and behavior [36, 37].

12.4.2 Drug Solubility

Drugs with high solubility have high concentrations in the stomach and intestinal fluids, implying that significant drug absorption is possible. Due to the aqueous nature of stomach and intestinal fluids, hydrophilic medications have high solubility in these fluids, resulting in reduced membrane permeability. Therefore, it is important to evaluate a therapeutic product's dissolution and permeation. The biopharmaceutics classification method (BCS), which divides medicines into four categories based on their water solubility, is based on this concept. Class, I (high permeability, high solubility), Class II (high permeability, low solubility), Class III (low permeability, high solubility), and Class IV (low permeability, high solubility) are the groups that are separated (low permeability, low solubility). The number of H-bond donors/acceptors, which would be a component of the well-known Lipinski rule of five for determining drug likeness, is one implicit measure of dispersion. The "like dissolves like" rule, which stipulates that molecules are given a "solubility parameter," and are deemed soluble when their solubility findings differ very little, is an alternative philosophical explanation of solubility. Though solubility limitations can be deduced from existing investigational data or determined using a group-contribution technique from the molecular structure, molecular simulations can provide a more exact conclusion, especially with new drug candidates. Furthermore, only molecular simulations can offer a comprehensive knowledge of medication solubility variations caused by deviations in pH and salt concentration in the GIT [38, 39].

12.4.3 Carrier-Drug Miscibility

To improve the solubility of not that many soluble medications or sustain drug supersaturation, several carrier molecules (excipients) are utilized. Molecular simulations, as well as approaches for quantifying drug–excipient interactions, can aid in the selection of such excipients. The best excipients are inert components that interact structurally with drug molecules, but the drug–excipient binding should not be too robust to inhibit later drug molecule liberation. Via Flory–Huggins interactions characteristics, which may be obtained utilizing molecular simulations, is one method for assessing drug–excipient interactions, particularly for polymeric active ingredients. More importantly, molecular simulations can disclose the phase behavior of excipient–drug combinations, including alterations in excipient–drug mixtures and complexation that depend on composition. Jha and Larson (2014) used atomistic MD simulations to assess the interaction and chemistries between phenytoin, a medicine, and several cellulosic polymers (hydroxypropyl-methylcellulose, HPMC, and hydroxypropyl-methylcellulose acetate succinate, HPMCAS). Researchers assessed drug–drug and polymer–drug radial statistical features to correlate polymer–drug complexation with drug aggregation inhibition; drug permeability via transporters was just employed to measure prescription outflow. They were also able to reverse the dissolving process by removing water molecules one at a time, then re-equilibrating, because the ultimate solvent-free state matches the dose form. pH-responsive excipients, which store pharmaceutical molecules in the stomach at a lower pH and discharge them at a progressive pH in the intestine, have recently sparked a lot of attention. As recently proven for the doxorubicin–graphene and doxorubicin–nanodiamond systems [39, 40], molecular simulations could be a useful method for developing such inert ingredients.

12.4.4 Drug Crystallization

Numerous oral dose forms chromatographic retention drug molecules, and poorly soluble medicines can crystallize in the body as well. Because crystalline versions of medicine have lower absorption than amorphous forms, excipients that impede crystallization are frequently used to decrease crystallization. Crystallization models are highly computationally costly since nucleation is an uncommon occurrence and crystals have long-range correlation coefficients (needful simulations of superior system dimensions). Crystal growth and dissolution simulations have advanced significantly over the past 10 years, even though crystal germination simulations of big pharmaceutical compounds from molecular simulations are still a long way off. These simulations might be used to discover crystal expressions with accelerated development rates, predict the crystal habit of novel pharmaceutical substances, and analyze the capacity of different medications to prevent crystallization. The meticulous yet vigorous coarse-graining techniques [27] that successfully mimic crystal formation with the right parameterization obtained from atomistic simulations [38, 39] should be specifically mentioned.

12.4.5 Drug Loading and Release

When discussing carrier-mediated drug delivery, the terms "drug loading" and "drug release" relate to the drug's incorporation into the transporter and succeeding release to the body. Researchers should change the environment to reverse the trajectory of drug flow since the drug loading and releasing activities are triggered by a difference in osmotic pressures within and external the carrier. Because the ultimate rate of drug release takes place in an aqueous medium inside the body, drug content is often done in a non-aqueous solvent. Although homogeneous methods are mainly based on the Higuchi equation and have been very effective in the education of drug loading and announcement from carrier matrices, these methods frequently disregard the structural features of the carriers. Particularly when it comes to the creation of pharmacological disintegrant nanostructures, structural simulations can offer a precise explanation of the drug loading/release process [40]. Due to the significant discrepancy seen between modeled and actual length scales, molecular simulation, particularly continuous techniques, does not allow for a direct comparison with empirically obtained loading/release data of pharmaceutical formulations. The drug diffusion coefficient can be calculated using the carrier and then coupled to experimental data or utilized as an estimation of parameters in a continuous model. Despite the fact that these links are typically straightforward and helpful for evaluating the effectiveness of various carriers, medication diffusivity cannot be accurately assessed since it depends on network size. Another more convenient approach is the usage of systematic coarse-graining methods that parameterize coarse-grained simulations utilizing atomistic simulations. For instance, researchers analyzed the drug concentration and release behavior of pH-sensitive amphiphilic copolymers using atomistic MD and coarse-grained dissipative particle dynamics (DPD) simulations [1, 41].

12.5 Molecular Simulation of Drug Delivery Strategies with Dendrimer

Aptamers are globular macromolecules that radiate from a central structure on many massively branched arms. In the rapidly expanding field of nanoscale academics and technology, dendrimer nanostructures are a unique class of macromolecular architecture and precision production modules. As nanoscale building blocks for high-performance materials, dendrimers and dendritic materials are quickly gaining appeal. The most versatile and structurally and compositionally regulated molecular building blocks to date are generally considered dendrimers. Dendrimers may be customized to a specific database's requirements by altering the chemical composition of the core, shells, and especially, the surface layer.

Technologies based on dendrimers allow for exciting new connections in chemistry, biology, and complex materials. Due to the distinct, exact, and predictable dendritic polymer topologies, dendrimers can serve as suitable delivery vehicles, both *in vitro* and *in vivo*. Dendrimers are a system that makes a variety of diagnostic and therapeutic products possible by acting as the vehicle—the targeting and delivery mechanisms. Due to their exact and programmable composition, variable solubility, effective membrane transport, maximum loading capability, high homogeneity and cleanliness, little toxicity and immunogenicity, and bioattachment capability, aptamers are prospective drug delivery vehicles [42].

Dendrimers' structural characteristics affect how they work. Knowing the dendrimer structure and function, and also logical dendrimer design, have proved unachievable in the absence of molecular modeling.

A fourth generation (dansyl terminated) propylene amine dendrimer was immersed in CH_2Cl_2, and MD simulations were utilized to investigate the system's immediate architecture before and after the encapsulation of several eosin Y dyes. Calculations show one of the guest molecules' impacts is to "close" the box architecture in a cubic box containing up to 3500 molecules and 12 eosins. First, before excess eosins are completely expelled and their number is reduced to the research's limit of six, many entrances–exits of guest molecules in the polymer backbone are visible in a little over a millisecond. The visitor molecules are separated from the dendrimer's core by two different distances, and their surroundings may be anything. Eosins move around the inside of the hyperbranched monomer in the same way that solvent molecules do, and they sometimes mix up [43, 44].

12.6 Molecular Simulation of Drug Delivery Strategies with Polymer

A crucial area for drug delivery applications is polymer, which has recently been the subject of multiple computer simulation studies. Computer simulations were utilized in 1997 by Li *et al.* [45] to examine the molecular mobility in amorphous polymers. In four amorphous polymers, Li *et al.* looked at the diffusion coefficients of small molecules with molecular weights ranging from 16.04 (methane) to 452.50 (Fluocinolone acetonide). The study's results and the predicted distribution coefficient were compared. The numbers for hydrophobic contact diffusing across hydrophobic polymers that were seen and those that were predicted are equal, according to the researchers. Sadly, the testing results do not match the estimated values for hydrophilic diffusions and/or a hydrophilic polymer. Hofmann *et al.* [46] examined the structure, dynamics, and transport processes of dense amorphous membrane polymers using atomistic molecular modeling approaches. Through extensive atomistic MD simulations, the motion of many tiny molecules in behaviors that can range from rubbery to rigid chain glassy polymers was studied. To analyze the transportation, bulk polymer models and interface models for liquid feed mixtures at contact with the upstream left layer were employed. Simulated flexible chain rubbery polymers and stiff chain glassy polymers exhibit significant difference that is significant for virtually perceptible variances in particular molecule diffusion when the stationary architecture and dynamic characteristics of the charge density are compared. The contact models have been developed to capture essential characteristics of pervaporation processes that had hitherto been overlooked.

A "smart" nanoporous scheme that can be exposed and adjacent in answer to changes in the environment was investigated using MD simulations. The comb polymer molecules linked to the pore induced the rodglobule transition, which explains the dynamic pore size ability. The simulation results show that modifying the solvent excellence can result in a substantial alteration in the pores aperture, which can be exploited to preferentially trap big molecules by letting smaller molecules flow through to the hole without restriction. The simulation results show that these kinds of nanoporous procedures could be created to adapt to alterations in environmental factors like pH, temperature, and chemical content, with a diversity of potential application areas counting process of chemical discharge, molecular sieves, smart drug development, and ultra-filtration [45].

NMR and MD simulations were performed by Mathias *et al.* [47] to examine the idea of drug-loaded fluorocarbon-based micelles. The IPDU (isophorone diurethane) serves as a bridge, joining the PEG chains to a fluoroalkyl group on each end, making Rf-IPDU-PEGs a particular form of fluoroalkyl-blended poly(ethylene glycol) polymer (Rf-PEG). In freshwater, the Rf-IPDU-PEGs create hydrogels with favorable sol–gel coexistence characteristics. They can be used to dispense medications. The scientists found that the middle IPDU shells of the Rf-IPDU-PEG micelles had been loaded with the hydrophobic drug molecules. To produce personalized drug-loading characteristics, the linkage between the Rf group and the PEG chain must be altered.

A brand-new hybrid polymeric micelle for drug delivery was proposed by Lin *et al.* [48]. PLGA and PEG-linked corona strands are the building blocks of these micelles. The creation of PEG channels and micelle corona microphase separation are both caused by the transformation of PLGA from water-soluble randomized spirals to a water-insoluble α-helix once it is subjected to acid. Medication diffusion from micelles is expedited thanks to these channels, which link the internal core to the outside environment. The models of the hybrid micelle structure were tested using MD simulations. Hybrid micelle discharge can be managed using micelle structures, such as the mixture's arrangement, or outside influences, including pH.

Kuramochi *et al.* [49] investigated the micelle structure of n-acetylated poly(ethylene glycol)-poly(benzyl L-glutamate) block copolymers using MD simulations. The hydrophobic PBLG inner core and hydrophilic PEG outer shell of the micelle structure are expected to be present in the equally two-thirds majority, according to researchers. The PEG blocks formed a close, strong connection, while the PBLG blocks formed an α-helix, maintaining the micelle structure's dynamic stability throughout the simulation. Numerous hydrogen connections between the water molecules in the solution and the PEG blocks kept their helical shape.

12.7 Liposome-Based Molecular Simulation of Drug Delivery Strategies

The medication's capacity to enter cell membranes is essential for medicinal administration. The extent to which route of administration, dosage schedule, delivery device architecture, candidate selection or chemical modification, and narcotics trafficking may all have an impact on drug dispersion and efficacy of pharmaceutical delivery. Computational techniques for evaluating lipid bilayer susceptibility can be useful for drug delivery applications that are controlled by membrane transport, such as liposomal delivery [48].

The medication's capacity to cross cell membranes is essential for therapeutic administration. The amount to which transmembrane penetration and narcotics trafficking may be specifically altered by the selection of candidates or synthesis method, administration route, dosage schedule, or devices, has an impact on drug delivery effectiveness. Delivery systems controlled by membrane transport, such as liposomal delivery, may be designed and optimized using computational approaches for estimating membrane lipid permeability [48].

It has been demonstrated that simulations may confidently predict solute absorption percentages across the phospholipid bilayer and can shed light on a number of pervasion possessions that are challenging to test (e.g., locations for favorable solute partitioning).

MD simulations have been utilized to educate a variety of solute molecules, as well as the diffusion coefficients for these solute molecules ranging from 10^5 to 10^7 cm^2/s. The amount of deformation is often higher for bigger solutes or for solutes that penetrate more densely packed headgroup/acyl chain areas. On the other hand, it depends on a fine equilibrium of their hydrophobicity, communications with specialized cellular membranes, and supplemental entropic belongings brought on by the decision to order cellular membranes, there can be a wide range in the partition coefficient (PC) of these solutes from water into different areas of a lipid bilayer. Apart from n-alkanes, almost all solutes examined to date colocalize preferentially in regions that also include polar headgroups [48, 50, 51].

Bassolino-Klimas *et al.* [52] investigated benzene molecules in a 36-lipid-molecule all-atom fully hydrated dimyristoylphosphatidylcholine (DMPC) bilayer using MD simulations. It was shown that the simulation period was sufficient to achieve regional benzene diffusivity in different parts of the bilayer for simulations inside the region of 1ns and to produce structural parameter agreement. The torsional movements of the phospholipid chains that act as "gates to entrance" cause hydrocarbon molecules to "rattle" in holes for extended periods of time before actually bouncing up to six to eight times to another gap, the researchers revealed using mean-squared displacement plots.

In order to explore the thermodynamic characteristics of methanol and propanol in DMPC bilayers at 311K, Jin and Hopfinger [53] employed MD and discovered strong agreement with experimentation. Also with shorter simulation periods, it may well be possible to estimate the diffusion coefficient of extremely minor molecules since the dispersion of these molecules seemed to behave in an Einsteinian manner despite the fact that the simulation durations were just 100 ps.

MD simulations have been confirmed to be beneficial in solving questions about small molecule partitioning in membranes and medicinal molecule binding in membranes.

The impact of various absorption substituents on the free energy of transmission from H_2O into the ordered-chain area of an L-phase dipalmitoylphosphatidylcholine (DPPC) bilayer was determined by utilizing the thermodynamic addition technique. The effects of changes in chemical structure on solute partition ratios in bulk liquid and membrane/water have been widely studied in experiments using the semi-empirical functional group contribution approaches [48]. Xiang and Anderson used a joint thermodynamic integration and perturbation approach to calculate the decision to the order area of DPPC and the comparative free energy in water by changing the -H atom in p-toluic acid to various substituents (-Cl, $-CH_3$, $-OCH_3$, -CN, -OH, -COOH). It was demonstrated that the organized acyl chain area had the same biochemical specificity as the obstacle domain for the transportation of these polar solvents by the mobility study results and the estimated free energy of transferring for these functional groups (0.5–5.1 kcal/mol).

Oversized pharmaceutical compounds that are possible candidates for liposomal delivery might benefit from MD simulations to forecast their membrane binding and loading competences and to understand storage and releasing dynamics. A protonated 20(S)-4-aminobutyrate relieved prodrug and a lipophilicsilate can camptothecin analogue (DB-67), an anticancer drug presently in preclinical education, were dispersed in a hydrated DMPC bilayer (DB67-AB). To enhance the ability of the hydrophobic segments of the chemical compound to the acyl chain area and the hydrophobic interactions of the polar functional groups situated on the together tops of the molecules either with water or polar atoms inside the OH groups, the rigid part of the chemical compound is tilted just under the plane of the bilayer outer walls. Researchers found that both compounds' centers of mass appeared to be near the polar carbonyl groups of DMPC. These findings, which were corroborated by stability studies [48], imply that the presence of DMPC vesicles may limit lactone hydrolysis.

Saito *et al.* [54] previously assessed the effect of partly fluorinated phospholipid bilayers on membrane structure and water permeability using MD modeling. Compared to regular benzene rings, fluorocarbon chains have an all-trans conformation, which creates a tightly organized approach in the transmembrane core. A modified hole implantation Widom technique was utilized to compute the water's free energy distributions throughout the bilayers. The fluorinated bilayer membrane had a lower water permeability because it had a 1.2 kcal/mol greater free energy hurdle than the non-fluorinated lipid bilayer. A hollow dispersion experiment revealed that the fluorinated membrane had a lower free capacity as a result of the closely packed chains, which explained the greater free energy barriers.

12.8 Molecular Simulation of Drug Delivery Strategies with Carbon Nanotube

Carbon atoms form the backbone of SWCNTs (single-walled carbon nanotubes), which have an open-ended, hollow cylindrical structure.

As a consequence, they can make internal holes large enough to store biomolecules, which have huge ramifications for drug delivery and innovation. In biological situations, they are noncytotoxic, allowing them to carry numerous molecules across cellular membranes through endocytosis deprived of producing cell death or harm. Fullerenes (C60), non-fullerene molecules, DNA oligonucleotides, peptides, protein fragments, and amylase have all been shown to be encased in carbon nanotubes (CNT). Surfaces made of carbon nanotubes have the potential to be useful in the administration of medications. The results of MD simulations on the peptide medicine Zadaxin's insertion into carbon nanotubes in a water solvent were published by Liu and Wang [54]. Single-walled carbon nanotubes, or nanocrystals, were a kind of carbon nanotube. Thymosin, a molecule that has been shown in several clinical trials to be a benign, gentle, and successful therapy for chronic hepatitis B, is the basis for the synthetic drug Zadaxin. The author shows that Zadaxin could be enclosed in CNTs big enough and long enough to hold it in a mixture of water. In this transplantation, it is crucial to have a good CNT-Zadaxin-water appealing contact. It was proposed that the (14) CNT's diameter was indeed the size that was essential for the implanted Zadaxin [47].

Some other crucial areas for computer simulation of carbon nanotubes for medicinal purposes is the customized waterfall via carbon nanotubes. MD simulations look at the features of limited transport of water and the few carbonyl-based oxygenation catalyst surfaces. At lower hydration levels, it was found that water diffusion was restricted. At somewhat high hydration levels, the diffusion appeared to be of the Fickian type; it increases as the concentration of limited water molecules rises, peaks, and then decreases as H_2O expands the nanotubes. In order to respond to the conclusions, the author employs two competing dynamics processes—"cluster-breakage" and "cluster-libration"—that were identified through the study of many simulation snapshots [55].

MD modeling of carbon nanotube-based medicine administration and releasing devices was discussed by Cheng *et al.* [56]. The researchers claim that a peptide trapped non-bonded or bonded to a nanofiber's outer surface may be released by another nanotube through a procedure termed competing replacements. The potential between two nanotubes pushes competitive substitutions, making the van der Waals contact essential to the operation.

12.9 Conclusion

Newton's laws of statistical mechanics and motion will guide future medicine delivery methods. This seemingly unachievable ambition may one day become a reality, thanks to the phenomenal growth of the molecular modeling field. However, two challenges must always be addressed: the first is creating computers and computer programs that are quick enough even to run these simulations, and the second is figuring out the chemical reaction underlying pharmacological activity in the body, which must be mimicked. There is still a long way to go until this objective is reached, even though each of these components of progress has recently shown promise. In particular, the system under consideration and time scales of simulated and real-world systems continue to differ greatly, notwithstanding huge increases in information processing in the past few decades. Therefore, unlike *in vitro* and *in vivo* research, *in silico* methodologies created on the idea of molecular simulations have been, up to this point, utilized by federal departments and agencies. They offer a way to rate the efficiency of suggested pharmacological compositions and also a way to get molecular knowledge about certain systems. The majority of the time, simulation predictions cannot provide a real comparison to a real observation, but they do offer a helpful correlation between observed and expected values. For example, the drug seen experimentally exactly resembles the drug diffusion coefficient measured in simulations. Molecular simulations offer a huge amount of potential as a computational tool for drug carrier testing and might aid in the creation of design criteria for drug delivery formulations, notwithstanding their shortcomings. Particularly when compared to continuum techniques, which neglect the chemical and structural characteristics of pharmacological formulations, molecular simulations offer a clear benefit. These variables affect drug solubility, carrier-drug miscibility, drug crystallization, drug loading, and release of the drug, in addition to drug transport and membrane damage [1, 33].

Multiscale simulations, particularly those that use systematic coarse-graining, are also another important opportunity for a research study. They might be open to expanding the scope of atomistic simulations' application to magnitudes lengthier scales, and they might provide a methodical strategy for gathering hyperparameters for continuum approaches applied at physiological length scale dimensions. Given that their effect is predicted to decrease as duration scales increase, it is difficult to decide which degrees of freedom should be averaged out. The creation and application of effective sampling methods (like metadynamics) to quicken atomistic simulations are another crucial subject. These come in quite handy for performing free-energy calculations and researching slow or unusual phenomena (e.g., crystal nucleation). There is a real opportunity for advancement in efficient Monte Carlo parallelization methodologies and the development of specialized, low-cost hardware for molecular simulations from a computational standpoint. Each research endeavor eventually has to win the confidence of stakeholders to provide a significant influence, in this instance the pharmaceutical business, physicians, and academic researchers. The notion that stochastic mechanics theories should be used to describe the behavior of drug formulations may seem logical to physicists, but not to others who prefer (and often correctly) to trust practical data over model expectations. Mathematical modeling has long been utilized in the field of medicine administration to explain experimental findings and extend results to bigger populations. This is due to the fact that the model parameters of continuous approaches are frequently established by the user. As a result, accepting and testing the premise that simulation forecasts could precede or even influence systematic trials is required. Most critically, industry-computational scientist partnerships in this field can expedite work in fields where molecular simulations can have a significant influence [42, 43].

In the subject of medicine delivery, significant progress has recently been accomplished. This is because drug delivery now uses computational approaches and technologies, leading to enhanced drug delivery tactics. As a means of reducing costs in medication design and production, computational approaches to anticipate the binding and dynamics between therapeutic compounds and their transporters are gaining popularity. Drug delivery approaches using dendrimers, polymer micelles, liposomes, and carbon nanotubes are the four categories of computational investigations in this review. Substantial advancements in computer modeling have made it possible to comprehend several types of pharmaceutical distribution processes. For issues like dissemination and dynamics that are challenging to investigate in a laboratory setting, computation simulations are very helpful. Notwithstanding major advancements in the delivery of drug molecular modeling, a number of significant obstacles still exist. One of the main difficulties is accurately calculating the extraction efficiency for medication delivery systems. Since the way the drug delivery reacts varies because of the environment, a technique to recover the characterization of several ecological characteristics for simulation, such as pH, solvent effect, and temperature, is required. Forecasts of loading capacity and drug particle dispersion are especially important since they are linked to the rational development and optimization of drug delivery systems [38].

Acknowledgment

Authors are thankful to the management of Acropolis group of Institutes, Indore Madhya Pradesh and Department of Pharmaceutical Sciences, Mohan Lal Sukhadia University, Udaipur, Rajasthan.

Funding

There is no funding issued.

References

1. Katiyar, R.S. and Jha, P.K., Molecular simulations in drug delivery: Opportunities and challenges. *Wiley Interdiscip. Rev.: Comput. Mol. Sci.*, vol. 8, p. e1358, 2018.
2. Leach, A.R., *Molecular modeling: principles and applications*, Pearson education, Upper Saddle River, 2001.
3. Adcock, S.A. and McCammon, J.A., Molecular dynamics: survey of methods for simulating the activity of proteins. *Chem. Rev.*, 106, 1589–615, 2006.
4. Merz, K.M., Using quantum mechanical approaches to study biological systems. *Acc. Chem. Res.*, 47, 2804–2811, 2014.
5. Mucs, D. and Bryce, R.A., The application of quantum mechanics in structure-based drug design. *Expert Opin. Drug Discov.*, 8, 263–276, 2013.
6. Meunier, M., Goupil, A., Lienard, P., Predicting drug loading in PLA-PEG nanoparticles. *Int. J. Pharm.*, 526, 157–166, 2017.
7. Pogodin, S., Werner, M., Sommer, J.U., Baulin, V.A., Nanoparticle-induced permeability of lipid membranes. *ACS Nano*, 6, 10555–10561, 2012.
8. Martínez, L., Villalobos, R., Sánchez, M., Cruz, J., Ganem, A., Melgoza, L.M., Monte Carlo simulations for the study of drug release from cylindrical matrix systems with an inert nucleus. *Int. J. Pharm.*, 369, 38–46, 2009.
9. De Vivo, M., Masetti, M., Bottegoni, G., Cavalli, A., Role of molecular dynamics and related methods in drug discovery. *J. Med. Chem.*, 59, 4035–4061, 2016.
10. Thota, N., Hu, Z., Jiang, J., Ibuprofen loading and release in amphiphilic peptide FA32 and its derivatives: a coarse-grained molecular dynamics simulation study. *Mol. Simul.*, 42, 679–687, 2016.

11. Gubbins, K.E. and Moore, J.D., Molecular modeling of matter: Impact and prospects in engineering. *Ind. Eng. Chem. Res.*, 49, 3026–3046, 2010.
12. Peppas, N.A. and Narasimhan, B., Mathematical models in drug delivery: How modeling has shaped the way we design new drug delivery systems. *J. Control. Release*, 190, 75–81, 2014.
13. Steuperaert, M., Falvo D'Urso Labate, G., Debbaut, C., De Wever, O., Vanhove, C., Ceelen, W., Segers, P., Mathematical modeling of intraperitoneal drug delivery: simulation of drug distribution in a single tumor nodule. *Drug Deliv.*, 24, 491–501, 2017.
14. Cherkasov, A., Muratov, E.N., Fourches, D., Varnek, A., Baskin, I.I., Cronin, M., Dearden, J., Gramatica, P., Martin, Y.C., Todeschini, R., Consonni, V., QSAR modeling: where have you been? Where are you going to? *J. Med. Chem.*, 57, 4977–5010, 2014.
15. Sterling, T. and Irwin, J.J., ZINC 15–ligand discovery for everyone. *JCIM*, 55, 2324–2337, 20152015.
16. Walsh, M.R., Koh, C.A., Sloan, E.D., Sum, A.K., Wu, D.T., Microsecond simulations of spontaneous methane hydrate nucleation and growth. *Science*, 326, 1095–1098, 2009.
17. Cho, K. and Joannopoulos, J.D., Ergodicity and dynamical properties of constant-temperature molecular dynamics. *Phys. Rev. A: At. Mol. Opt. Phys.*, 45, 7089, 1992.
18. Huang, W., Dalal, I.S., Larson, R.G., Analysis of solvation and gelation behavior of methylcellulose using atomistic molecular dynamics simulations. *J. Phys. Chem. B*, 118, 13992–14008, 2014.
19. Jha, P.K. and Larson, R.G., Assessing the efficiency of polymeric excipients by atomistic molecular dynamics simulations. *Mol. Pharmaceutics*, 11, 1676–1686, 2014.
20. Marrink, S.J., Risselada, H.J., Yefimov, S., Tieleman, D.P., De Vries, A.H., The MARTINI force field: coarse grained model for biomolecular simulations. *J. Phys. Chem. B*, 111, 7812–7824, 20072007.
21. Qi, Y., Ingólfsson, H.I., Cheng, X., Lee, J., Marrink, S.J. and Im, W., CHARMM-GUI martini maker for coarse-grained simulations with the martini force field. *J. Chem. Theory Comput.*, 11, 4486–4494, 2015.
22. Huang, W., Ramesh, R., Jha, P.K., Larson, R.G., A systematic coarse-grained model for methylcellulose polymers: Spontaneous ring formation at elevated temperature. *Macromolecules*, 49, 1490–1503, 2016.
23. Pan, A.C., Borhani, D.W., Dror, R.O., Shaw, D.E., Molecular determinants of drug–receptor binding kinetics. *Drug Discov. Today*, 18, 667–673, 2013.
24. Filion, L., Hermes, M., Ni, R., Dijkstra, M., Crystal nucleation of hard spheres using molecular dynamics, umbrella sampling, and forward flux sampling: A comparison of simulation techniques. *J. Chem. Phys.*, 133, 244115, 2010.
25. Birru, W.A., Warren, D.B., Han, S., Benameur, H., Porter, C.J., Pouton, C.W., Chalmers, D.K., Computational models of the gastrointestinal environment.

2. Phase behavior and drug solubilization capacity of a type I lipid-based drug formulation after digestion. *Mol. Pharmaceutics*, 14, 580–592, 2017.
26. Mandal, T., Marson, R.L., Larson, R.G., Coarse-grained modeling of crystal growth and polymorphism of a model pharmaceutical molecule. *Soft Matter*, 12, 8246–8255, 2016.
27. Durrant, J.D. and McCammon, J.A., Molecular dynamics simulations and drug discovery. *BMC Biol.*, 9, 1–9, 2011.
28. Chodera, J.D., Mobley, D.L., Shirts, M.R., Dixon, R.W., Branson, K., Pande, V.S., Alchemical free energy methods for drug discovery: progress and challenges. *Curr. Opin. Struct. Biol.*, 21, 150–160, 2011.
29. Hong, G., Cornish, A.J., Hegg, E.L., Pachter, R., On understanding proton transfer to the biocatalytic [Fe—Fe] H sub-cluster in [Fe—Fe] H2ases: QM/MM MD simulations. *Biochim. Biophys. Acta, Bioenerg.*, 1807, 510–517, 2011.
30. Jorgensen, W.L., Special issue on polarization. *J. Chem. Theory Comput.*, 3, 1877–1877, 2007.
31. Cieplak, P., Dupradeau, F.Y., Duan, Y., Wang, J., Polarization effects in molecular mechanical force fields. *J. Phys.: Condens. Matter*, 21, 333102, 2009.
32. Arnarez, C., Uusitalo, J.J., Masman, M.F., Ingólfsson, H.I., De Jong, D.H., Melo, M.N., Periole, X., De Vries, A.H., Marrink, S.J., Dry Martini, a coarse-grained force field for lipid membrane simulations with implicit solvent. *J. Chem. Theory Comput.*, 11, 260–275, 2015.
33. Aminpour, M., Montemagno, C., Tuszynski, J.A., An overview of molecular modeling for drug discovery with specific illustrative examples of applications. *Molecules*, 24, 1693, 2019.
34. Lee, H., Molecular simulations of PEGylated biomolecules, liposomes, and nanoparticles for drug delivery applications. *Pharmaceutics*, 12, 533, 2020.
35. Troendle, E.P., Khan, A., Searson, P.C., Ulmschneider, M.B., Predicting drug delivery efficiency into tumor tissues through molecular simulation of transport in complex vascular networks. *J. Control. Release*, 292, 221–234, 2018.
36. Casalini, T., Rosolen, A., Henriques, C.Y.H., Perale, G., Permeation of Biopolymers Across the Cell Membrane: A Computational Comparative Study on Polylactic Acid and Polyhydroxyalkanoate. *Front. Bioeng. Biotechnol.*, 8, 718, 2020.
37. Wadhwa, R., Yadav, N.S., Katiyar, S.P., Yaguchi, T., Lee, C., Ahn, H., Yun, C.O., Kaul, S.C., Sundar, D., Molecular dynamics simulations and experimental studies reveal differential permeability of withaferin-A and withanone across the model cell membrane. *Sci. Rep.*, 11, 1–15, 2021.
38. Hossain, S., Kabedev, A., Parrow, A., Bergström, C.A., Larsson, P., Molecular simulation as a computational pharmaceutics tool to predict drug solubility, solubilization processes and partitioning. *Eur. J. Pharm. Biopharm.*, 137, 46–55, 2019.

39. Bunker, A. and Róg, T., Mechanistic understanding from molecular dynamics simulation in pharmaceutical research 1: drug delivery. *Front. Mol. Biosci.*, 7, 604770, 2020.
40. Walden, D.M., Bundey, Y., Jagarapu, A., Antontsev, V., Chakravarty, K., Varshney, J., Molecular simulation and statistical learning methods toward predicting drug–polymer amorphous solid dispersion miscibility, stability, and formulation design. *Molecules*, 26, 182, 2021.
41. Salo-Ahen, O.M., Alanko, I., Bhadane, R., Bonvin, A.M., Honorato, R.V., Hossain, S., Juffer, A.H., Kabedev, A., Lahtela-Kakkonen, M., Larsen, A.S., Lescrinier, E., Molecular dynamics simulations in drug discovery and pharmaceutical development. *Processes*, 9, 71, 2020.
42. Zeng, S., Ji, Y., Shen, Y., Zhu, R., Wang, X., Chen, L., Chen, J., Molecular dynamics simulations of loading and unloading of drug molecule bortezomib on graphene nanosheets. *RSC Adv.*, 10, 8744–8750, 2020.
43. Sandoval-Yañez, C. and Castro Rodriguez, C., Dendrimers: Amazing platforms for bioactive molecule delivery systems. *Materials*, 13, 570, 2020.
44. Chawla, M., Kaushik, R.D., Singh, J., Optimization and computational studies evaluating molecular dynamics of EDA cored polymeric dendrimer. *Sci. Rep.*, 10, 1–10, 2020.
45. Li, Y. and Hou, T., Computational simulation of drug delivery at molecular level. *Curr. Med. Chem.*, 17, 4482–4491, 2010.
46. Hofmann, D., Fritz, L., Ulbrich, J., Schepers, C., Böhning, M., Detailed-atomistic molecular modeling of small molecule diffusion and solution processes in polymeric membrane materials. *Macromol. Theory Simul.*, 9, 293–327, 2000.
47. Mathias, E.V., Liu, X., Franco, O., Khan, I., Ba, Y., Kornfield, J.A., Model of drug-loaded fluorocarbon-based micelles studied by electron-spin induced 19F relaxation NMR and molecular dynamics simulation. *Langmuir*, 24, 692–700, 2008.
48. Lin, J., Zhu, J., Chen, T., Lin, S., Cai, C., Zhang, L., Zhuang, Y., Wang, X.S., Drug releasing behavior of hybrid micelles containing polypeptide triblock copolymer. *Biomaterials*, 30, 108–117, 2009.
49. Kuramochi, H., Andoh, Y., Yoshii, N., Okazaki, S., All-Atom molecular dynamics study of a spherical micelle composed of N-acetylated poly (ethylene glycol)–Poly (γ-benzyl l-glutamate) block copolymers: A potential carrier of drug delivery systems for cancer. *J. Phys. Chem. B*, 113, 15181–15188, 2009.
50. Parchekani, J., Allahverdi, A., Taghdir, M., Naderi-Manesh, H., Design and simulation of the liposomal model by using a coarse-grained molecular dynamics approach towards drug delivery goals. *Sci. Rep.*, 12, 1–15, 2022.
51. Hashemzadeh, H., Javadi, H., Darvishi, M.H., Study of Structural stability and formation mechanisms in DSPC and DPSM liposomes: A coarse-grained molecular dynamics simulation. *Sci. Rep.*, 10, 1872, 2020

52. Bassolino-Klimas, D., Alper, H.E., Stouch, T.R., Solute diffusion in lipid bilayer membranes: an atomic level study by molecular dynamics simulation. *Biochemistry*, 32, 12624–12637, 1993.
53. Jin, B. and Hopfinger, J., Characterization of lipid membrane dynamics by simulation: 3. Probing molecular transport across the phospholipid bilayer. *Pharm. Res.*, 13, 1786–1794, 1996.
54. Saito, H., Shinoda, W., Mikami, M., Enhanced hydrophobicity of fluorinated lipid bilayer: A molecular dynamics study. *J. Phys. Chem. B*, 112, 11305–11309, 2008.
55. Liu, Y.-C. and Wang, Q., Dynamic behaviors on zadaxin getting into carbon nanotubes. *J. Chem. Phys.*, 126, 03B614, 2007.
56. Cheng, Y., Pei, Q.X., Gao, H., Molecular-dynamics studies of competitive replacement in peptide–nanotube assembly for control of drug release. *Nanotechnology*, 20, 145101, 2009.

13

Molecular Docking: An Essential Step in Computer-Aided Drug Design

Lata Potey[1]*, Suchita Waghmare[2], Anshu Chaudhary Dudhe[3], Rupesh Dudhe[4] and Prafulla Sabale[2]

[1]Shree Sainath College of Pharmacy, Nagpur, India
[2]Rashtrasant Tukdoji Maharaj Nagpur University, Nagpur, MS, India
[3]Nagpur College of Pharmacy, Nagpur, Maharashtra, India
[4]Adarsh Institute of Pharmacy, Nandanvan, Nagpur, Maharashtra, India

Abstract

Medicinal chemistry is concerned with the discovery, creation, interpretation of physiologically active chemical mode of action, and identification at the molecular level. Medicinal plants have historically proven their value as a source of molecules with therapeutic potential, and nowadays still represent an important pool for the identification of novel drug leads. Many potent compounds found in plants can be employed for the design, synthesis, and semi-synthesis of novel medications for the treatment of diseases. Finding and creating new medications are a laborious process that entails hunting for promising hits, turning those hits into leads, and then finally, validating those leads into therapeutic candidates in clinical trials. Drug discovery is a process that involves the identification of candidates, synthesis, characterization, validation, optimization, screening, and assays for therapeutic efficacy. Once a compound has shown its significance in these investigations, it will initiate the process of drug development earlier to clinical trials. The new drug development process must continue through several stages to make medicine that is safe, effective, and meets all regulatory requirements. Introducing a new drug to the market is a complex and time-consuming process that can cost pharmaceutical companies an average of $ 1 trillion by 2030 and 10 years of research and development. Against this complex background, this chapter examines the role of the computer-aided drug design potential of the remarkable developments in the basic biomedical sciences for global health care.

**Corresponding author*: latapotey@rediffmail.com; ORCID: orcid.org/0000-0001-9537-5347

Rishabha Malviya, Sonali Sundram and Dhanalekshmi Unnikrishnan Meenakshi (eds.)
Drug Delivery Systems Using Quantum Computing, (389–406) © 2024 Scrivener Publishing LLC

Keywords: Drug design, drug discovery, new drug, health care

13.1 Introduction

13.1.1 Medicinal Chemistry

The field of research known as medicinal chemistry is concerned with the discovery, creation, interpretation of physiologically active chemical mode of action, and identification at the molecular level. In addition to studying, identifying, and synthesizing different metabolic products of these medications and related molecules, medicinal chemistry also focuses on drugs [1].

Many potent compounds found in plants can be employed for the design, synthesis, and semi-synthesis of novel medications for the treatment of diseases. Morphine is the first pure natural product that was utilized for therapeutic use in 1826. Salicilin, a natural molecule that Bayer discovered from *Salix alba* in 1899, was the inspiration for the first semi-synthetic drug, aspirin, which is well-known for this distinction [2]. The first drugs were discovered by sampling higher plants such as opium, belladonna, ephedrine, codeine, digitoxin, and pilocarpine at random. However, a significant class of natural and synthetic products, known as heterocyclic compounds, demonstrates wide biological activity, is of synthetic interest, and serves as incredibly adaptable building blocks for the production of bioactive chemicals. Thus, research in the domain of life sciences is still very interested in creating new heterocyclic compounds with desirable medicinal activities [3].

A new drug's development takes at least 12 to 15 years and more than $ 1 trillion, from an original concept to the finished product's commercialization. According to industry data, for every drug that has been given the go-ahead for medical use, up to 10,000 compounds may be produced, tested, and evaluated for safety [4]. However, in humans, only 10 compounds may be tested clinically. Even after a new medication enters the market, its success is not guaranteed [5].

To overcome obstacles, medicinal chemistry has experienced dramatic modifications in recent years. The rapid progress in biological sciences has enabled a deeper comprehension of what happens at the molecular level in the human body. This strategy needs to comprehend the target's structure, function, and mechanism of action by which it interacts with the active medication [6].

The key phases of drug research, development, and design include: (Figure 13.1)

- Drug discovery: locating a lead molecule:
  - ➢ Choosing a disease
  - ➢ Deciding on a pharmacological target
  - ➢ Identifying a bioassay
  - ➢ Locating the lead molecule
  - ➢ Isolating and purifying the lead molecule, as necessary
  - ➢ Locating the structure of the lead chemical, as necessary
- Drug design:
  - ➢ Structure–activity relationships (SARs)
  - ➢ Pharmacophore identification
  - ➢ Target interaction improvement (pharmacodynamics)
  - ➢ Pharmacokinetic characteristics improvement
- Drug development includes the following steps:
  - ➢ Preclinical research
  - ➢ Design and manufacture
  - ➢ Development of chemicals and processes
  - ➢ Clinical trial studies
  - ➢ Registration and marketing
  - ➢ Financial gain

13.1.2 Computer-Aided Drug Design (CADD) and Molecular Docking

Finding and creating new medications are a laborious process that entails hunting for promising hits, turning those hits into leads, and then finally validating those leads into therapeutic candidates in clinical trials. The amount of money spent on drug development and research has significantly increased during the past few decades [7]. However, it was discovered from the existing situation that there is a high failure rate and low productivity due to ineffective medication discovery and development processes (Figure 13.1). One of the most recent and efficient approaches is

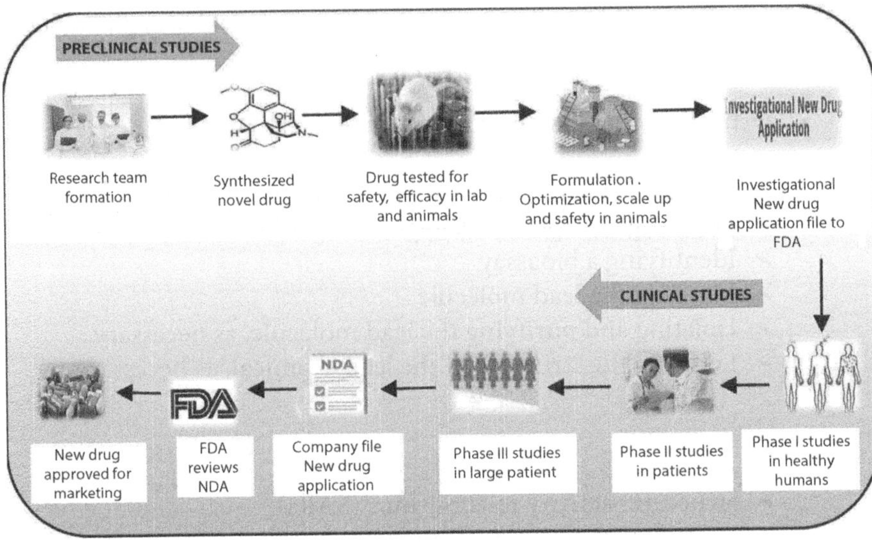

Figure 13.1 Drug discovery and development process.

computer-aided drug design (CADD) and drug development, which also conserves time, money, and resources [8].

13.1.3 History and Progression of CADD

The CADD discovery notion first appeared in a cover piece of Fortune magazine in 1981, with the title "Next Industrial Revolution: Designing Drugs by Computer at Merck." In the same era, high-throughput screening (HTS) was also introduced for users to find novel therapeutics [9]. The number of lead molecules discovered after the screening was also lower when using the conventional HTS approach, which required screening a huge number of molecules. Hence, the use of CADD before the HTS study provides a significant increase in the result and also helps to identify the more specific molecule [10]. With the help of CADD, the molecules possessing less activity or binding affinity towards the binding site can be neglected. Numerous success stories are based on the usage of CADD, combining carbonic anhydrase inhibitors and dorzolamide in the treatment of cystoid macular edema [11]. The HIV protease inhibitors drug like Saquinavir, Indinavir, and Ritonavir are used to treat HIV. The renin inhibitor Aliskiren is used to treat essential hypertension. The ACE inhibitor Captopril is used in the treatment of hypertension. Virtual screening is one of the most important methods for CADD, often known as virtual library screening. In 2003, a group at Biogen

Idec used virtual screening to search for inhibitors that would focus on the transforming growth factor 1 receptor kinase [12]. The Biogen Idec team found 87 hits, one of which had the same structure as the lead found by Eli Lilly using conventional HTS. This example showed how virtual screening in CADD may produce the same molecules as a full HTS operation while costing much less and requiring much less work. Through virtual screening, a significant portion of hits for kinases, and over time, G-protein-coupled receptors (GPCRs) have been identified [13].

13.1.4 Classification of CADD

CADD can be divided into the following two groups:

A. Structure-based drug design (SBDD)
B. Ligand-based drug design (LBDD)

The structure-based technique uses the 3D structure of the receptor or protein acting as the target molecule to screen potential ligand compounds. It is subsequently followed by the methods of synthesis, biological testing, and optimization. Contrarily, ligand-based drug design entails gathering a sizable number of ligand molecules with various structures, known potencies, and biological activity [14]. The potency of these compounds is then increased, and new chemical molecules are discovered by virtual screening of a sizable chemical library. Utilizing both drug design approaches yields more successful results than using just one since they can complement each other's strengths and weaknesses [15].

13.1.4.1 Structure-Based Drug Design (SBDD)

In SBDD, information from the 3D macromolecule, such as a protein or receptor structure, is obtained from the area where the drug is intended to bind, and this information is utilized to design and assess ligands based on how they are expected to interact with the protein binding site. The protein structure was aided in SBDD by using X-ray crystallography, cryo-electron microscopy, Nuclear Magnetic Resonance (NMR), homology modeling, and molecular dynamic simulation. Virtual screening strategy and *de novo* approach are the two primary subcategories of SBDD. The *de novo* drug design checks information from the protein or receptor structure to find a little fragment that closely resembles the binding site. To ensure synthetic accessibility, this fragment should be connected based on the connection rule, producing a structurally unique ligand that can be used for further screening. Virtual screening, on the other side, employs a database of small compounds that are already available to find specific activities that can replace ligands that are already present in the protein structure or target regions [16].

13.1.4.2 Ligand-Based Drug Design (LBDD)

LBDD is used only when the structure of the 3D target protein is not known. It involves consideration of the ligands having known biological activity and can be used to identify the physicochemical and structural properties of the ligands and predict biological activity on the basis of these properties. Generally, it is assumed that the compounds possessing similar structures have similar biological activities. Quantitative structure-activity relationships (QSAR) and pharmacophore-based methods are essential steps involved in LBDD [17].

13.1.5 Application of CADD in Drug Discovery and Development

Chemical biology and computational drug design methods are used in drug development to quickly identify and improve lead compounds (Figure 13.2). The main goals of chemical biology are to clarify the biological function of a target and the mechanism of action of chemical

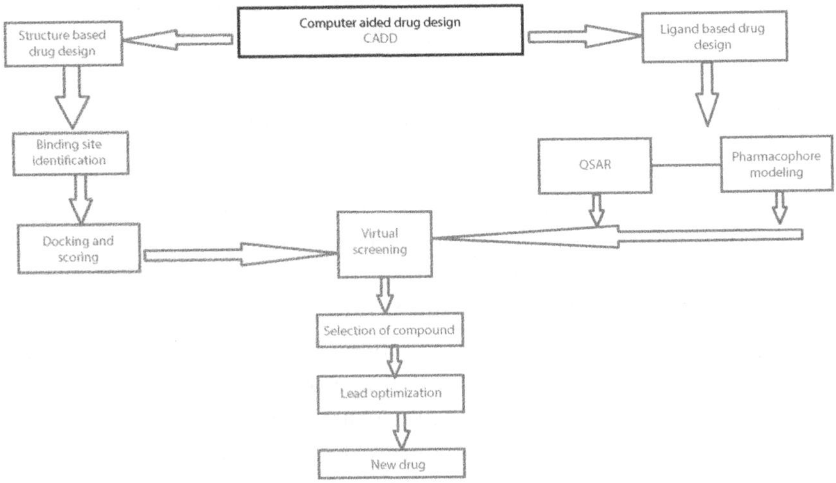

Figure 13.2 Drug design based on structure and ligand.

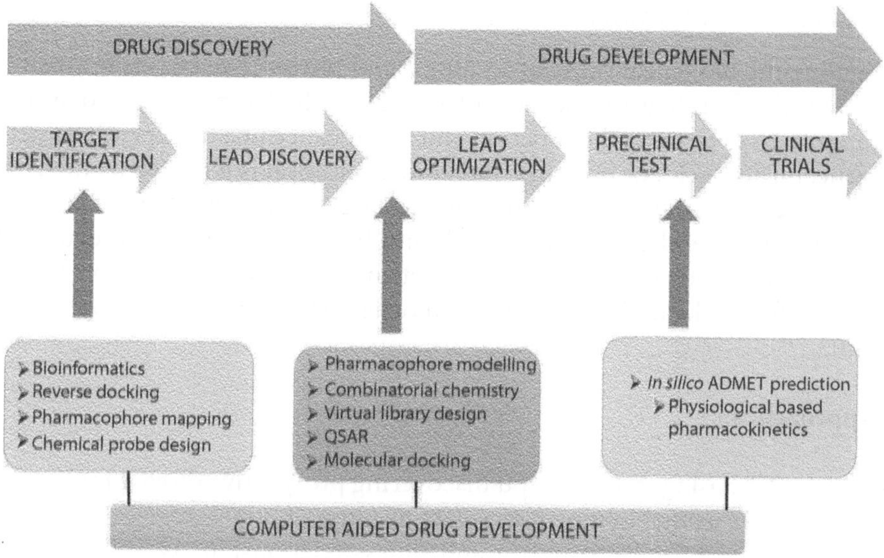

Figure 13.3 Drug discovery approaches.

modulators. Meanwhile, CADD (Figure 13.3) searches for potential drug candidates based on the structural information of either the target molecule or known ligands with biological activity [18].

13.1.6 Stages of Drug Discovery

Drug development is a vast area that encompasses many different disciplines, including pharmacology, computational chemistry, organic synthesis, and chemical and structural biology. As a result, it was divided into several stages, including [19]:

- Identification of target
- Validation of target
- Identification of lead
- Optimization of lead
- Characterization of product
- Formulation and development
- Preclinical research
- Investigational new drug
- Clinical trials
- New drug application
- Approval

13.1.6.1 Identification of Target

Finding the biological cause of an illness and prospective targets for treatment is the first stage in discovery of a drug. The first step in target identification is to determine the function and significance of a therapeutic target (gene, nucleic acid, or protein). Target identification is followed by molecular mechanism characterization of the targets. A perfect target would satisfy clinical and commercial needs and be safe, effective, and "druggable." The methods used to identify targets may be based on concepts from various fields like molecular biology, biochemistry, genetics, or biophysics [20].

Approaches:

- Choosing, ranking, and discovering prospective disease targets using bioinformatics data mining
- Genetic polymorphism and illness relationship, as well as genetic association
- Changes in mRNA and protein levels in the expression profile
- Mechanistic studies using *in vitro* cells that analyze pathways and phenotypes
- Functional screening employs technologies that are particular to a target or by knockdown or knockout [21].

13.1.6.2 Target Validation

Target validation is the procedure used to certify the intended molecular target of a tiny molecule, such as a gene, protein, or nucleic acid. It involves identification of the structure–activity relationship (SAR) of the tiny chemical analogs, the development of a drug-resistant mutant of the hypothesized target, knocking down or overexpressing the suspected target, and monitoring of the known signaling systems downstream of the presumed target. Target validation is the process of demonstrating the functional importance of the selected target in the sickness manifestation. The ultimate test is whether a medicine is effective in a clinical setting, even if it is immensely valuable to validate a drug's efficacy and toxicity in a number of disease-relevant cell models and animal models [22]. Target validation consists of two essential steps:

- ➢ Reproducibility: The initial step after identifying a drug target, whether using a particular technology or by a review of the literature, is to repeat the experiment to ensure that it can be effectively replicated. The target validation method includes affinity chromatography, expression cloning, protein microarray, reverse transfected cell microarray, biochemical suppression, siRNA, DNA microarray, system biology, and evaluation of already existing medications.

- ➢ Changing the environment where the ligand (drug) interacts with the target [23]:

    - Genetic modification of the target gene (*in vitro*); gene editing techniques include shRNA, siRNA, miRNA, CRISPR, and knocking out genes (viral transfection of mutant genes)
    - High-affinity antibodies that bind to the target and prevent subsequent interactions
    - Chemistry of genomes chemical methods for proteins encoded in the genome

13.1.6.3 Identification of Lead

Chemical leads are drug-like molecules that are synthetically stable, useful, and active in primary and secondary assays. They also have sufficient specificity, affinities, and selectivities for the target receptor. For this, the definition of the structure–activity link, the viability of the synthetic process, and primary evidence of the effectiveness and target engagement are required. The characteristics of a chemical lead are the following [24]:

- SAR explained
- Ability of the drug (preliminary toxicity, hERG)
- Synthetic feasibility
- Selection of mechanistic assays
- *In vitro* assessment of drug resistance and efflux potential
- Evidence of chemical class effectiveness *in vivo*
- PK/Toxicity of the chemical class known based on preliminary toxicity or *in silico* studies

A chemical lead is a stable, practicable, drug-like molecule that can be synthesized, is active in primary and secondary assays, and has appropriate specificity, affinities, and selectivities for the target receptor. To do this, the structure–activity relationship must be defined, the viability of a synthetic process must be determined, and there must be some preliminary evidence of *in vivo* efficacy and target engagement [25].

13.1.6.4 Lead Optimization

After identifying a key lead chemical, a drug candidate is generated through a process known as lead optimization. A potential drug is put through an iterative process of synthesis and characterization to see how its chemical structure and activity relate to the interactions with its targets and metabolism. In the previous stages of drug discovery, lead optimization is used to identify prospective compounds among the hits from hit-to-lead high-throughput screening assays. Lead optimization, the penultimate stage of early-stage drug development, involves evaluating putative leads for a variety of characteristics, such as binding processes and selectivity. Lead optimization aims to preserve advantageous features in lead compounds while addressing structural shortcomings. Lead compounds (small molecules or biologics) must have their chemical structures changed in order to increase target specificity and selectivity to create a preclinical therapeutic candidate. Additionally assessed are toxicological characteristics, pharmacodynamic, and pharmacokinetic factors. Laboratories must gather data on the toxicity, stability, efficacy, and bioavailability of leads in order to properly describe the molecule and choose the optimal path of optimization [26]. For the sake of this downstream selectivity profiling and additional research, researchers working on drug discovery need rapid strategies to choose a reduced number of drug candidates. In order to understand and predict *in vivo* pharmacokinetics utilizing *in vitro* studies, lead optimization now entails high-throughput DMPK (drug metabolism and pharmacokinetics) screening. Chemical changes to prospective pharmaceuticals' structures are made through optimization in order to create new medications with greater potency and safety characteristics. Pharmaceutical and biopharmaceutical drug development labs are increasingly using automated screening technologies. The identification and measurement of metabolites are done using mass spectrometry. A crucial method for quickly and reliably assessing drug candidates and their metabolites in tissue structure is matrix assisted laser desorption/ionization (MALDI) imaging [27]. For the discovery and optimization of lead molecules in focused screening programs, NMR fragment-based screening (FBS) has also become a widely employed approach in the pharmaceutical sector.

13.1.6.5 Product Characterization

Each unique drug molecule with therapeutic potential has certain properties, such as size, form, power, vulnerability, application, toxicity, and biological activity. To describe the compound's mode of action, early pharmacological investigations are useful. The creation and formulation of active pharmaceutical ingredients (APIs) are characterized physicochemically throughout the pharmaceutical formulation stage of drug development to offer the best bioavailable, efficient, and stable dosage form for a certain delivery route. During pre-formulation, the following parameters are evaluated:

- Solubility in different media and solvents
- Dissolution of the active pharmaceutical ingredient (API)
- Accelerated stability services under various conditions
- Solid state properties (polymorphs, particle size, particle shape, etc.)
- Formulation services and capabilities
- Formulation development of new chemical entities (NCE)
- Optimization of existing formulations
- Process development for selected dosage forms
- Novel formulations for the improved delivery of existing dosage forms
- Controlled release and sustained release formulations
- Self-emulsifying drug delivery systems
- Colloidal drug delivery systems
- Sub-micron and nano-emulsions

13.1.6.6 Preclinical Testing

Preclinical studies evaluate a drug's effectiveness and safety in animal models with an eye toward prospective effects on humans. Additionally, the relevant regulatory authorities must approve the preclinical trials. The regulatory agencies are responsible for ensuring that clinical trials are conducted ethically and safely, and they will only approve medicines that are both effective and safe. A fundamental set of specifications for acceptable technology needs for preclinical drug development has been developed by The International Council for Harmonisation of Technical Requirements for Pharmaceuticals for Human Use (ICH). Preclinical trials can be conducted in two ways: pharmacology and toxicology, in general [28]. The pharmacokinetic and pharmacodynamic characteristics of drugs are the focus of pharmacology. Exploring adverse pharmacological effects in appropriate animal models and keeping an eye on them in toxicological investigations are crucial. Pharmacokinetic studies are essential for determining the safety and effectiveness of absorption, distribution, metabolism, and excretion-related variables. These studies provide data on the rate of absorption for various routes of administration, which aids in dosage form selection and the regulation of the drug's half-life through distribution, metabolism, and excretion rates. The medicine's half-life explains its safety profile, which is a requirement for a drug to receive regulatory agency approval. The drug's bioavailability and affinity are determined by using a drug distribution mechanism, which explains the therapeutic efficacy of the drug. The possibility of going through stages of the biotransformation process and producing drug metabolites is provided by drug metabolism. Understanding the processes and enzymes involved in biotransformation is also aided by this. Both *in vitro* and *in vivo* drug toxicological studies can be used to evaluate a drug's toxicological effects. *In vitro* investigations can be done to evaluate the immediate effect on cell proliferation and phenotypic. Toxicological effects can be quantified and qualitatively analyzed via *in vivo* research. Selecting the proper animal species for toxicity studies is crucial because many medications are species-specific. Typically, the suggested use of the chemical in clinical research is based on *in vivo* investigations to assess pharmacological and toxicological activities, including the mode of action [29].

13.1.6.7 The Investigational New Drug (IND) Process

Before beginning clinical trials, drug developers must submit an Investigational New Drug application to the FDA. Developers must include the following in the IND application:

- Data from preclinical and toxicology studies
- Data on drug production and clinical research protocols for the planned studies
- Data from earlier clinical research (if any)
- Information regarding the researcher/developer

13.1.6.8 Clinical Research

Clinical trials are performed on volunteers and are designed to provide specific answers concerning the security and effectiveness of medicines, vaccines, other therapies, or novel ways to administer existing medications.

Clinical trials adhere to a particular research protocol that has been created by the manufacturer, investigator, or researcher. Before beginning clinical research, developers must begin the IND process. As they organize the clinical study, they will consider what they aim to achieve for each of the numerous clinical research phases. Research questions and objectives are created before a clinical study starts by reviewing earlier data regarding the medication [30, 31]. They then choose:

- The criteria for participant selection
- The number of participants in the trial
- The length of the study
- The dosage and method of administration of the dosage form
- The evaluation of parameters; gathering and analyzing data

13.1.6.8.1 Phase 0 Clinical Trial

Phase 0 consists of first-in-human (FIH) exploratory trials carried out in accordance with FDA regulations. In human microdose studies, commonly referred to as Phase 0 trials, 10 to 15 volunteers receive single sub-therapeutic doses. These trials provide pharmacokinetic data or aid in the imaging of particular targets without causing pharmacological effects. Research organization conduct Phase 0 investigations to determine which of their drug candidates has the best pharmacokinetic properties in people.

13.1.6.8.2 Phase 1: Safety and Dosage

The initial evaluation of medicine is done in Phase I involves 20 to 80 healthy volunteers. Patients are often only utilized when a drug's mechanism of action indicates that healthy individuals will not tolerate it. However, researchers undertake Phase 1 trials in individuals with a particular kind of disease if a novel medicine is recommended for use in patients with such a disease. Studies in the first phase are attentively watched to gather data on pharmacodynamics in the human body. To determine what dosage of a drug the body can handle and what its immediate side effects are, researchers modify the dosing schedule based on information from animal tests. Scientists gain knowledge regarding the drug's efficacy, side effects associated with dosage increases, and mechanism of action as a Phase I trial goes on. Information collected by the clinical trials Phase I must be considered in the Phase II study design. About 70% of the time, drugs progress to the next stage [32].

13.1.6.8.3 Phase II: Effectiveness and Adverse Effects

The purpose of Phase II trials, which involve larger patient populations (a few hundred), is to confirm the results of Phase I safety evaluations and examine the drug's efficacy. These tests are insufficient to determine if the medication will have therapeutic effects. Phase II trials give researchers more safety information. These data are used by researchers to establish new Phase III research procedures, develop research methods, and revise their research topics. Drugs move on to the subsequent phase in about 33% of cases. Clinical Phase II studies play a significant role in determining therapeutic dosages for the extensive Phase III investigations [33].

13.1.6.8.4 Phase III: Monitoring of Medicine Effectiveness and Unfavorable Side Effects

Phase III research is conducted to demonstrate whether a product offers a specific group of people an action benefit or not. These trials, which occasionally go by the name of pivotal studies, involve 300 to 3,000 individuals. Most safety information comes from Phase III trials, as the earlier study could not have been able to find less frequent adverse effects. However, because Phase III studies involve a greater number of volunteers and last longer, the findings are more likely to reveal persistent or unusual adverse effects. Drugs move on to the following stage of clinical study in about 25%–30% of cases. An application to sell the drug may be submitted by the pharmaceutical industry if a drug's safety and efficacy for the intended use have been established through prior testing, preclinical research, and clinical trials. All of the drug's submitted data are carefully reviewed by the FDA review team before a decision is made on its approval.

13.1.6.8.4.1 New Drug Application

The entire history of a therapeutic molecule is told in a New Drug Application (NDA). Its goal is to confirm that the medicine is both safe and effective for the population being researched. Preclinical data through Phase III trial data must all be included by the drug developer in the NDA. Reports on all investigations, data, and analyses must be included by developers. The results of clinical trials, developers must also do the following:

- Proposed labeling
- Safety updates
- Information on drug abuse

- Patent information
- Institutional review board compliance details
- Usage instructions

13.1.6.8.4.2 FDA Review

The FDA panel of reviewers may take between 6 and 10 months to decide whether to approve the NDA after a thorough evaluation. The FDA review panel will reject an incomplete NDA if it is submitted to them. Working with the developer to update the prescribing information is crucial if the FDA decides that a medication has proven to be reliable and efficient for the purpose it was designed. Labeling is employed to explain this. The grounds for approval and instructions on how to take the drug are clearly stated on the label. However, before the medicine can be licensed for commercialization, there are still problems that need to be resolved. In other circumstances, the FDA requires more research. The developer has the option of moving forward with the development or not at this point. A formal appeal procedure is available for developers who are upset by an FDA decision.

13.1.6.8.5 Phase IV: Post-Market Drug Safety Monitoring

After the FDA has approved the pharmaceutical or device, Phase IV trials are conducted. These investigations are also recognized as post-approval or post-marketing surveillance, requiring continuing technical assistance and pharmacovigilance.

In Phase IV studies, a range of observational approaches and evaluation patterns are used to analyze the effectiveness, viability, and safety of an engagement in actual circumstances. Regulatory agencies may require Phase IV studies (such as a modification in labeling or an action plan for risk management or minimization), or the sponsoring corporation may carry them out for reasons such as competition. Therefore, over the days, weeks, months, and frequently, even years that make up a drug's shelf life, genuine safety must be proven. The FDA evaluates complaints about prescription and over-the-counter medications and may decide to add warnings to dosing instructions or other practice recommendations, as well as other events for more severe adverse drug reactions [34].

13.2 Conclusion

Finding and creating new medications is a laborious process that entails hunting for promising hits, turning those hits into leads, and then finally validating those leads into therapeutic candidates in clinical trials. Drug discovery is a process that involves the identification of candidates, synthesis, characterization, validation, optimization, screening, and assays for therapeutic efficacy. Once a compound has shown its significance in these investigations, the process of drug development will be initiated prior to clinical trials. The new drug development process must continue through several stages to make medicine that is safe, effective, and that meets all regulatory requirements.

References

1. Fernandes, J.P.S., The Importance of Medicinal Chemistry Knowledge in the Clinical Pharmacist's Education. *Am. J. Pharm. Educ.*, 82, 2, 6083, 2018, doi: 10.5688/ajpe6083.
2. Bokhari, F.F., Albukhari, A., Design and Implementation of High Throughput Screening Assays for Drug Discoveries, in: *High-Throughput Screening for Drug Discovery*, Saxena, S.K. (Ed.), **2021**.
3. Newman, D.J., Natural Products as Sources of New Drugs over the Nearly Four Decades from 01/1981 to 09/2019. *J. Nat. Prod.*, 83, 3, 770–803, 2020. https://doi.org/10.1021/acs.jnatprod.9b01285.
4. Horizny, C., The Drug Discovery Process. *Taconic Biosci.*, **2019**, available at https://www.taconic.com/resources/drug-development-process
5. Deore, A.B., Dhumane, J.R., Sonawane, R., The Stages of Drug Discovery and Development Process. *Asian J. Pharm. Res. Dev.*, 7, 6, 62–67, **2019**, doi: http://dx.doi.org/10. 22270/ajprd.v7i.
6. Li, T., Guo, R., Zong, Q., Ling, G., Application of molecular docking in elaborating molecular mechanisms and interactions of supramolecular cyclodextrin. *Carbohydr. Polym.*, 276, 118644, **2022**, https://doi.org/10.1016/j.carbpol.2021.118644.
7. Fan, J., Fu, A., Zhang, L., Progress in molecular docking. *Quant. Biol.*, 7, 83–89, **2019**.
8. Grinter, S.Z. and Zou, X., Challenges, applications, and recent advances of proteinligand docking in structure-based drug design. *Molecules*, 19, 7, 10150–10176, **2014**.
9. Pedro, H.M., Torres, A.C.R., Sodero, P.J., Floriano, P., Key Topics in Molecular Docking for Drug Design. *Int. J. Mol. Sci.*, 20, 18, 4574, **2019**. https://doi.org/10.3390/ijms20184574.
10. Lavecchia, A. and Giovanni, C., Virtual screening strategies in drug discovery: a critical review. *Curr. Med. Chem.*, 20, 13, 2839–60, **2013**.
11. Lionta, E., Spyrou, G., Vassilatis, D.K., Cournia, Z., Structure-Based Virtual Screening for Drug Discovery: Principles, Applications and Recent Advances. *Curr. Top. Med. Chem.*, 14, 16, 1923–1938, **2014**.

12. Kaur, T., Madgulkar, A., Bhalekar, M., Asgaonkar, K., Molecular Docking in Formulation and Development. *Curr. Drug Discov. Technol.*, 16, 1, 30–39(10), **2019**. https://doi.org/10.2174 /1570163815666180219112421.
13. Prathipati, P., Dixit, A., Saxena, A.K., Computer-Aided Drug Design: Integration of Structure-Based and Ligand-Based Approaches in Drug Design. *Curr. Comput. Aided Drug Des.*, 3, 133–148, **2007**.
14. Joy, S., Macalino, Y., Gosu, V., Hong, S., Choi, S., Role of computer-aided drug design in modern drug discovery. *Arch. Pharm. Res.*, 38, 1686–1701, **2015**.
15. Pawar, S.S. and Rohane, S.H., Review on Discovery Studio: An important Tool for Molecular Docking. *Asian J. Res. Chem.*, 14, 1, 86–88, **2021**, doi: DOI: 10.5958/0974-4150.2021.00014.6.
16. Song, S., Zhang, Q., Yang, H., Guo, J., Xu, M., Yang, N., Yi, J., Wang, Z., A combined application of molecular docking technology and indirect ELISA for the serodiagnosis of bovine tuberculosis. *J. Vet. Sci.*, 23, 3, e50, **2022**, doi: 10.4142/jvs.21270.
17. Li, Y., Li, Y., Ning, C., Yue, J., Zhang, C., He, X., Discovering inhibitors of TEAD palmitate binding pocket through virtual screening and molecular dynamics simulation. *Comput. Biol. Chem.*, 98, 107648, **2022**.
18. Tao, X., Huang, Y.K., Wang, C., Chen, F., Yang, L.L., Ling, L., Recent developments in molecular docking technology applied in food science: a review. *Int. J. Food Sci. Technol.*, 55, 1, 33–45, **2020**.
19. Ripphausen, P., Nisius, B., Peltason, L., Bajorath, J., Vadis, Q., Virtual Screening? A Comprehensive Survey of Prospective Applications. *J. Med. Chem.*, 53, 8461–8467, **2010**.
20. Mohanty, M., Mohanty, P.S., Molecular docking in organic, inorganic, and hybrid systems: A tutorial review. *Monatshefte für Chemie-Chemical Monthly*, 154, 7, 683–707, 2023.
21. Muhammed, M.T., Aki-Yalcin, E., Molecular docking: principles, advances, and its applications in drug discovery. *Lett. Drug Des. Discov.*, 21, 3, 480–495, 2024.
22. Sahu, D., Rathor, L.S., Dwivedi, S.D., Shah, K., Chauhan, N.S., Singh, M. R., Singh, D., A review on molecular docking as an interpretative tool for molecular targets in disease management. *ASSAY Drug Dev. Techn.*, 22, 1, 40–50, 2024.
23. Stocks, C.J., Schembri, M.A., Sweet, M.J., Kapetanovic, R., For when bacterial infections persist: Toll-like receptor-inducible direct antimicrobial pathways in macrophages. *J. Leukoc. Biol.*, 103, 1, 35–51, **2018**.
24. Ciddi, V., Natural products derived from plants as a source of drugs. *J. Adv. Pharm. Technol. Res.*, 3, 4, 200–201, **2012**.
25. Hughes, J.P., Rees, S., Kalindjian, S.B., Philpott, K.L., Principles of early drug discovery. *Br. J. Pharmacol.*, 162, 1239–1249, **2011**.
26. Kutchukian, P.S. and Shakhnovich, E.I., De novo design: balancing novelty and confined chemical space. *Expert Opin. Drug Discov.*, 5, 8, 789–812, **2010**.

27. Patrick, G.L., *An introduction to medicinal chemistry*, Third edition, pp. 161–162, Oxford University Press, Oxford, United Kingdom, **2006**.
28. Sethi, A., Joshi, K., Sasikala, K., Alvala, M., Molecular Docking in Modern Drug Discovery: Principles and Recent Applications, in: *Drug Discovery and Development*, Gaitonde, V., Karmakar, P., Trivedi, A. (Eds.), **2019**.
29. Malviya, R., Sharma, P.K., Sundram, S., Dhanaraj B. Balusamy, R.K., *Bioinformatics tools and big data analytics for patient care*, vol. 1, p. 240, Routledge Taylor & Francis Group, Chapman & Hall, **2022**.
30. Deore, A.B., Dhumane, J.R., Wagh, R., Sonawane, R., The stages of drug discovery and development process. *Asian Journal of Pharmaceutical Research and Development (AJPRD)*, 15, 7, 62-7, 2019.
31. Kinders R., Parchment, R.E., Ji, J., Kummar, S., Murgo, A.J., Gutierrez, M., Collins, J., Rubinstein, L., Pickeral, O., Steinberg, S.M., Yang, S., Phase 0 clinical trials in cancer drug development: from FDA guidance to clinical practice. *Mol. Interv.*, 1, 7, 325, 2007.
32. Chen, J., Luo, X., Qiu, H., Mackey, V., Sun, L., Ouyang, X., Drug discovery and drug marketing with the critical roles of modern administration. *Am. J. Transl. Res*, 10, 12, 4302, 2018
33. Bhattaram,V.A., Booth, B.P., Ramchandani, R.P., Beasley, B.N., Wang,Y., Tandon, V., Duan, J.Z., Baweja, R.K., Marroum, P.J., Uppoor, R.S., Rahman, N.A., Impact of pharmacometrics on drug approval and labeling decisions: a survey of 42 new drug applications. *The AAPS Journal*, 7, E503-12, 2005.
34. Zhou, S.F., Zhong, W.Z., Drug design and discovery: principles and applications. *Molecules*, 22, 2, 279, 2017.

14

Challenges and Emerging Problems in CADD

Akshita Arora[†]**, Simranjeet Kaur**[†] **and Amandeep Singh***

Department of Pharmaceutics, ISF College of Pharmacy, Punjab, India

Abstract

Pharmaceutical synthesis has traditionally been a lengthy, expensive, and labor-intensive process that requires many years of experience and considerable personnel. However, with the introduction of information technology, it can be done more efficiently. During the last few decades, the area of drug discovery, which leads to the discovery of novel ligands, has evolved into a modern science that employs both computational and experimental methods. Computer-aided drug design (CADD) has emerged as a striking feature of numerous programming languages in a variety of contexts and research scenarios. To uncover, reinforce, or analyze medications and other physiologically active substances, computational chemistry is employed in CADD. It involves creating novel chemicals, docking them to a target protein, examining molecular interactions, and assessing binding strength along with therapeutic properties. To advance to further stages of the drug research pipeline or the market, hit compounds have to be identified and optimized, which have been streamlined by CADD. Despite CADD's contributions at several points in the drug discovery process, challenges remain. To realize the promise of computational methodologies, research teams must improve communication. Other problems that must be addressed include proper scientific dissemination, data exchange, and education to minimize false expectations and enhance CADD production. This chapter's goal is to provide a thorough review of CADD, together with an analysis of its challenging issues and promising future directions.

***Keywords*:** CADD, problem, patent, CAD, future

**Corresponding author*: ad4singh@gmail.com

[†]Contributed equally

Rishabha Malviya, Sonali Sundram and Dhanalekshmi Unnikrishnan Meenakshi (eds.)
Drug Delivery Systems Using Quantum Computing, (407–442) © 2024 Scrivener Publishing LLC

14.1 Introduction

14.1.1 The Antiquity of CADD

In the early 1900s, German scientists Emiel Fischer (in 1894) and Paul Ehrlich (in 1909) popularized the lock-and-key analogy, a paradigm-defining drug associated with a specific receptor, culminating in biological activity. From synthesizing through testing, scientists in the early 1960s did everything by hand, relying on their knowledge and imagination. However, to develop a drug by trial and error would necessitate vigorous effort and time. Eli Lilly was the first to achieve success in computational drug research, establishing the antibacterial properties of cephalosporins, also known as penicillin. The IBM650 used by Lilly's research group was the early computer to contain stored scientific programme.

In the 1970s, the Cambridge University Press hurled the Cambridge Structural Database (CSD) and the PDB (Protein Data Bank). Computational chemists were pleased by the evident possibility that these databases could yield a greater number of compounds with medicinal potential. Following the 1980s, multiple technological breakthroughs were made, such as the arrival of Apple Macintosh computers, which were advantageous for data entry, visual effects, and were capable of handling laboratory datasets. Furthermore, there were software advancements. A few examples include electronic mail (for communication), ChemDraw (to make two-dimensional diagrams), and ball-and-stick models (to view 3D structures of compounds). This also led to the introduction of approaches like QSAR (Quantitative Structure–Activity Relationships), quantum chemistry, molecular mechanics, and molecular simulations.

Since efforts to develop new medications in the computerized form in the 1980s resulted in an incredible influx of novel chemical entities reaching the pharmaceutical industry, the 1990s were a prosperous decade. In reality, in the 1990s, QSAR emerged as one of the most sophisticated methods of aiding medicinal chemists.

14.2 Preamble of CADD

CADD is an essential approach for collecting, analyzing, and evaluating molecular data for drug discovery and development, saving considerable hours and expenditures. CADD's Principle targets are to advance and start producing therapeutically active medicines and perhaps to gain deeper expertise in the field of molecular physiological functions. It can screen

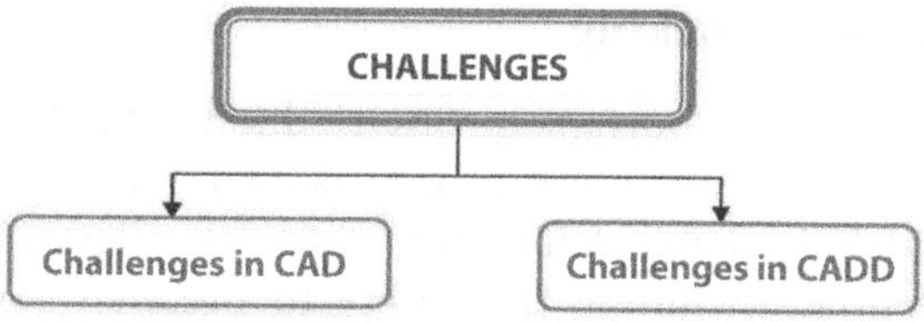

Figure 14.1 Classification of challenges.

big compounds quickly and circumvent the time-consuming stages of discovering significant medicinal chemistry, making it ideal for targeting and creating these bioactive medicines [1].

CADD approaches are primarily divided into two groups as shown in Figure 14.1:

1. Structure-based drug design
2. Ligand-based drug design

14.2.1 Structure-Based Drug Design

Structure-based drug design (SBDD) is a providential operation for discovering the lead and optimizing it more profitably in the fullness of time. It works scrupulously by assessing the ligand association with the target proteins for therapeutic and pharmacological effects using 3D structures of numerous accessible targets, such as enzymes or proteins [1]. SBDD is a multistep procedure involving protein structure preparation, binding site identification, ligand library preparation, docking, and scoring function.

14.2.2 Protein Structure Preparation

It is the preparatory phase enclosing the preparation of the target structure, thereby forming protein–ligand complex structures determined by structure biology techniques like nuclear magnetic resonance (NMR), X-ray crystallography, cryo-electron microscopy, or computational methods such as homology modeling and *de novo* folding [2].

14.2.3 Binding Site Identification

Binding sites are perhaps accentuated by hit-and-miss, as an illustration, through site-directed mutagenesis or X-ray crystallography, or by using online tools such as the DoGSite Scorer server, Meta pocket, DEPTH, LISE, and MS pocket, which assess protein–ligand interactions [2].

14.2.4 Molecular Docking

Docking is extremely important in SBDD due to its ability to find the perfect combination of molecules, as well as owing to its predicting capabilities. It is widely in demand for drug design. It offers a broad array of prediction potential, including the ability to precisely entangle ligands to their ultimate destination. Tasked with investigating crucial chemical processes like ligand-binding states and intermolecular interactions that might help ensure complex stability, the docking algorithms use various scoring functions to assess binding energies and rank the ligands [3].

14.2.5 Scoring Functions

Scoring functions are used to quantify the binding affinity of complex molecules. Force field, empirical, knowledge-based, and machine learning are the four broad categories of scoring functions [3].

14.3 Ligand-Based Drug Design

Ligand-based drug design is a dynamic approach in which ligands ligand-based drug design (LBDD), ligands that cling to the therapeutic targets are used rather than the drug target's 3D structure. According to the similarity principle, extremely similar compounds are prerequisites for having comparable bioactivities. Descriptors emphasizing chemical processes, geometry, electrostatics, and points of convergence are employed to analyze resemblance (e.g., pharmacophore points) [4]. Molecular descriptors, which can either be 1D, 2D, or 3D, allow for fast comparison of the structural and/or physicochemical aspects of different molecules. The two different LBDD techniques that are most frequently employed are pharmacophore modeling and quantitative structure–activity relationship (QSAR) [5].

14.4 Challenges

Computer-aided design has several challenges, which can be broadly grouped into the following two categories:

14.4.1 Challenges in CAD (Computer-Aided Design)

These challenges are grouped into three different categories:

- Geometry
- Interactive techniques
- Scale

14.4.2 Geometry

In any CAD application, geometry is the fundamental data type that designers utilize to define physical products. The basis for determining how to manufacture the entire product and how the product will behave is its geometric form, architecture, and system elements. Computational geometry algorithms are only implemented in a few commonly used libraries. Geometry is the essential building block of all CAD (computer-aided design), CAM (computer-aided manufacturing), and CAE (computer-aided engineering) software. Geometry encompasses both the mathematical forms and algorithms used to construct, combine, alter, and examine an object's geometrical characteristics [6].

14.4.2.1 Controlling Shape

The accomplishment of graphics design has been to give everything a precise mathematical representation. The new polynomial and piecewise polynomial methods (such as B-splines and splines) did not have implicit control over shape, in contrast to the hand-drawn and hand-constructed procedures, including liming. Consequently, it became essential to develop algorithms that supported both point fitting and shape control. The occurrence and location of inflection points are determined by shape control. Both engineering and manufacturing optimization require control. Different shape control strategies have been put forth by researchers. The majority of attempts failed because the approaches frequently did not correctly preserve shape. Shape anomalies can be found after the fact using a variety of techniques, and they must be manually corrected. We realized the fundamental tenets of the approaches were flawed when attempts to modify them to account for shape control generally failed to function. Although nonlinear approaches have demonstrated their effectiveness in numerous form control scenarios, the majority of these methods have not found a place in commercial CAD systems [7].

14.4.2.2 Interoperability

CAE analysis and current CAD systems do not work together properly. The geometry must be promptly and accurately challenged, for example, in computational fluid dynamics (CFD). The geometry is used to create a computational grid in most CFD programs. Building the grid dependably entails that there must be no unintentional gaps in the geometry, or what CFD experts describe as "watertight" geometry. Real geometry from CAD systems is rarely completely accurate. Tolerances and floating-point arithmetic are the two main causes of gaps, translational issues, and other problems. Although theoretically possible, floating-point arithmetic cannot be used in practice. To reduce the consequent inaccuracies, we could enforce a higher level of precision. Alternatively, we may switch to a rational arithmetic system and make floating points unnecessary. The precision of computed results is governed by tolerances, which are an inevitable part of modern CAD systems. Measuring the curve of an intersection between two surfaces is a straightforward example. When the geometry is represented by straightforward algebraic equations, there is a closed-form response for the intersection. For operations on equations of a high enough degree, closed-form solutions do not generally occur. If greater accuracy is required, certain CAD programs will recompute intersection curves. This does not resolve the issue, especially if more geometry is generated using the intersection curve. To regulate the estimation, tolerances are required. A permissive tolerance can produce quick but inaccurate findings. Poor performance or convergence can be caused by a tolerance that is too tight. Because of tolerance selection, even straightforward junctions become challenging. The excellence of downstream engineering, fabrication, and quality assurance analysis is determined by tolerances. The geometry generator must be aware of the types of analyses to be used to choose a tolerance that ensures a high chance of success. In conclusion, present math theory and digital arithmetic are inadequate to consistently manage complicated geometry procedures and to effectively communicate with downstream analysis tools. For downstream applications, the geometry should be as watertight as feasible, and algorithms should not produce topological abnormalities like self-intersections. Finding solutions to deal with inferior outcomes is a difficult task [8, 9].

14.4.2.3 Design Exploration

The difficulty lies in automated design discovery through multidisciplinary optimization. Geometry must maintain topological validity for design optimization. Two things need to be taken into account: how to structure geometry algorithms, which are fundamental to every optimization process; and how to parameterize geometry for downstream analysis. CAD systems are unable to satisfy the necessity of morphing. It is challenging to design and implement geometry systems that ensure the consistency of morphing operations. There are differences between geometric and morphological continuity. Geometry frequently contains discontinuities that must be kept during morphing. The geometry will not change rapidly like parameters do when there is morphing continuity. To guarantee accurate geometry for analysis and optimization, values for all parameters must be simultaneously set to appropriate levels. A geometry system that automatically changes parameters without user input while maintaining the integrity and intent of the design is necessary for design optimization. We must re-evaluate both the algorithms and the geometric design process in multidisciplinary design [10].

14.4.3 Interactive Techniques

Making more effective menus will not result in a significant improvement in interaction. Instead, improvement will rely on reconsidering the objectives of the process.

14.4.3.1 Reverse Engineering

The same actions done more quickly or cheaply do not lead to significant changes. They appear when you reverse techniques and procedures. For example, in the field of aerodynamics, designing for efficiency and safety is crucial. Government organizations like NASA invest millions of dollars in equipment to evaluate these systems, including wind tunnels and CFD analysis. This testing, however, is in the final stages of design. Now consider this reality in conjunction with one of the guiding design's principles: the more expensive a mistake is to repair, the later in the process it is discovered. Bad designs could be identified far earlier in the process by designers. Additionally, this would enable them to identify, investigate, improve, and, as early in the design process as feasible, comprehend the most feasible designs. Many other CAD workflow processes could use this CFD methodology. Consider the possibility of including tasks, like calculations for volume or strength, early in the workflow. As a result, designers could take into account the connectivity and compatibility of elements considerably earlier in the design process [11].

14.4.3.2 Everything in Its Place

Previously, a person's position in a CAD system could be used to determine what their current work was. For instance, at an automotive design studio, interiors are frequently handled in one area, full-sized clay models are kept in another, and color exploration takes place in yet another. Anyone passing through such a room typically will not be able to discern whether they are in the engineering or accounting department. A general-purpose workstation's grip on the engineer will be broken by ubiquitous computing. This transformation will force designers to transition from one specialized field to another in a way that harkens back to their greatest prior practices. The workspace will be divided into specialized regions, each with its own set of tools; and these tools will combine both private and public displays and technology. Mobilization increases the potential for collaboration and raises the visibility of certain activities. Future systems will face difficulties in both new forms of human-to-human contact and human-to-machine interaction.

14.4.3.3 What We Do

CAD firms may see themselves as primary suppliers of equipment for building excellent 3D models. As a result, they might include that they need to create a better, more practical modeler if they want to expand their firm. Additionally, it appears that this line of reasoning is both incorrect and irrational. Focus on the following queries:

- Is there a significant untapped resource for 3D models seeking an intuitive modeling tool with a little learning curve?
- Is there a lack of qualified individuals to meet the demand for 3D model creation?
- Are there any items that need to be simulated that current users cannot create with their current knowledge and resources?

The basic answer to each of these queries is "no," in general. Simply having a product or part manufactured is what modeling is, and it is not always the ideal approach. The CAD procedure will be divided between experts who build and design components and those who manufacture larger assemblies out of them as product complexity rises. Future CAD systems components will not be permanent items created from declarative data, as it is known in the computer science community; instead, they will have a big procedural element to them. The core problem is that things are challenging to scale, even the easiest ones. The CAD engineer's tools enable the designer to find the right components and alter them in such a way as to maintain the intended attributes [12].

14.4.4 Scale

While scientists and engineers are advancing the frontiers of macro and nanotechnology, still scale remains a difficulty for them. We can generate larger amounts of data thanks to advanced hardware. The scale challenges that are covered in this part are based on Boeing's experience, a sizable and varied aircraft company. The issues are directly related to changes made to the fundamental design-analyze-build-maintain process.

14.4.4.1 Understanding Vast Quantities of Data

Generally, the processes associated with designing, engineering, production, and management have existed mostly independently of one another. Data management presents a challenge, particularly as products get more complicated. Another section concerns people since the graphical terms used in the design phase differ significantly from those utilized throughout the product's manufacturing and assembly. Initial remedies for problems like maintainability and manufacturability include an emphasis on integrated design-build teams. The members of the team utilize several programs, as well as store their information in various locations using various configuration management strategies. Product lifecycle management systems enable businesses to centrally store information such as client specifications, architectural geometry, production plans, engineering analysis results, factory process designs, monitoring designs, and so forth in a centralized file that uses configuration management. Selling product life cycle management (PLM) systems can be difficult. With pre-facto forced integration, improved configuration management, and greater configuration transparency, the essential benefits increase costs in each area individually while lowering overall product prices. PLM technologies interact with other important business systems for managing activities like supply chain management (SCM), enterprise resource planning (ERP), and staffing. After product integration is institutionalized, the actual issues arise. Integrated products need to be looked at from a variety of perspectives [13].

14.4.4.2 Suitable Layout for Users

This challenge concentrates on the massive number of individuals whose duties at work depend on understanding the description of the digital product but who are not involved in data creation and analysis. These people take part in product operation, certification, fabrication, assembly, maintenance, and sales. Governmental bodies certify that the finished product performs following rules and specifications that have been made public. Since various audiences require different information, several groups develop their graphic vocabularies. Consider the following example.

Images are frequently used by designers and engineers. Specific elements' architecture and layout are impacted by the colors, but production and maintenance are unaffected. The engineering drawing continues to be the authoritative image for certification, production, and maintenance. A dramatic change occurs when an engineering drawing is used to communicate with a maintenance engineer. The basic 3D model can frequently be worked with using tools, such as the capacity to rotate, scale, and transform, as well as making anything in a scene transparent except for an item of curiosity. Some of these tools may even produce engineering drawings, but they fundamentally change the way that graphic representation is generated. Since communication norms and methods differ greatly from community to community and from corporation to corporation, the difficult part of graphic design is figuring out how to communicate with a certain system of users. As a result, businesses invest a lot of time and effort into editing or redoing images [14].

14.4.4.3 Retrieving Data Years Later

The requirements for tolerance, reliability, and other factors for products nowadays differ significantly from those over 40 years ago. The manufacturing, certification, and design authorities are all represented by the engineering drawing. Even though we are aware that engineering drawings have drawbacks, they also have a unique benefit: They may be stored for a very long time and still make sense. Consider the difficulty if the documents were kept digitally, since they are frequently copied to fresh media, and the media is not a concern in and of itself. Updating software versions is the main challenge. The majority of users often anticipate that part of their data will not successfully transition from one version to the next. Therefore, even years after the first release, businesses cannot afford to update to the latest version of the system made by the same company manufacturer. The data migration challenge gets more difficult as there is more configuration-managed data [15].

14.4.4.4 Constricted by the Light Speed

In a geographically dispersed workplace, it becomes impossible for users to transverse the hall, which is the central concern of this issue. The globalization of both large and small enterprises is a topic of ongoing controversy.

Companies are hiring people from all around the world for design, production, assembly, maintenance, and customer support services despite distinct national interests and boundaries. The incentives can include labor at a lower cost, trade offsets, or specialized technical skills. Effective distance collaboration for 3D design is a brief research and cultural problem because it loses the instant and comprehensive indications of a face-to-face session. Collaboration across a variety of stakeholders is the major productivity factor for reduced design cycle time. Computer-supported collaborative work is the subject of considerable research and a wide range of collaboration models. Collaborative task analysis is becoming more popular among researchers [16].

14.5 Challenges in CADD (Computer-Aided Drug Design)

In CADD, docking has progressed to a certain extent, although it is far from ideal. The majority of docking programs on the market can often estimate known binding poses. One of the docking's main drawbacks is its inability to calculate precise binding energies [6]. The most significant flaw in docking is probably the absence of an efficient scoring system and a discovery algorithm that can combine quality and precision. This section focuses on a few of the areas where the constraints listed above are most likely to have an effect [17].

14.5.1 Metal-Binding Docking

The significance of metal co-factors in several protein families is well known. Identical to covalent docking, ligand-metal interaction modeling needs significant advancements. Modeling interacts with calcium, zinc, and magnesium in particular (which are important metal ions for finding new drugs). Zinc has a mainly tetrahedral shape and is found on druggable targets. It receives a lot of attention when it comes to proper parameterization. Since it is possible for ligand-zinc interactions to be quasi-covalent, modeling of this interaction must take several factors into account; for example, the acidic behavior of this interaction permits "special" hydrogen bonding. As a result, AutoDock introduced force field reparameterization. When measured against other programs, it offered a more precise answer to the question and improved scores. Metal recognition is available in the following applications: AutoDock, FITTED, Ligand Fit, FlexX, Glide, DOCK, MOE, and MPs Dock [18].

14.5.2 Protein–Protein Docking

The critical analysis of protein interaction represents some of the early simulations of protein–protein interaction. It served as a contested space to test the abilities of various human teams, computer programs, and servers to precisely forecast the shape of chosen targets for protein–protein interactions. The complex's experimental construction is unknown to the competitors, but it is known to the judges. Interesting observations have come as a result of the numerous Critical Assessment of PRediction of Interactions (CAPRI) cycles. Even with minor adjustments, the interaction phenomena between proteins are highly complex. Protein–protein docking can be conceptualized as an entire complex estimation employing local searches for the binding sites to minimize each protein using coarse-grained models. Thus, the flexibility of the backbone poses the biggest obstacle to protein–protein docking. This has been accomplished with success using Monte Carlo methods. Due to improper sampling, the bound state of proteins still presents difficulty. Therefore, it is necessary to carry out extensive computer investigations to correctly distinguish between actual complexes and irrational predictions [19, 20].

14.5.3 Evaluation of Protein–Ligand Flexibility and Interaction

This is one of the biggest issues as it could be argued that assessing a protein–ligand complex's behavior in a changing environment is the right strategy. With the advent of Emil Fischer's binding model and the development of more effective tools and techniques, it has become evident that protein behavior is subtle. After all, ligand adaptability has been a central issue in medicinal chemistry and pharmacology. The exponential development of databases like the Protein Data Bank has led to the discovery of novel folding patterns and structural configurations. With this knowledge, it is possible to shed some light on the inner workings of proteins by looking for commonalities between binding sites and protein pockets. Five classes of "potential pocket dynamics" have been identified, all of which might coexist simultaneously in a single protein. This would allow bivalent binding, which might be exploited to increase selectivity [21].

14.5.4 Protein–Ligand Docking

It is possible to think of structure-based design as an "old-school" but incredibly effective strategy for druggable targets. In computational approaches, docking is the most effective since it allows for the input of several ligand changes and instantaneous results. Despite being effective, induced fit docking has limited simple space because only nearby residues are taken into account as agile, indicating that certain pocket dynamics and even all steric effects cannot be properly considered. Protein energy landscape exploration (PELE) is one technique for evaluating flexibility [22]. This Monte Carlo method combines perturbations to proteins and ligands. There are three distinct stages:

I. Protein and ligand perturbation
II. Side-chain testing
III. Reduction and acceptance using the Metropolis criterion

This serves as a helpful refinement, especially when a protein binding site is unknown. A recent study demonstrated the effective utilization of protein energy landscape exploration for the reliable assessment of poses and binding locations for extremely supple proteins. Machine learning is one method of approaching the issue. It is conceivable to use tools to evaluate the compatibility of protein and ligand structures using techniques like SOMs (self-organizing maps) or k-means using an ensemble of molecular dynamics snapshots. The most successful and widely recognized methodology for simple flexibility is ensemble docking. Much like Monte-Carlo techniques, protein energy landscape exploration techniques seemed to be used to create trustworthy groups. An illustration is MTflex, which determines free energy via Monte Carlo integration. The distinction between this method and others (such as rotamer libraries) is that when creating rotamers for interacting compounds depending on low energy values along the free energy surface, the molecules are introduced depending on a local energy criterion rather than to satisfy a conformational question. Using reference atom positions can lead to better outcomes. This technique produced more effective ensembles [23].

There are various methods for evaluating the flexibility of protein ligands. In MD simulations, metadynamics is one of the enhanced sampling strategies. To delve further into the configurational space of the system, biased potentials are applied to a certain subset of the system's variables, also referred to as collective variables. To create a sum of Gaussian potentials, metadynamics investigates conformations outside of local minima utilizing restraints. These potentials have already been exploited in docking by algorithms like taboo search and are repulsive in nature. The distinction is that the program's free energy surface can be recreated because of metadynamics. As a result, a binding mode may be more thoroughly validated using metadynamics [24].

14.5.5 Docking Peptides or Peptide-Like Ligands

Peptide sampling is very unpredictable and essentially unrepeatable owing to its considerable freedom of movement and adaptability. Peptides are currently observing a renaissance in medicine, with examples including romidepsin and dolastatin. Demonstrating that despite being versatile, remarkable polypharmacological effects may result from them and are appealing as protein–protein interaction modulators. The great degree of flexibility makes peptide docking more complex in terms of computations. There have been various attempts to overcome these obstacles [25]:

- The utilization of monomeric ensembles' derived templates and energy-based filtering;
- The use of a local search, application of constraints, and optimization of an orientation;
- Anchor-spot identification on the binding site and simulated-annealing molecular dynamics; and
- The use of the Latin squares method, selecting rotatable bonds systematically to find energy minima.

14.5.6 Water Solvation and Docking

A common criticism of docking is the removal of the explicit solvent (more precisely, the removal of the water molecules). This idea is frequently based on methods for limiting the degrees of freedom of a particular process that involve the scoring function and computing time. For proteins like HIV-1 protease that have highly conserved structural water molecules, this method is impossible. The issue extends beyond merely accounting for water molecules in computations and encompasses accurately assessing water binding contribution and its effects [26].

Attempts to include water molecules in docking measurements revealed little to no improvement in scores. Some ligands, however, are made specifically to dispense with water. With the proper handling of water molecules in these situations, docking simulations are more accurate. Currently, the majority of docking software can detect water molecules during calculations. Some difficulties include coring reparametrization, pose comparisons including various interacting waters, and accurate distinctions between displaced and non-displaced waters. The initial step is to identify significant water molecules and their functions. Methods involving molecular dynamics of water on the binding site, the Monte Carlo probability, free energy perturbation techniques, and water displacement have improved the accuracy of the modeling of water molecules in proteins. Additional techniques include the grid in the homogeneous salvation theory used by AutoDock, protein-centric, water docking utilizing Vina, WScore, and "hydrated docking" scripts for AutoDock [27].

14.5.7 Covalent Docking

Due to the toxicity and probable off-target consequences of irreversible covalent medicines, non-covalent pharmaceuticals have historically received the majority of attention in drug discovery. The COVID-19 outbreak and ensuing years, however, have seen a revival of covalent medications. Covalent pharmaceuticals are highly effective as compared to non-covalent endogenous substrates. They may provide additional benefits over non-covalent medications, such as improved efficacy. They may provide a lesser client burden and lesser drug resistance due to the shorter, less frequent dose and improved target specificity. For infectious disorders, covalent modifiers may be more focused, efficient, and specific [28].

The primary flaw in the covalent interaction models that are now in use is that they just simulate the development of bonds between instances that are known, whereas the covalent bonding process itself needs to be accurately assessed by quantum mechanics. Numerous manual processes and ad hoc fixes have been historically employed to overcome these limitations.

The binding site—is it known? Is the reactivity of the targeted amino acid known? The answers to these questions will determine the circumstances and requirements that apply. Particular nucleophilic amino acids present in the binding site can be aimed in the simplest scenario where the target site is well-known; often, the nucleophilic amino acid to be targeted is also known in advance. In these situations, the main goal of docking is simply to clarify the binding modalities and/or to order the molecule of choice following how strongly they bind to the protein's active site after being covalently attached to it [17].

More complex strategies are needed for complex activities, such as the design of covalent ligands for processes where little or no prior knowledge is available. By first describing covalent bond breakdown and then adding a new terminology to the non-covalent grade, to describe the covalent interaction of ligands, a scoring system based on quantum mechanics has been developed. This phrase refers to the difference in 'free' energy between covalent and non-covalent compounds. Such a strategy aids in overcoming the widespread denial of the energy contributions from the creation of covalent bonds [29]. The use of quantum mechanics/molecular mechanics (QM/MM) approaches for simulating enzyme reactions has steadily improved over the past few decades. These are some difficulties that could arise in the upcoming years.

14.5.7.1 Bigger QM Regions

From a conceptual standpoint, the use of QM/MM methods to examine enzyme reactions presupposes that all of the atoms involved are included inside the QM domain. Indirect yet crucial roles in enzyme processes are also played by several residues of amino acids and atoms surrounding the active site. Using extremely large QM regions has long been a target for every QM/MM modeler. The decision of the QM area has been constrained by the CPU available power because the time required for a QM/MM measurement is highly dependent on the time required for a QM measurement. Early QM/MM enzymology research frequently used QM areas with 20 atoms or more that had undergone the density functional theory (DFT). At this time, the majority of research already uses DFT QM areas of nearly 200 atoms. The desire to enlarge the QM zone remains strong, especially as more intricate processes involve the investigation of significant structural change at the active site [30].

14.5.7.2 Utilization of Polarizable Force Fields and Polarized Embedding

One of several greatest current limitations in the most popular QM/MM research on enzymatic reactions is the problem of trying to implement a polarizable embedding process to contribute to the polarization in the MM region caused by the difference in charge arising from the chemical reaction occurring in the QM region. Many reactions involve the formation of charges along the reaction pathway, with intermediates or products exhibiting considerable charge modification relative to the reactants. This charge difference should naturally be mostly balanced out by interactions with molecules that are present in the QM region, such as amino acid residues.

The strength of the action, the QM zone's size, and naturally, the special characteristics of enzymes all play a significant role in this condition. The majority of QM/MM research still uses the same biomolecular force fields that were widely used 20 years ago. The availability of better versions of AMBER, CHARMM, GROMACS, and other programs has led to advancements in electrostatics and several other structural aspects. Consider the fact that when reacting with other charged and polar molecules or the solution, charged and polar molecules in enzymes can become quite polarized. The geometry may be affected by this polarization. However, the environment around amino acid residues affects their charges, which vary based on their location within the protein, orientation, and activity with some other molecules, among other characteristics. Over the past few decades, polarizable force fields have become better. There are currently several great options for treating biomolecular systems. AMOEBA, DRUDE, CHARMM variable charge, SIBFA, and GEM are notable examples [31, 32].

Accounting for these effects would be greatly enhanced by the introduction of polarizable force fields in QM/MM techniques. In recent years, some alternatives have become accessible. In the next 10 years, programs like LICHEM are projected to be crucial in helping in the modeling of enzyme reactions [33].

14.5.7.3 Improvements on the QM Level

Density functionals have undergone significant advancements over the past decades. Grimme's density functionals, the Minnesota density functionals, and the development of dispersion correction are notable examples. The majority of research tests for uniformity in the activation or reaction-free energy limits using a variety of different options. The use of techniques other than DFT in QM/MM optimization of enzymatic reactions is currently uncommon in contrast. However, it would be extremely important when treating enzymes, especially in circumstances where the existence of excited states is essential [33].

14.5.8 Pose vs. Scoring

One of the docking's ongoing issues is hit selection. As they were created from a test set, the scoring functions are constrained. As a result, unless the score of a given ligand is compared to a reference, it does not tell us much. The outcomes, such as symmetric ligands and enantiomer pairings, may be unclear in some situations [34]. So the issue is how to get good possibilities from a scoring system. Correct structure preparation would be the first step. The use of molecular dynamics to prepare the structures of molecular targets, such as by conducting energy minimization and assigning charges, has significantly improved docking [35]. The use of force field optimization, molecular mechanics, Poisson–Boltzmann/surface area optimization, and optimization using semi-empirical charges have all been shown to improve docking outcomes for ligand synthesis. Furthermore, the ability of force field optimization with explicit solvent to forecast complicated energies has been demonstrated more accurately than more sophisticated semi-empirical approaches. Additionally, it has been proposed that enhancing model representation may help predict binding more accurately. To this purpose, work has been done to create additional binding models. Models, however, do not significantly enhance binding affinity prediction. Some systems allow changing the scoring parameters and weight assignment in addition to changing the scoring functions, such as Vina. Further refining is taken into account for rotational terms and solvation. Machine learning has been used in recent studies to enhance docking and scoring. Machine learning is the study of the creation of adaptable programs that can advance themselves without user input. These techniques fall into two categories: feature-based and similarity-based. One method that has optimized scores is PostDOCK, which takes into account 3D elements that are not reflected in scoring [36].

14.5.9 Chemical Space

Chemical space analysis and visualization have played a crucial role in decision-making and tracking the development of the drug discovery process. The searchable chemical universe offers a plethora of opportunities to locate hit molecules, especially with the advent of approaches to scale them up quickly. The big challenge query could be phrased as follows in this situation: What areas of the chemical landscape need to be looked into to find possible therapeutic candidates? There have not been many novel chemical entities discovered in the drug-like chemical realm that are medically important. Finding new starting points may require exploring new areas. Expanding the searchable drug-like chemical space, revisiting and repurposing current chemical libraries, particularly significant corporate collections, and the SAR of in-house data are some of the challenges in this approach. These challenges include recovering lost hits, using novel tools to search classic chemical spaces, and revisiting "neglected chemical spaces," such as poorly researched types of compounds like peptides, peptidomimetics, and biologics. These kinds of compounds may provide encouraging beginning points for the development of therapeutic candidates for novel or challenging targets [29, 37]. The challenges involved in chemical space are shown in Figure 14.2.

14.5.10 Biological Space

Finding and confirming molecular targets with therapeutic applications is a vital task in drug discovery. The main challenge is to figure out and

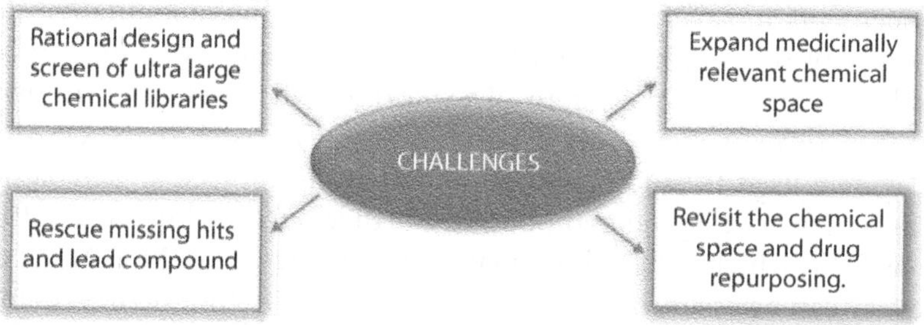

Figure 14.2 Showing challenges in chemical space.

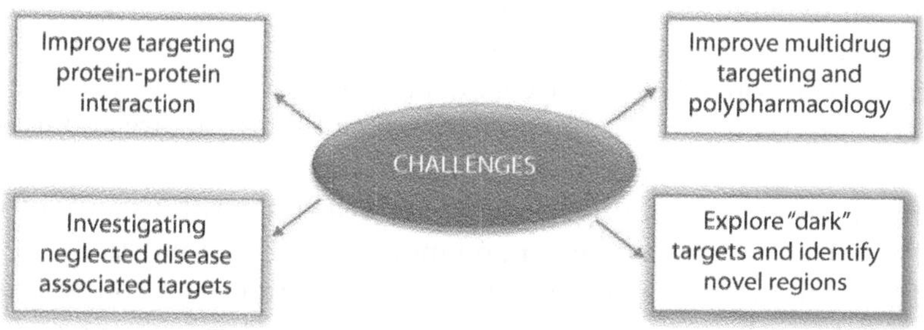

Figure 14.3 Showing challenges in biological space.

search for the areas of the physiologically relevant chemical space. A recent thorough analysis has focused on identifying the therapeutically important targets. Studying polypharmacology and attempting to predict how a drug candidate would interact with each of its potential target partners is also crucial. Improvements in multi-target drug design and multi-target property prediction are among the challenges, as are finding suitable strategies to address protein–protein interactions and continuing to investigate the molecular targets linked to neglected diseases. The finding of druggable binding pockets, allosteric sites, and transient binding sites that can be used for drug discovery can also be assisted by computational approaches [37, 38]. The challenges involved in biological space are shown in Figure 14.3.

14.6 Future Prospects of CADD

CADD appears to be a promising tool with a year of expertise in assisting in the development of therapeutically active molecules. Considering advancements in several techniques, such as docking, quantitative structure–activity relationship (QSAR), molecular dynamics simulation, and virtual screening, the paradigm of exploring and developing innovative pharmaceuticals is constantly expanding. For the last 3 years, QSAR has been widely used to study compound activity in molecular databases, as well as in newly designed compounds. Furthermore, by boosting software such as DFT, which is based on quantum strategies (QM/MM/MD), drug discovery can be assumed to contribute toward milestone achievements.

14.6.1 Molecular Dynamics Simulations

Molecular dynamics is a sophisticated cybernated tool whose prominence has skyrocketed in recent years due to its versatility and tremendous scientific breakthroughs. These are generally used to better understand the biology of molecules (typically proteins), including structure and dynamics and their precise placement using mechanics, and force fields [39]. Improvements in computing hardware, as well as algorithms and software used in MD, have enabled longer and more accessible simulations.

MD has both merits and drawbacks. Concentrating on constraints for future panorama such as latency resulted in the development of GPUs (graphics processing units) for atomistic performance. GPU's simulation can now be performed quickly for the creation of multiple arithmetic logic units (ALUs) and automictically for MD. GPUs have evolved into fully programmable, high-performance processors to generate molecular representation and network topology. Due to these characteristics, no descriptor engineering is needed [40]. GPUs have specific MD codes, each with a distinctive performance, such as AMBER, LAMMPS, GROMACS, NAMD, ACEMD, Desmond, Espresso, Folding Home, and HOOMD-blue. HOOMD-blue and GROMACS were the only GPUs used exclusively, while NAMD was the least used. Among these codes, HOOMD-blue is dynamic. It employs entirely device-resident data structures and provides a comprehensive feature set for general-purpose particle-based simulation. Another issue is that covalent bonds are fixed and must be properly specified at the commencement of a simulation because they do not disintegrate or generate in conventional MD simulations. Many software applications were employed for simulations including ACEMD, DOCK, GROMACS, NAMD, and YASARA. The automatic setup feature of GROMACS, for instance, is available when it is used with a user-friendly interface (such as Gromita and jSimMacs).

Emphasizing advantages and specific applications: (1) It leads to the discovery of ligands or targets, such as G-protein coupled receptor (GPCRs).

(2) Predicting how a biomolecular system will react to deformation, for instance, determining the impact of various proteins' associations with ligands (with or without them). (3) Different biomolecules' flexibility can be evaluated, and new structures, such as X-ray crystal structures, can be generated. (4) To overcome grammatical errors. (5) To create allosteric pharmaceuticals that bind to a target at a different place than the native ligand. Therefore, simulation is helpful and useful for discovering the structural origins of diseases, elucidating the functional mechanisms of proteins and other biomolecules, and designing and optimizing small molecules, peptides, and proteins [41].

14.6.2 Virtual Screening

A prominent computerized method of establishing and identifying novel medicinal compounds is virtual screening (VS). It effectively operates by optimizing and locating possibly active chemicals from virtual databases, typically against a protein of interest. The two types of VS that are most used are ligand-based and structure-based methods. Both share a variety of traits that are effective in finding novel drugs. The success of these methods spikes continuously because of the advancements made in certain algorithms, computational power, and structural properties of both databases, as well as protein molecules.

Future improvements in the effectiveness of virtual screenings should come from a variety of sources, for example, the discovery of numerous screening databases that are designed specifically for screening the compound. ZINC, PubChem, ChEMBL, NCI, DUD, Asine, Maybridge, and ChemDiv are a few of these databases, each with a particular application. Among them, ZINC is a prevalently utilized dataset with a spectrum of uses, including the accessibility of various file formats (2D, 3D SDF, and SMILES) and physicochemical parameters (pH range, xlogP, number of H-bond donors and acceptors, net charge, molecular weight), while being made up of numerically and alphanumerically enlightened compounds that are represented in regioisomeric forms and tautomeric form [42]. Another crucial factor to take into account when looking for promising candidates for tentative inquiries is that, for instance, SBDD does not take into account pharmacokinetics aspects. To eliminate compounds that might have permeability and solubility problems, the estimation of parameters like the partition coefficient (logP) and polar surface area (PSA) might be employed as a preliminary filter.

Furthermore, many well-established virtual screening software applications were specified, executing multiple activities to screen the compounds quickly, such as AutoDock, GLIDE, GOLD, LibDocK, Molecular Operating Environment (MOE), Molegro Virtual Docker (MVD), and Surflex-Dock. GLIDE and perhaps even AutoDock were the most widely used programs, followed by GOLD when compared to other programs.

AutoDock is an automated method that is frequently used to forecast how tiny molecules, such as therapeutic candidates, associate with targets of known 3D structures. Aside from that, AutoDock might be used for virtual screening, structure-based design, protein–ligand docking, and X-ray crystallography. In contrast to LigandScout, PyRx, PyMol, and YASARA, AutoDock can be integrated into other software packages. The two software generations, AutoDock 4 and AutoDock Vina, are their types. A more current version of AutoDock 4 that is GPU-accelerated is called AutoDock-GPU. One of the most frequently referenced molecular docking software tools is AutoDock Vina, which uses the Monte Carlo algorithm to investigate the compounds and optimize the particle size. Grid maps for all-atom kinds are precalculated instantly by AutoDock Vina. Platforms like MAC OS X, Linux, Unix, and Windows are the only ones that support AutoDock. Conversely, a VS tool called GOLD uses a genetic algorithm and a scoring system based on force fields. It can run on multiple operating systems, including AIX, Linux, Solaris, Sun, Unix, and Windows [43].

14.6.3 Density Functional Theory (DFT)

DFT is a very well-liked cornerstone of quantum mechanics and is becoming increasingly important for the drug-designing process since it produces successful results, which makes it a preferred tool to use for medicinal studies. It effectively determines the ground state energy and other molecular variables by relying on electron density as a less complex alternative to wave function formalism. The use of this method has surged in recent years as a result of several breakthroughs, including a significant degree of efficacy and precision in forecasting the geometries of medicinal molecules. Furthermore, to explain the drug molecule's response processes, DFT studies are frequently used. For medication design, it can accurately compute the transition state. The hybrid functional B3LYP has provided significant insight into the mode of pharmacological action. In addition, it can forecast complex interactions between medicine and its target, including covalent bonds, hydrogen bonds, dipole–dipole interactions, ionic interactions, and interactions involving hydrophobic molecules. To improve medication design, DFT can be used to predict how an enzyme will act [44].

14.6.4 Drug Design-Related Web Servers

Computer-aided drug design is made possible by a wide range of software programs and online services. Big data in pharmaceutical research tend to be more complex because most drugs, perhaps fine molecules, have an abundance of statistics, including biological activity, drug–drug interactions, and protein–ligand interactions. In this regard, the web server's main objective is to help and encourage medicinal researchers in their efforts to improve lead compounds. Numerous web servers for developing novel drugs are included in both structure- and ligand-based drug design.

In structure-based design and development of compounds, web servers involved are Caver Web, DOCK Blaster, Pharmit, e-LEA3D, dockNmine, and MedusaDock. Pharmit is a popular web server that uses pharmacophores and molecular shapes to automate the screening of large compounds in large compound databases.

(1) It accepts the input, initializes it using PDB Code accession, then allows the user to choose the "Load receptor" and prepare it in any file format (for example, Open Babel) after selecting the ligand from the PDB file and uploading it by pressing "ENTER Pharmit Search."
(2) After input, pharmacophores are identified and initialized using pharmacophore files, particularly from LigBuilder, LigandScout, PharmaGist, or Pharmer.
(3) Then, molecules are perfectly related using the shape search method known as the Volumetric Aligned Molecular Shapes (VAMS), which locates precise matches by employing inclusive and exclusive shape limits. The final shape can be changed in a modern interactive interface by adopting 3Dmol.js, which enables improved 3D molecular graphics to refrain from resorting to plugins or Java.
(4) Results are given from lowest to highest depending on variables like RMSD and similarity score for various queries (such as pharmacophore or shape searches), which are computed using search modalities. Pharmacophore searches are often conducted using RMSD, whereas a similarity score is typically used for shape searches.

A complimentary Web molecular screening test, DOCK Blaster, is similar to Pharmit in that it creates a comprehensive collection screening using a PDB file and ligand settings. The graphical user interface is an established architecture that enables both learners and specialists to achieve worthwhile results [45].

FragRep, which uses fragment replacement to create structure-based ligand designs quickly, has recently gained popularity and computational reliability. To locate appropriate pieces for their projects, FragRep effectively explores the fragment space. The creation of efficient fragment-based computer algorithms has a significant positive impact on the search for novel medications. A system for automated ligand optimization for speedy drug design can help research teams. Some tools for automatic fragment replacement include LUDI, ReCore, Scaffold Replacement, SHOP, ParaFrag, and Core Hoping. SHOP produces presumed frameworks from databases by comparing the 3D structures of the scaffolds in the database to the query scaffold, while ParaFrag generates unique components comparing the surface-based similarity of molecular fragments. ReCore, on the other hand, presents an algorithm for quick queries and creates a fragment database using small molecule crystal structure conformations obtained from the Cambridge Crystallographic Structure Database [46].

Another popular website that aims to gather both private and public data with a single tool is DockNmine. If necessary, contemporary connection studies, like ROC curves or enhancement analysis, are performed on the docking data. The selection of actives and the non-selection of decoys are both represented by ROC curves. The number of molecules successfully docked is indicated by the receiver operating characteristic curve, or ROC AUC.

TargetNet, SwissTarget Prediction, Multi-fingerprint Similarity Search Algorithm (MuSSel), MolTarPre, RFQSAR, and SuperPred are among the ligand-based drug design web servers that are available. Countless sites for each molecule can be predicted using TargetNet, an automated web server. The use of contemporary chemogenomics datasets is extremely significant in a considerable section of QSAR models for enhanced prognosis. The service predicts user molecular activity for 623 human proteins by upgrading QSAR models for every protein domain and developing a drug-target interaction (DTI) profile that can be used in application areas. A tiny, bioactive molecule's most likely macromolecular targets are predicted by SwissTarget Prediction, in contrast. The assessment relies on the merger of 2D and 3D similarity on more than 3000 proteins of three different types with a library of 370,000 verified active participants using RFQSAR. This web server allows for the recognition of targets utilizing the chemical structure–activity (SAR) relationship. Researchers generated a total of 1288 human targets employing random forest methods and bioactivity data from the ChEMBL database.

14.6.5 Big Data

Huge quantities of information with complex data structures are considered "big data." Big data are predominantly exploited in all fields of study and engineering, including pharmaceutical innovation, despite being used principally in the field of informatics. Machine learning (ML) methods have proven useful in drug discovery and development. Numerous data sets are swiftly procreated by methods like combinatorial chemistry and high-throughput screening (HTS) [47]. The information gathered could significantly speed up the search for new drugs within big data, providing insights into key chemical structures that can be used to predict outcomes *in vitro*, *in vivo*, and in the clinic. The databases for drug candidates have expanded quickly, and new modeling techniques involve some classes, including records of chemical repositories (e.g., Enamine REAL Database, PubChem, and ChEMBL); chemical records originally tailored for drugs and substances that resemble drugs (e.g., DrugBank and eDrug3D); libraries of bioactive molecules, along with genome sequencing and proteomics data (e.g., BindingDB); datasets retrieving biological parameters acquired from assay screening and mechanistic databases.

Drug research and development have benefited from the use of machine learning techniques. For instance, QSAR approaches are used to swiftly forecast a large number of novel compounds for multiple endpoints, including physicochemical attributes and biological activities. It is a quick and effective way to determine if there is a clinically meaningful linkage among chemical structures, their features, and distinct actions like the creation of new chemical descriptors, topological descriptors, and the use of new nonlinear modeling techniques. Through the virtual screening of drug candidates using QSAR modeling, the efficiency of drug discovery has greatly improved. Prioritizing drug candidates over chemical products can help them have the required biological activities (such as therapeutic ones) and fewer negative side effects. Testing targeted chemicals chosen from a validated model could greatly increase hit rates by implementing QSAR models. To create novel biomarkers for personalized therapy, data from the 'omics' (genomics, transcriptomics, proteomics, and metabolomics) that were recently publicly revealed were indeed exploited with ML. These findings might help us comprehend the intricacy of disease development better. As a result, in the age of big data, CADD may generate significant value by using ML models for a variety of data, including public databases [48].

14.6.6 Quantum Mechanics (QM)

In the last century, the field of pharmaceutical research has seen considerable interest in quantum mechanics (QM) as a different approach. To better describe complexation and predict binding affinities, it is being researched gradually. QM approaches are methodically improvable, offering a significant level of transferability. The QM/MM/MD approach, which combines the QM/MM approach with molecular dynamics (MD) simulation, is a potent and propitious tool for examining the dynamics of chemical reactions in complex molecular systems, drug delivery, and molecular device features. Additionally, at least some of the free energy contributions, such as the effects of electronic polarization, metal coordination, and covalent bonding, are calculated explicitly using QM, making them systematically better and more accurate [49].

QM/MM approaches are currently an ineradicable alternative tool for the assessment of enzyme reaction processes and are cropping up in new and more useful applications as computer power and technical advancement expand. This approach often involves the use of enzymes that can be separated into two regions: an electrical region that is addressed by quantum mechanics (the QM region) and a surrounding region that is treated by molecular mechanics (the MM region). The segment of a chemical system that is currently engaged in a chemical process is positioned in the QM zone, while other components of the system are found in the MM region. Thus, the significant region is said to only be indirectly influenced by the MM region. The underlying idea behind QM/MM hybrid approaches is that an enzymatic reaction is viewed as a metamorphosis that otherwise incorporates a restricted number of atoms of each element at the active site (the QM region) but still has an impression on the remainder of the enzyme and its proximity (the MM region).

By assessing the intensity of the associations created including the changes in electron density, spin, and other electronic properties of the target and inhibitor, QM/MM concepts can be used to fine-tune specific interactions at specific places in enzymes. Computational enzymatic catalysis is where QM/MM approaches are most frequently applied in the study of enzymes. To confirm or contradict various mechanistic theories on the catalytic pathway of a particular enzyme reaction, QM/MM approaches are frequently used. This method has been effective in therapeutic applications due to its wider benefits. Additionally, a variety of models, such as the ONIOM [33] (our Own N-layer Integrated molecular Orbital molecular Mechanics) model, are accessible for effective modeling. Various patents in CADD related to diseases are listed in Tables 14.1 and 14.2.

14.7 Recent Patents in CADD

Table 14.1 Patents associated with diseases in CADD.

Patents	Patent no.	Outcomes	References
Method, apparatus, and system for computer-aided tracking, navigation, and motion teaching	US9082319B2	The invention relates to tracking devices, monitoring a patient's position, orientation, and movement of their anatomy during therapy. The present invention's computer-aided navigation system is used to reduce the issues of arthroscopic hip surgery. They can also take the form of computer software, which, when used by a processor, enables the processor to do particular tasks.	[50]
Computer-assisted tumor response assessment and evaluation of the vascular tumor burden	US11350900B2	The innovation includes a computational method for locating target lesions within the tumor and using cross-sectional pictures to assess the realistic tumor response to anti-cancer therapy. The technique further entails gathering data to confirm the prevalence or absence of metastases, assessing changes in lesion metrics, and obtaining the response based on the tumor response criteria.	[51]
Compositions, methods, and computer systems related to making and administering modified T cells	US9499855B2	The techniques disclosed relate to computational methods and tools for creating T cell clones with a collection of synthetic chimeric antigen receptors and delivering the T cell clone cell line(s) to a subject in need of it thereby, T CAR cells, used as an immunotherapy tool for a person suffering from a disease.	[52]

(*Continued*)

Table 14.1 Patents associated with diseases in CADD. (*Continued*)

Patents	Patent no.	Outcomes	References
Congenital heart disease monitor	US9687160B2	A congenital heart disease monitor uses a sensor emitting optical radiation into a tissue site to detect these radiations after it has been attenuated by the pulsating blood flowing through the tissue site, followed by receiving a signal and calculating the physiological parameter in response at a baseline site and a comparison site.	[53]
A surgical catheter having side exiting medical instrument and related systems and methods for four-dimensional soft tissue navigation	US1024903B2	The surgical instrument, as claimed, such as a steerable surgical catheter with a biopsy device or a surgical catheter with a side-exiting medical instrument, visually simulates the fluid flow scene of the patient's body cavity from the point of view of a surgical instrument located in the cavity of the patient.	[54]
Device and method for electroporation-based treatment of stenosis of a tubular body part	US10702326B2	The invention relates to non-thermal irreversible electroporation-based medical devices and procedures for treating or preventing stenosis through electrodes included in variants of the invention's balloon catheter devices. These devices permanently electroporate a section of the inner circumference of a body's tubular structure.	[55]

(*Continued*)

Table 14.1 Patents associated with diseases in CADD. (*Continued*)

Patents	Patent no.	Outcomes	References
Method for tracking, predicting, and proactively correcting malocclusion and related issues	US11147652B2	The innovation offers computerized tools and techniques for forecasting a patient's futurity dental or orthodontic issues, as well as for projecting the position of an intraoral object into the patient's intraoral cavity. It involves receiving digital data points that indicate the intraoral cavity's actual state and based on the velocity, one can predict the intraoral object's position.	[56]

Table 14.2 Recent general patents in CADD.

Patents	Patent no.	Outcomes	References
Generative machine learning systems for drug design	US10776712B2	The generative models are used to create chemical molecules satisfying several criteria and having desired properties, such as activity against a target. Machine learning techniques are used to train these models, with training sets made up of chemical compounds and biological or chemical data related to the chemical compounds.	[31]

(*Continued*)

Table 14.2 Recent general patents in CADD. (*Continued*)

Patents	Patent no.	Outcomes	References
Computer-aided analysis of medical images	US9642586B2	The innovation includes an apparatus and a procedure for computational analysis of medical images to identify calcified plaques in vessels and to perform a diagnosis to identify arterial disorders based on the discovered calcified plaques. A semiautomatic approach is used for this purpose.	[32]
Medical device disposables information systems and method	US11322239B2	This invention relates to medical devices with disposable components and offers a method for controlling the usage of a disposable with an infusion pump, identifying information from a disposable, and involves figuring out if the disposable is suitable for use, and also involves deciding whether or not to deliver an infusate.	[33]

(*Continued*)

Table 14.2 Recent general patents in CADD. (*Continued*)

Patents	Patent no.	Outcomes	References
Method and apparatus for the analysis of molecular configurations and combinations	US8036867B2	The invention relates to bioinformatics, proteomics, molecular modeling, drug design software, and computer-aided molecular design and computational modeling of molecular combinations. Functions of molecules are computed for molecular docking. The computations might be applied to model and/or calculate the probabilities of molecular interactions.	[34]

14.8 Conclusion

In the field of drug research and discovery, computer-aided drug design is a useful tool. It allows us to quickly and cheaply identify the drug candidate that has the most potential. It constantly offers the possibility of advancement in the field of drug discovery. Computer-aided drug design has helped advance scientific knowledge in recent years, and it will be crucial in the coming years. Several strategies are still being developed and worked upon. Additionally, several innovative ideas and methods are propelling CADD into the future. The latter is evident in "big data," quantum mechanics, DFT, and other fields, as well. In addition to this, it also faces challenges, some of which are connected to computer-aided design and others to computer-aided drug design. A further challenge is to keep enhancing conventional methods of drug discovery to collaborate with, mutually enhance, and benefit from other conventional and proven methods of drug discovery.

References

1. Sabe, V.T. *et al.*, Current trends in computer aided drug design and a highlight of drugs discovered via computational techniques: A review. *Eur. J. Med. Chem.*, 224, 113705, 2021.
2. Wang, X. *et al.*, Structure-based drug design strategies and challenges. *Curr. Top. Med. Chem.*, 18, 12, 998–1006, 2018.
3. Batool, M., Ahmad, B., Choi, S., A structure-based drug discovery paradigm. *Int. J. Mol. Sci.*, 20, 11, 2019, 2783.
4. Surabhi, S. and Singh, B., Computer aided drug design: an overview. *J. Drug Deliv. Ther.*, 8, 5, 504–509, 2018.
5. Sharma, V., Wakode, S., Kumar, H., Structure-and ligand-based drug design: concepts, approaches, and challenges, in: *Chemoinformatics and Bioinformatics in the Pharmaceutical Sciences*, pp. 27–53, 2021.
6. Wang, X. and Bi, Z., New CAD/CAM course framework in digital manufacturing. *Comput. Appl. Eng. Educ.*, 27, 1, 128–144, 2019.
7. Vukašinović, N. and Duhovnik, J., Controlling CAD Curves by Equations, in: *Advanced CAD Modeling*, pp. 135–163, Springer, 2019.
8. Ramnath, S. *et al.*, Interoperability of CAD geometry and product manufacturing information for computer integrated manufacturing. *Int. J. Comput. Integr. Manuf.*, 33, 2, 116–132, 2020.
9. Hadj, R.B. *et al.*, An interoperability process between CAD system and CAE applications based on CAD data. *Int. J. Interact. Des. Manuf.*, 12, 3, 1039–1058, 2018.
10. Mukkavaara, J. and Sandberg, M., Architectural design exploration using generative design: framework development and case study of a residential block. *Buildings*, 10, 11, 201, 2020.
11. Lee, H.-J. *et al.*, Digital Workflow to Fabricate Complete Dentures for Edentulous Patients Using a Reversing and Superimposing Technique. *Appl. Sci.*, 11, 13, 5786, 2021.
12. Kim, H. *et al.*, Deep-learning-based retrieval of piping component catalogs for plant 3D CAD model reconstruction. *Comput. Ind.*, 123, 103320, 2020.
13. Kafle, K. *et al.*, Dvqa: Understanding data visualizations via question answering, in: *Proceedings of the IEEE conference on computer vision and pattern recognition*, 2018.
14. Desai, R., McCann, J., Coros, S., Assembly-aware design of printable electromechanical devices, in: *Proceedings of the 31st Annual ACM Symposium on User Interface Software and Technology*, 2018.
15. Mazandu, G.K. *et al.*, Designing data-driven learning algorithms: A necessity to ensure effective post-genomic medicine and biomedical research,

in: *Artificial Intelligence-Applications in Medicine and Biology*, vol. 5, pp. 3–18, 2019.
16. Pinheiro, H.M. and da Costa, R.M., Pupillary light reflex as a diagnostic aid from computational viewpoint: A systematic literature review. *J. Biomed. Inf.*, 117, 103757, 2021.
17. Stanzione, F., Giangreco, I., Cole, J.C., Use of molecular docking computational tools in drug discovery. *Prog. Med. Chem.*, 60, 273–343, 2021.
18. Lin, Y.-F. *et al.*, MIB: metal ion-binding site prediction and docking server. *J. Chem. Inf. Model.*, 56, 12, 2287–2291, 2016.
19. Vakser, I.A., Protein-protein docking: From interaction to interactome. *Biophys. J.*, 107, 8, 1785–1793, 2014.
20. Sunny, S. and Jayaraj, P., Protein–protein docking: Past, present, and future. *Protein J.*, 41, 1, 1–26, 2022.
21. Du, X. *et al.*, Insights into protein–ligand interactions: mechanisms, models, and methods. *Int. J. Mol. Sci.*, 17, 2, 144, 2016.
22. Grinter, S.Z. and Zou, X., Challenges, applications, and recent advances of protein-ligand docking in structure-based drug design. *Molecules*, 19, 7, 10150–10176, 2014.
23. Grebner, C. *et al.*, Binding mode and induced fit predictions for prospective computational drug design. *J. Chem. Inf. Model.*, 56, 4, 774–787, 2016.
24. Zhang, W. *et al.*, EDock: blind protein–ligand docking by replica-exchange monte carlo simulation. *J. Cheminf.*, 12, 1, 1–17, 2020.
25. Agrawal, P. *et al.*, Benchmarking of different molecular docking methods for protein-peptide docking. *BMC Bioinf.*, 19, 13, 105–124, 2019.
26. Lie, M.A. *et al.*, Molecular docking with ligand attached water molecules. *J. Chem. Inf. Model.*, 51, 4, 909–917, 2011.
27. Prieto-Martínez, F.D., Arciniega, M., Medina-Franco, J.L., Molecular docking: current advances and challenges, in: *TIP. Revista especializada en ciencias químico-biológicas*, vol. 21, 2018.
28. Li, Q. *et al.*, Potential clinical drugs as covalent inhibitors of the priming proteases of the spike protein of SARS-CoV-2. *Comput. Struct. Biotechnol. J.*, 18, 2200–2208, 2020.
29. De Cesco, S. *et al.*, Covalent inhibitors design and discovery. *Eur. J. Med. Chem.*, 138, 96–114, 2017.
30. Mera-Adasme, R., Domínguez, M., Denis-Alpizar, O., A benchmark for the size of the QM system required for accurate hybrid QM/MM calculations on the metal site of the protein copper, zinc superoxide dismutase. *J. Mol. Model.*, 25, 6, 1–7, 2019.
31. Zhang, C. *et al.*, AMOEBA polarizable atomic multipole force field for nucleic acids. *J. Chem. Theory Comput.*, 14, 4, 2084–2108, 2018.
32. Aytenfisu, A.H., Yang, M., MacKerell Jr., A.D., CHARMM Drude polarizable force field for glycosidic linkages involving pyranoses and furanoses. *J. Chem. Theory Comput.*, 14, 6, 3132–3143, 2018.

33. Magalhães, R.P., Fernandes, H.S., Sousa, S.F., Modelling enzymatic mechanisms with QM/MM approaches: current status and future challenges. *Isr. J. Chem.*, 60, 7, 655–666, 2020.
34. Ramírez, D. and Caballero, J., Is it reliable to use common molecular docking methods for comparing the binding affinities of enantiomer pairs for their protein target? *Int. J. Mol. Sci.*, 17, 4, 525, 2016.
35. Ash, J. and Fourches, D., Characterizing the chemical space of ERK2 kinase inhibitors using descriptors computed from molecular dynamics trajectories. *J. Chem. Inf. Model.*, 57, 6, 1286–1299, 2017.
36. Anighoro, A. and Bajorath, J.R., Three-dimensional similarity in molecular docking: prioritizing ligand poses on the basis of experimental binding modes. *J. Chem. Inf. Model.*, 56, 3, 580–587, 2016.
37. Medina-Franco, J.L., Grand challenges of computer-aided drug design: The road ahead, *Front. Drug Discov.*, p. 728551, Frontiers Media SA, 2021.
38. P., W., W.P., Krutoholow, E., Kohler, M., Blaney, J., Funatsu, K., Luebkemann, C., Schneider, G. *et al.*, Rethinking Drug Design in the Artificial Intelligence Era. *Nat. Rev. Drug Discov.*, 19, 5, 353–364, S2020.
39. Hospital, A. *et al.*, Molecular dynamics simulations: advances and applications. *Adv. Appl. Bioinform. Chem.*, 8, 37, 2015.
40. Hollingsworth, S.A. and Dror, R.O., Molecular dynamics simulation for all. *Neuron*, 99, 6, 1129–1143, 2018.
41. Śledź, P. and Caflisch, A., Protein structure-based drug design: from docking to molecular dynamics. *Curr. Opin. Struct. Biol.*, 48, 93–102, 2018.
42. Kumar, S., Online resource and tools for the development of drugs against novel coronavirus, *In Silico Modeling of Drugs Against Coronaviruses*, pp. 735–759, Springer, 2021.
43. van Hilten, N., Chevillard, F., Kolb, P., Virtual compound libraries in computer-assisted drug discovery. *JCIM*, 59, 2, 644–651, 2019.
44. Tandon, H., Chakraborty, T., Suhag, V., A brief review on importance of DFT in drug design. *Res. Med. Eng. Stud.*, 39, 46, 2019.
45. Sunseri, J. and Koes, D.R., Pharmit: interactive exploration of chemical space. *Nucleic Acids Res.*, 44, W1, W442–W448, 2016.
46. Shan, J. *et al.*, FragRep: A Web Server for Structure-Based Drug Design by Fragment Replacement. *J. Chem. Inf. Model.*, 60, 12, 5900–5906, 2020.
47. Prieto-Martínez, F.D. *et al.*, Computational drug design methods—current and future perspectives. *In silico Drug design*, Kunal Roy (Ed.), pp. 19–44, Elsevier, 2019.
48. Zhao, L. *et al.*, Advancing computer-aided drug discovery (CADD) by big data and data-driven machine learning modeling. *Drug Discov. Today*, 25, 9, 1624–1638, 2020.
49. Cavasotto, C.N., Adler, N.S., Aucar, M.G., Quantum chemical approaches in structure-based virtual screening and lead optimization. *Front. Chem.*, 6, 188, 2018.

50. Dolan, B., *Generation of computer-based discovery avatars based on tokenization of prioritized source data*. US patent 9740987, assigned to GCP IP, 2017.
51. Smith, A.D., *Computer-assisted tumor response assessment and evaluation of the vascular tumor burden*, 2022, Google Patents.
52. Hyde, R.A., Kindsvogel, W.R., McKnight, G.L., *Compositions, methods, and computer systems related to making and administering modified T cells*. US patent 9499855, assigned to Elwha LLC, 2016.
53. Kiani, M.J.E., *Congenital heart disease monitor*, 2017, Google Patents.
54. Holsing, T.L., Hunter, M., Stern, E., *Surgical catheter having side exiting medical instrument and related systems and methods for four dimensional soft tissue navigation*. US patent 10249036B2, assigned to Veran Medical Technologies, 2019.
55. Neal, I.R.E. *et al.*, *Device and method for electroporation based treatment of stenosis of a tubular body part*, 2020, Google Patents.
56. Mason, D. and Brandt, D., *Method for tracking, predicting, and proactively correcting malocclusion and related issues*. US patent 11147652B2, assigned to Align technology Inc, 2021.

Index

Printed and bound by CPI Group (UK) Ltd, Croydon, CR0 4YY

17/09/2026

14964276-0007